Ethical Innovation for Global Health

Chieko Kurihara • Dirceu Greco • Ames Dhai
Editors

Ethical Innovation for Global Health

Pandemic, Democracy and Ethics in Research

Springer

Editors
Chieko Kurihara
Specially-appointed Professor
Kanagawa Dental University
Yokosuka, Kanagawa, Japan

Dirceu Greco
Department of Internal Medicine, School of Medicine
Federal University of Minas Gerais
Belo Horizonte, Minas Gerais, Brazil

Ames Dhai
School of Clinical Medicine
University of the Witwatersrand
Johannesburg, Gauteng, South Africa

ISBN 978-981-99-6162-7 ISBN 978-981-99-6163-4 (eBook)
https://doi.org/10.1007/978-981-99-6163-4

This Springer imprint is published by the registered company Springer Nature Singapore Pte Ltd.
The registered company address is: 152 Beach Road, #21-01/04 Gateway East, Singapore 189721, Singapore

Paper in this product is recyclable.

Contents

Part II Historical and International Perspectives on the Development of Ethical Principles in Research Involving Humans

Part III Alternative Frameworks for Innovation and Drug Development Strategies

About the Editors

Introduction to Ethical Innovation for Global Health: Pandemic, Democracy, and Ethics in Research

Chieko Kurihara, BA Specially appointed Professor at Kanagawa Dental University, and Vice-Chair of the Certified Review Board/Senior Researcher at the National Institutes for Quantum Science and Technology, Japan. Editor of a Japanese medical journal *Clinical Evaluation*; a founding co-representative of the Center for Bioethics Policy Study, a non-profit academic study group. After graduation from the Department of Economics, School of Political Science and Economics, Waseda University, in 1983, she has been engaged in book/journal publications, editorial and theater works, and then in the study of bioethics, especially focusing on ethics of research involving humans. Not only working as a member of various types of ethics committees of Japanese academic societies, and of a Task Groups on GCP regulations designated by the Ministry of Health, Labour and Welfare, she has been engaged in developing international ethics documents or literatures as a member of several Task Groups of the International Commission of Radiological Protection (ICRP), as well as Ethics Working Group of the International Federation of Pharmaceutical Physicians and Pharmaceutical Medicines Development (IFAPP). Her main interests and responsibilities include bioethics studies, management of ethics committees, education and consultancy on research ethics, and quality assurance (audit) of academic research involving humans.

Dirceu Greco, MD, PhD Professor Emeritus of Infectious Diseases and Bioethics at the School of Medicine, Federal University of Minas Gerais (UFMG) in Belo Horizonte, Brazil. He received his MD degree and his PhD from UFMG, with further specialization in Clinical Immunology at State University of New York, Buffalo, and the University of London, England. Member (2018–2024), Brazilian National Health Council (CNS), representing the Brazilian Society of Bioethics (SBB). Member (2018–2024) and one of the vice-Chairs of UNESCO's (Paris) International Bioethics Commission (IBC). Chair (2018–2024), Brazilian Society of Bioethics (SBB). Member (2023 on), WHO Expert Group on ethics and governance of infectious disease outbreaks and other emergencies of public health importance. Member (2020–2023) WHO Access to COVID-19 Tools Ethics Working Group. Director (2010–2013) Department of STI, AIDS and Viral Hepatitis (Ministry of Health, Brazil). Member (2007–2010), Brazilian Research Ethics Commission (CONEP). Founding member and chair, UFMG Research Ethics Committee (1994–1998). Dean for Graduate Affairs (1994–1998), UFMG. Temporary consultant for many national/international institutions such as the Brazilian AIDS Programme, WHO, UNITAID, UNAIDS, UNICEF, UNESCO, CIOMS, World Medical Association (WMA), and the United States Presidential Commission for the Study of Bioethical Issues. Main topics of interest include infectious and parasitic illnesses, bioethics, public health, human rights, and clinical immunology.

Ames Dhai, MASSAf, PhD, MBChB, FCOG, LLM
Founder and Past Director of the Steve Biko Centre for Bioethics at the Wits Faculty of Health Sciences (2007–2019), Professor of Bioethics and Health Law at the Wits School of Clinical Medicine, Specialist Ethicist at the Office of the President and CEO of the South African Medical Research Council (SAMRC), member of the Academy of Science South Africa (ASSAf) and Chairperson of the UNESCO International Bioethics Committee. Some of her current activities include: Chairperson of the SAMRC Bioethics Advisory Panel, the SA Human Sciences Research Council Research Ethics Committee and the SA National Blood Services Research Ethics Committee, and the ASSAf Biosafety and Biosecurity Committee. She is a board member of the SA Medical Association; Vice-Chair on the Ministerial Advisory Committee for COVID-19 Vaccines; member of the Ministerial Advisory Committee on Unrelated Organ Transplants; and a member of the Advisory Committee to the SA Law Commission Project on Medical Litigation. She is editor-in-chief of the *South African Journal of Bioethics and Law* and Associate Editor of the *South African Medical Journal*. Professor Dhai started her career as a medical doctor, specialized as an obstetrician and gynecologist, and then moved into the disciplines of bioethics and health law.

About the Contributors

Part I Relevant Constructions from Global South and/or Asian Paradigms: Brazil, South Africa, Taiwan, South Korea, and Japan

Democracy Restoration in Brazil, *the Constitutional Guarantee of Health as a Right for All, Giving Rise to a Universal Health System* (SUS) and of a National Research Ethics Commission

Dirceu Greco (See "About the Editors")

Bernardo Galvão-Castro, MD, PhD Full Professor at the Bahiana School of Medicine in Salvador, Bahia, Brazil, and Emeritus Researcher at the Oswaldo Cruz Foundation (Fiocruz). He received his PhD in Immunology, University of Geneva, Switzerland (1977); MsC in Human Pathology, Federal University of Bahia (1974); Bachelor's Degree in Medicine, Federal University of Bahia (1969). Vice-Director of the Oswaldo Cruz Institute, FIOCRUZ, Rio de Janeiro (1981–1982); Vice-Director of the Gonçalo Moniz Research Center, FIOCRUZ, Bahia (2007–2008); Human Health and Medicine Postgraduate Course Coordinator, Bahiana School of Medicine and Public Health (2004–2015); Former International AIDS Society Governing Council Member; Former Member of the International Retrovirology Association (IRVA) Board; Member of the UNAIDS/WHO Steering Committee on HIV/AIDS Diagnostics and Reagent Development (1989–1991); Member of the UNAIDS/WHO Steering Committee on HIV/AIDS Vaccine (1998–2001); Member of the Brazilian HIV/AIDS National Committee (1985–2001); Member of the

Brazilian HIV/AIDS Vaccine Steering Committee (1991–2001); Member of the UNAIDS International Network for HIV-1 Isolation and Characterization (1991–2001); Coordinator of the Brazilian Network for HIV Isolation and Characterization (1993–2001); Member, Bahia Academy of Medicine; Member of the Academy of Sciences of Bahia; Member of the Brazilian Academy of Science.

Equitable Access to COVID-19 Vaccines, Vaccine Research, and Vaccine Apartheid on the African Continent: Challenges and Recommendations

Ames Dhai (See "About the Editors")

Response to COVID-19 Pandemic and Ethical Innovations in Taiwan

Ian Chen, MD, LLM, SJD A physician and a lawyer and is responsible for the human research protection program of the National Taiwan University Hospital, Taiwan. After graduation from Kaohsiung Medical School in 1991, he studied law and practiced both law and medicine in Taiwan, and later holds a Juries Science Doctor degree from Washington University in St. Louis, United States. He has also worked for the Ministry of Health and for the organization responsible for the accreditation of hospitals and other healthcare services in Taiwan. His research interests include research ethics and health law.

Yvonne Chen, PhD holds a PhD in Economics from the University of Wisconsin in the USA. She is the head of Shenandoah University Global MBA in Asia Pacific as well as the Chief Assessment Officer and Associate Dean for the School of Business in the USA. As Professor of Economics, Chen teaches healthcare economics. She regularly publishes in scholarly journals and has been a distinguished member of the Who's Who of American Women since 2005. In 2019 she was the recipient of the Albert Nelson Marquis Lifetime Achievement Award. In 2017 and 2020, she twice received the Wilkins Award, the highest recognition for faculty and administrators at Shenandoah University.

Daniel Fu-Chang Tsai, MD, PhD is a family physician and bioethicist. He graduated from the National Taiwan University College of Medicine (NTUCM) and earned his PhD in bioethics from the University of Manchester, UK. He is the founding professor of the Graduate Institute of Medical Education and Bioethics, a joint professor in the Department of Family Medicine and the Graduate Institute of Clinical Medicine, and the past Director of the Department of Social Medicine at the NTUCM. He is an attending physician in the Department of Medical Research, the director of the Ethics Center, the chairman of the Research Ethics Committee, and the executive secretary of the Clinical Ethics Committee at National Taiwan University Hospital. He is also the director of the Center for Biomedical Ethics at National Taiwan University, the past vice president of the International Association of Bioethics, and the past president of the Taiwan Association of Institutional Review Boards. He is a member of the Merk Ethics Advisory Panel and was awarded Honorary Membership by the UNESCO Chair of Bioethics in 2015, Goldman-Berland Lectureship in Palliative Medicine in 2019, and Hastings Center Fellow in 2021. His research interests include research ethics, clinical ethics, cross-cultural bioethics, and medical ethics education.

Between Truth and Profit: Scientific Misconduct Case of Human Cloned Embryonic Stem Cell and Revisiting Cases During COVID-19 Pandemic

Young-Joon Ryu, MD, PhD is Professor in the Department of Medical Humanities and Medical Ethics, as well as the Department of Pathology at Kangwon National University College of Medicine. He earned his MD degree from Kosin University and his PhD from Seoul National University. He witnessed research misconduct by his advisor, Hwang Woo-Suk, while studying at Seoul National University as a PhD graduate student, and reported it to MBC TV. Afterwards, he was forced out of the hospital where he worked and spent 2 years struggling on the streets. Even in oppression and hardship, he did not lose faith in science and continued his research. He continued his journey as a clinical specialist in pathology at Korea University and Asan Medical Center. He also obtained a degree in medical humanities, broadening his research interests to include ethics. He has served as an expert member of the fifth Presidential National Bioethics Committee, as well as the president of Korea Human Resources Bank and Korean Human Brain Bank. He is currently the research director of the Neuroethics Project of the National Research Foundation of Korea.

Therapeutic Misconception as the Basis for Vaccine Nationalism of Japan: A Historical Reflection and Perspectives for Global Public Health

Chieko Kurihara (See "About the Editors")

Takeo Saio, MD Department of Internal Medicine and Psychiatry, Fuji Toranomon Orthopedic Hospital; Chief, K&S Consulting Office for Occupational Mental Health. SMBC Nikko Securities Inc. Consultant in occupational health; Certified Specialist and Supervisor of the Japanese Society of Psychiatry and Neuropsychiatry (JSPN). Member of the JSPN Committees of Pharmaceutical Affairs; Suicide Prevention; and Occupational Health. Editorial Committee member of the Japanese journal *Clinical Evaluation*. After graduating from Gunma University (Japan) School of Medicine in 1989, he has practiced in psychiatry, gastroenterology, occupational health, and general internal medicine. He engaged in introducing the basic philosophy behind evidence-based medicine (EBM), public understanding of science (PUS), as well as ethics and regulatory science to Japanese people by way of publishing Japanese translation of books on background information of these areas, including *Evidence-Based Medicine Toolkit* (Badenoch D & Heneghan C, 2002.), *The Institutional Review Board Member Handbook* (Amdur RJ, et al., 2002), *The Resourceful Patient* (Gray JAM, 2002), *The Truth About the Drug Companies: How They Deceive Us and What to do About It* (Angell M, 2004), *Inside the FDA: The Business and Politics Behind the Drugs We Take and the Food We Eat* (Hawthorne F, 2005).

Part II Historical and International Perspectives on the Development of Ethical Principles in Research Involving Humans

The Declaration of Helsinki as a Living Document: Revisiting Its Principles in a Global Pandemic

Ramin W. Parsa-Parsi, MD, MPH Head of the Department for International Affairs at the German Medical Association (GMA) in Berlin, Germany. Prior to joining the GMA, he worked with Harvard Medical International in Boston, USA, as the Director of Health Policy. He has been a member of the Board of the European Doctors (CPME), the Council of the World Medical Association (WMA), and the WMA Medical Ethics Committee. He has chaired international work-groups for the 2013 revision of the WMA Declaration of Helsinki, the 2017 revision of the WMA Declaration of Geneva, and the 2022 revision of the WMA International Code of Medical Ethics. He holds a master's degree in public health from Harvard University, obtained a doctoral degree at the Institute of Clinical Pharmacology at the University of Heidelberg, Germany, and received his MD from the University of Cologne in Germany, where he also did his postgraduate medical education in Hematology/Oncology. He received additional training at medical schools and hospitals in the USA, France, and Switzerland.

Otmar Kloiber, MD, PhD has served as Secretary-General of the World Medical Association since 2005, after he left the German Medical Association as Deputy Secretary General. He holds an MD (1984) and PhD (1986) from the University of Cologne, was a postdoctoral fellow in the Department of Biochemistry at the University of Minnesota, Duluth, and a scientific research fellow at the Max Planck Institute for Neurological Research. He holds an honorary doctorate from the Victor Babes University, Timisoara, Romania, and was appointed Clinical Professor in Health Administration at the Brooks College of Health, University of North Florida, from 2009 to 2013. He is interested in the development of deontology under the influence of health system organization and its relation to the provision of medical care. He provided advice to numerous governments on medical ethics and socio-medical issues. His advocacy focus is on equitable access to quality healthcare for all people.

Hidden Medical War Crimes and the Emergence of Bioethics in Japan

Rihito Kimura, LLM, PhD is a faculty affiliate at the Kennedy Institute of Ethics, Georgetown University, and Professor Emeritus of Bioethics and Law at Waseda University, Tokyo, Japan. After graduating from Waseda University's Graduate School of Law, he taught at Chulalongkorn University, Thailand (1965–69), Saigon University, Vietnam (1970–72), then became an Associate Director and Professor at the Ecumenical Institute of the World Council of Churches and taught at the affiliated Graduate School of Ecumenical Studies, University of Geneva, Switzerland (1972–75). He became a visiting scholar at the Center for the Study of World Religions, Harvard University (1978–80). In 1980, he became a Director of International/Asian Bioethics Program at the Kennedy Institute of Ethics at Georgetown University. In 1987 he was appointed as a Professor of Bioethics and Law at Waseda University. He is on the Editorial Advisory Board of the *Journal of Medicine and Philosophy* (Oxford Univ. Press). He has served as an advisor to international organizations such as CIOMS, WHO, and UNESCO and chaired the Hospital Ethics Committee of Tokyo Metropolitan Government, and was a member of the Council for Welfare and Health Policy Science, Ministry of Health, Labor and Welfare, Government of Japan (1997–2003). He is also past president of Japan Association for Bioethics (2008–2011).

From Nuremberg to Helsinki: Historicizing the Codification of Post-War Research Ethics

Ulf Schmidt, DPhil Senior Professor of Modern History in the Faculty of Humanities at the University of Hamburg (UHH), founding-director of the Centre for the Study of Health, Ethics, and Society (CHES), and a Fellow of the Royal Historical Society. His research interests are in the area of the history of modern medical ethics, warfare, and policy in twentieth-century Europe and the United States. He is especially interested in the history of authoritarian regimes and modern dictatorships. He has published widely on the histories of Nazi Germany, human experimentation during the Cold War, the Nuremberg Doctors' Trial and the

Nuremberg Code, eugenics and euthanasia, chemical and biological warfare, and propaganda and persuasion. He is the author, among others, of *Justice at Nuremberg: Leo Alexander and the Nazi Doctors' Trial* (Palgrave Macmillan, 2004); *Karl Brand: The Nazi Doctor. Medicine and Power in the Third Reich* (Continuum, 2007); *Secret Science. A Century of Poison Warfare and Human Experiments* (OUP, 2015); *Propaganda and Conflict: War, Media and the Shaping of the Twentieth Century* (Bloomsbury, 2019), and *Ethical Research: The Declaration of Helsinki, and the Past, Present, and Future of Human Experimentation* (OUP, 2020). He has recently been awarded a six-year ERC-Synergy grant on "Taming the European Leviathan: The Legacy of Post-War Medicine and the Common Good."

CIOMS Research Guidelines: Considering the Needs of Developing Countries

Lembit Rägo, MD, PhD Secretary-General of the Council for International Organizations of Medical Sciences (CIOMS) since 2016. From 1983 to 1999 he was a Professor of Pharmacology and Clinical Pharmacology at Tartu University, Estonia. He was the founder and first Director-General of the Estonian regulator, State Agency of Medicines, during 1991–1999. In 1999 he joined the World Health Organization (WHO) in Geneva as Coordinator of the Quality Assurance and Safety of Medicines team. In 2013 he was appointed as the Head of the newly formed WHO unit of Regulation of Medicines and other Health Technologies, which combined all WHO's regulatory activities for medicines, vaccines, and diagnostics. He has also served as Secretary of the WHO's internal research ethics committee. During 2000–2016 he represented WHO as an observer in the Steering Committee of the International Council of Harmonization (ICH), and since 2016 he represents CIOMS as an ICH observer. He was also a member of the Lancet Commission for Essential Medicines (2015–2016), and of the Committee on Mutual Recognition Agreements in the Regulation of Medicines convened by the U.S. National Academies of Sciences, Engineering, and Medicine (2017–2018). Dr. Rägo has published numerous scientific articles including on several topics related to medicines regulation.

Monika Zweygarth, Dipl Translation DOZ is a technical editor at the Council for International Organizations of Medical Sciences (CIOMS). She has lived in various African countries and has over 25 years' experience in medical writing and data management. From 1993 to 2009 she worked at the School of Pharmacy of the Medical University of Southern Africa (now Sefako Makgatho Health Sciences University), where she was instrumental in completing a large number of publications and projects, some of which contributed to national progress in pharmaceutical services in South Africa. In 2010 she moved to Geneva, where she held various contracts with the Global Fund and the World Health Organization (WHO). In particular she provided continued support to the WHO unit of Regulation of Medicines and other Health Technologies. In 2018 she joined CIOMS, where she has supported the development of Working Group reports and other publications on topics related to bioethics, medical product development, and safety.

Ethics of Placebo-Controlled Trials: Historical Analysis Including Experiences During the COVID-19 Pandemic

Chieko Kurihara, Dirceu Greco, Ames Dhai (See "About the Editors")

Takeo Saio (See Part I)

Hiroe Tsubaki, PhD Director-General and Professor Emeritus of the Institute of Statistical Mathematics in Tokyo, Vice President of Transdisciplinary Federation of Science and Technology, and a member of International Academia for Quality. After he received his PhD degree from the University of Tokyo, as an applied statistician for quality management he worked in drafting statistical guidelines of clinical trials for the Ministry of Health and Welfare and licensing practices for new drug application as an expert member of the New Drugs Investigation Committee of the Central Pharmaceutical Affairs Council in the 1990s. He also conducted ethical monitoring of clinical trials as a member of the Central Ethics Committee for Strategic Research of the Ministry of Health, Labour and Welfare in the 2000s. He was awarded the Deming Prize Honor Award in 2021 for his social implementation of quality management concepts in various fields including clinical evaluation.

Post-Trial Access: Historical Analysis Considering the Experience of COVID-19 Pandemic

Chieko Kurihara, Dirceu Greco, Ames Dhai (See "About the Editors")

Our "WMA Declaration of Helsinki": Opinions and Proposals from Patient and Public for Research Ethics

Chieko Kurihara (See "About the Editors")

Keiko Inoue, Associate degree in Law is a breast cancer patient and a victim of medical malpractice. She serves as a board member of the Association for Medical Malpractice Victims, providing counseling and support for victims and raising issues for society. She has served as a member of Relief Operations Committee of PMDA (Pharmaceuticals and Medical Devices Agency), and several third-party committees on research misconduct. She currently serves on Ethics Review Committees and a Certified Review Board at a national university and a national research institute. From her own experience, she realized that the general public without a medical background needs basic knowledge to participate in various fields in medicine, and is currently an instructor at the Japanese Institute for Public Engagement (Ji4pe) study course where she studied.

Hiroto Kai is a patient of rare blood disease and is presently an undergraduate student at Aichi University of Education. The affliction has necessitated medical intervention for a duration exceeding 5 years. He is pursuing his academic interests in the field of bioethics, drug development processes, and science pedagogy. He has received fundamentals of the process of pharmaceutical medicine at the Japanese Institute for Public Engagement (Ji4pe). Furthermore, he has augmented his knowledge base via coursework offered in the EUPATI Open Classroom, where he focused on medical research and development. He holds memberships with the Chemical Society of Japan and the Japanese Society on Thrombosis and Hemostasis. In the spring of 2023, he commenced his studies at Aichi University of Education, where he is conducting research on bioethics and pharmaceuticals from an educational and young perspective.

Katsura Suzuki, Pharmacist PwP (Person with Parkinson), involved in drug idea generation and discovery research for 27.5 years in ASKA Pharma. Member of the Japan Institute for Medical Development (Ji4pe). Engaged in facilitation of shared decision-making. Member of an international patient network, PD Avengers, research committee. Founder of the Life Design Institute for PwP Japan, the aim of the hub in Japan of PD Avengers. Member of the Patient Council of the World Parkinson's Coalition from Jan 25th, 2022, trying to define common principles for clinical research, and those common principles will be written by this group. Examples might include how to handle data, how to involve people with Parkinson's in the design and conduct of studies, and how to reword studies so that people with Parkinson's are referred to as partners or participants rather than subjects. Started a doctorate course at Juntendo University Graduate School of Medicine, neurology, in April 2022. Pharmacist license holder.

Haruko Saeki, BA Director, My Informed Consent. After graduating from the Osaka University of foreign studies, Haruko Saeki worked for a medical congress organizer. When she lost her father to pancreatic cancer, she felt uncomfortable communicating with her father's doctor. She could not point out what the feeling was at that time. Then, with her husband's business, she left Japan and lived in Milan, Italy, for several years. There she became acquainted with palliative care staff and a volunteer-group leader. In Italy, telling the truth (diagnosis) is less important than caring the patient with dignity. The communication gap with doctors in Japan was caused by different ways of thinking about Quality of Life and dignity of patient. The patient's wishes were often disregarded. If medical staff make much of patient's own decision, there would be closer communication. Upon returning to Japan, she started a lay person's group specialized in communication training for medical staff. In addition to her role in ethics committees, she also serves as consumer representative for Cochrane Japan.

Yoshikazu Funabashi, BEng graduated from Keio University in 1976 with a Bachelor of Engineering. After graduation, I worked for the Japan branch of a major American IT company as an SE/IT Specialist and Marketing Specialist. However, during my employment, I had bladder cancer and left the company 2 years later. I also recently had prostate cancer. As a peer supporter, I joined the NPO Cancer Patients Support Organization in 2014 and received training as a peer supporter. Furthermore, in order to improve the quality of peer support, I studied to become a Japan Society of Clinical Oncology Cancer Network Navigator in 2020 and received certification as a result. In 2023, the Certified Network Navigator Committee appointed me to serve as an external member of the Navigator Working Group. Currently, as a Cancer Survivor, I provide peer support at hospitals in Tokyo and Yokohama, as well as at a hospital in Kanagawa Prefecture as a part-time employee.

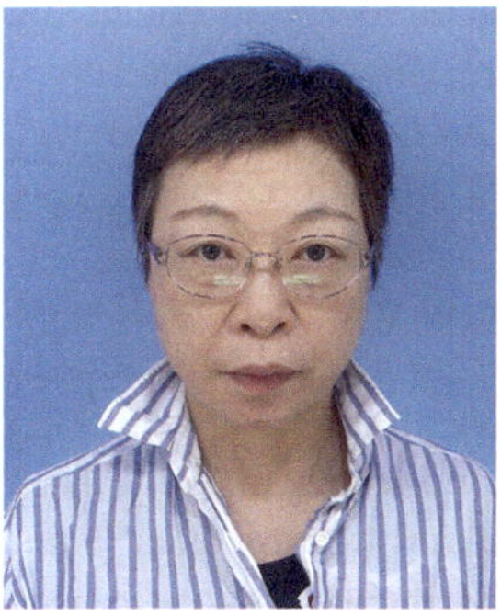

Noriko Kishi, BA Executive Director of CMT Association JAPAN (J-CMT). She was diagnosed with Charcot-Marie-Tooth disease in 1993 and joined the founding J-CMT in 2008. After graduating from the Department of Literature at Rikkyo University in 1983, she became the editor-in-chief at the Japan Finance News until 2014, then served as a senior researcher at the Japan Credit Information Reference Center Corp. from 2014 to 2020. Since 2020, she has been promoting project planning to improve the QOL of all intractable disease and disabled parties. Her academic affiliation includes the Japan Society for Disability Studies, the Japan Society of Social Design Studies, and the Japan Society for Business Ethics. The main organizations she belongs to as a member are the Japanese Institute for Public Engagement (Ji4pe), the Japan Fundraising Association (JFRA), the Stanford Social Innovation Review Japan (SSIR-J), and the Social Impact Management Initiative (SIMI).

Akemi Kuge, MA in Social Welfare Freelance social worker. After completing graduate school at Osaka Women's University, she worked as a law clerk and a writer for a local government women's policy public relations magazine and a consumer cooperative's bulletin. From her own experience, she became interested in the welfare of children who have not attended school. While working as a support worker for the mentally disabled and as a committee member of local government and social welfare councils, she obtained qualifications as a certified social worker and a mental health social worker. She has worked as a school social worker for 6.5 years. Participated in the establishment of a family peer support group for children and adolescents with mental disorders and served as its president for 5 years. She focuses on adolescent mental health welfare and suicide prevention education. Currently, she aims for the social restoration of all people with difficulties in life and their families, and the realization of empowerment, liberation, and well-being, including those who support them. She is also trying to promote networking and advocacy in education, welfare, healthcare, medical care, labor, justice, and community life as the environment for this, through activities that are not bound by position or affiliation.

Toshie Murakami, Bachelor of Psychology I work part-time as a peer supporter at Sagamihara Kyodo Hospital. As a survivor of cervical and breast cancer, I have been dedicated to peer support since 2007, grateful to be alive. In 2017, I obtained certification as a Senior Navigator from the Cancer Medical Network of the Japan Society of Clinical Oncology (JSCO), feeling the importance of ensuring the quality of peer support. Since then, I have served as an external committee member of Cancer Care Network Navigator Certification Committee of JSCO. I have also had the opportunity to present the need for peer support at JSCO annual meeting for three consecutive years. Related Committees: External Committee Member, Cancer Care Network Navigator Certification Committee of the Japan Society of Clinical Oncology (JSCO); Committee Member, Kanagawa Prefecture Cancer Control Review Board. Patient Groups: Representative of facilitators, Sagamihara Kyodo Hospital Cancer Patients' Association Fukiiso; Vice Representative, Cancer Support Sagamihara; Representative, Cancer Philosophy in Sagamihara F.

Yoshiko Saito, Master of Social Design is a breast cancer survivor and 2018 Master's graduate in Social Design at Rikkyo University. She completed her research on expanding peer support networks for cancer survivors in hospitals throughout Japan, and now dedicates her time to helping others in her community. After suffering from breast cancer in 2015, Saito participated in the Breast Cancer Survivor Coordinator Program sponsored by NPO Cancer Net Japan to learn more about the medical care system and the mental implications of breast cancer to serve others in similar situations. In 2016, she completed a training course at NPO Cancer Patients Support Organization. With her two qualifications, she now works as a peer supporter at both private and government-funded medical institutions in the greater Tokyo area. Over the last 10 years, Saito has moved around the globe, teaching Japanese in Asia before returning to Japan. Saito graduated from Rikkyo University in 1986 with a BA in Christian Ethics and Middle and High School Education in Social Studies.

Eiko Uchida President, Breast Cancer Patient Support Group BOUGAINVILLEA (in charge of breast cancer), Facilitator, Patient Voice Council (in charge of Basic Act on Medical Care), Vice President, Nangan Net (in charge of all cancer and intractable disease patients), Town Net Cafe (engaged in Local Civic Activities). In January 1994, while in Singapore, I experienced total breast cancer resection, chemotherapy, and breast reconstruction. Having encountered such medical care in Singapore that emphasizes patient self-determination and information disclosure, promoting informed consent, I established the Breast Cancer Patients' Association in January 1998 after returning back to Japan. It was out of gratitude to and requital for medical providers who saved my life. In our activities, we set the goal of the International Federation of Patient Organizations as the five principles for measuring patient centricity in healthcare, supporting the final outcome of patient and family to be their satisfaction of feeling of "living my own life even if having cancer." I am also working for the enactment of the "Basic Act on Medical Care," which should be the parent law of the healthcare system. Award: Received a letter of appreciation from the 23rd Annual Meeting of the Japan Breast Cancer Society. Received the Excellence Award at the Tokyo Metropolitan Women's Advancement Awards.

Naoki Tsutsumi, PhD is CEO of Tutti Quality Assurance Network, a consulting firm specializing in clinical research. He has more than 30 years of experience working for pharmaceutical companies as CRA and Study Manager in Bayer and as Global Auditor in AstraZeneca. He was a Project researcher at the University of Tokyo from 2020 to 2021 and a research fellow at the National Center of Neurology and Psychiatry from 2021 to 2023. He received a PhD from the University of Tokyo for his research on paradigm shifts in clinical research. He has published a series of academic papers regarding Patient and Public Involvement (PPI) and Remote Quality Assurance Method. He is also one of the authors of a widely used reference book in Japan named *Pocket Handbook for Multinational Clinical Research*. He has been involved in program development and implementation as a board member of the Japanese Institute for Public Engagement (Ji4pe), which offers a variety of educational programs for patients, citizens, academia, and the pharmaceutical industry. He was a committee member of Academy Global Committee of Association of Clinical Research Professionals (ACRP) from 2012 through 2018 and served as a member of Academy Board of Trustees in ACRP headquarter from 2020 to 2022.

Kyoko Imamura, MD, DrMedSci, PhD is the past president of IFAPP, after completed her presidency in 2018–2020. She is an orthopedic surgeon and holds MD, Dr of Medical Science (clinical pharmacology), and PhD (outcomes research). Since joining the pharmaceutical industry in 1995, she has built up her career in R&D and Medical Affairs and held management positions in international pharmaceutical companies. At the same time, she joined Japanese Association of Pharmaceutical Medicine (JAPhMed), the Japanese member association of IFAPP, and served as the president from 2009 to 2016. Since 2017, she has been engaged in academic career, first as the project professor of Pharmaco-Business Innovation, and later as the professor of Social Cooperation Program of IT Healthcare, at Graduate School of Pharmaceutical Sciences, The University of Tokyo. She has led research projects to promote innovative clinical trials including virtual (site-less) clinical trials, promoting pharmacists'

leadership in community, patient and public involvement, and real-world data studies. In 2020, she developed Japanese Institute for Public Engagement (Ji4pe) as the non-profit organization and has promoted social engagement in drug development and health policy projects as the founding President.

Part III Alternative Frameworks for Innovation and Drug Development Strategies

Medicines Development for Global Health: Learning from COVID-19 Vaccines R&D

Varvara Baroutsou, MD, PhD, GFMD, EMAUD President IFAPP. With over 35 years of experience as Consultant in Internal Medicine, and Pharmaceutical Medicine Expert, Varvara Baroutsou led Research & Development practice where she focused on improving medicines development, translational research, clinical trial operations, and evidence generation for innovative treatments. She has started her career as a Hospital Physician & Clinical Investigator and then moved on at senior leadership roles in Clinical Development and Medical Affairs of several large pharmaceutical and biotechnology companies in Europe. Currently she is affiliated with not-for-profit scientific organizations, academia, patients' associations, and multinational organizations to implement complex change management initiatives to pursue excellence in clinical research, patient engagement, and responsible ethical conduct with her latest role being President of International Federation of Associations of Pharmaceutical Physicians and Pharmaceutical Medicine (IFAPP). Varvara's academic records include scientific journal reviewer and editor experience, publications, book chapters and advisory contributions to educational, technical, and regulatory standards committees. She holds a Doctor of Medicine degree from the University of Athens, Greece, is a licensed certified Consultant, a Global Fellow in Medicines Development with a postgraduate degree in Health Economics from Stockholm School of Economics, and holds a European Market Access Diploma from the University of Lyon.

Development of Portfolios and Pipelines of Drugs for the Treatment, Prevention and Control of Neglected Tropical Diseases

Kiyoshi Kita, PhD was a professor (1998–2016) at the University of Tokyo and is Dean of Nagasaki University, School of Tropical Medicine and Global Health (2015-). He was educated at the Department of Biological Sciences, the University of Tokyo, and graduated in 1980. He moved to the Department of Parasitology, Juntendo University, School of Medicine (1983). Then, promoted to associate professor of the Department of Parasitology, the Institute of Medical Science, the University of Tokyo (1991–1998). Prof. Kita has been studying bacterial and mitochondrial respiratory chains from the viewpoint of oxygen homeostasis and found that mitochondrial fumarate reductase plays an important role in the parasitic adaptation and cancer cells. Furthermore, he developed several promising anti-helmintics, trypanocidal, and anti-COVID-19 drugs. He has been dispatched by JICA as a team leader of medical cooperation project to Paraguay (1984–1985). Thus, his research has contributed not only to basic biology but also directly to human health. He was President of Japanese Society of Parasitologist (2003–2006) and was President of Japanese Biochemical Society (2009–2011). He was awarded the 17th Award of Duke of Edinburgh by the Japan Academy in 2020.

Haruki Yamada, PhD is a Professor Emeritus in Kitasato University/the Kitasato Institute and was Chair of the Board, DND*i* Japan (2009–2023, now Board member). He is currently also Chief Director of Kitasato Research Center for Environmental Science and Visiting Prof. of Tokyo Univ of Pharmacy and Life Sciences. He worked for 31 years in the Kitasato Institute under his mentor Prof. Satoshi Ōmura (2015 Nobel laureate in Physiology and Medicine) as Directors of Research Division at Oriental Medicine Research Centre and WHO Collaborating Centre for Traditional Medicine in the Kitasato Institute, and Prof. and Director at Kitasato Institute for Life Sciences and Dean of Graduate School of Infection Control Sciences in Kitasato Univ. He also involved in DND*i* as the Scientific Advisory Committee Member for 7 years when DND*i* was founded in 2003.

Fumiko Hirabayashi, RPh joined the Drugs for Neglected Diseases *initiative* (DND*i*) in 2004 and led its operations in the country as DND*i*'s Japan Representative, with responsibility for building research collaborations and platforms for the development of new drugs against neglected diseases. She is a board member of DND*i* Japan since 2016. Currently, she is also an affiliated researcher of the School of Tropical Medicine and Global Health, Nagasaki University. Before joining DND*i*, she worked for Médecins Sans Frontières (MSF) in the field and served as the Japan Coordinator of MSF's Access to Essential Medicines Campaign when she supported the establishment of DND*i*. Initially, she spent 8 years with Roussel Uclaf and Kaken Pharmaceutical in Tokyo, and also worked in Southeast Asia, including Hong Kong, Thailand, and Malaysia. Ms. Hirabayashi graduated in pharmaceutical science from Hoshi University.

Simon L. Croft, BSc, PGCE, PhD, FRSB, DSc is an Emeritus Professor in the Faculty of Infectious and Tropical Diseases at the London School of Hygiene & Tropical Medicine (LSHTM) and an Advisor to programs on leishmaniasis and Skin NTDs. He worked on the discovery and development of anti-infective drugs for over 40 years in academia, industry, and public-private partnerships (PPPs). Funding from WHO, EU, MRC, UK DFID, Medicines for Malaria Venture, and the Bill & Melinda Gates Foundation gave him the opportunity to engage in research from the discovery and development of novel drugs and formulations for the treatment of leishmaniasis, malaria, human African trypanosomiasis (sleeping sickness) and American trypanosomiasis (Chagas disease), to operational research on disease control and implementation. Projects on miltefosine, AmBisome®, and topical paromomycin reached clinical trials for the treatment of leishmaniasis. More recent research interests included PK-PD relationships, predictive models for drugs and vaccines, topical formulations, and drug resistance. From 2004 to 2007 he was the first R & D Director of the Drugs for Neglected Diseases *initiative* (DND*i*), Geneva, and from 2008 to 2014 he was Dean of Faculty of Infectious and Tropical Diseases at the LSHTM.

Patient and Public Involvement (PPI) and Pharmaceutical Development Through Open Innovation Processes: Recent Activities

Kotone Matsuyama, BPharm, RPh Chair of the Ethics Working Group, IFAPP, since June 2021; Community Lead of Bioethics, Drug Information Association (DIA) Japan, since 2022. She is Professor at the Department of Health Policy and Management, and Vice President of the Center for Strategic Research Initiative, at Nippon Medical School since 2017. She graduated from Kyoto University Faculty of Pharmaceutical Science and obtained a license as a pharmacist. She started her Research and Development (R&D) career as a data manager in 2003 and has worked as a project manager since 2005. She spent 12 years at the Translational Research Informatics Center, Foundation for Biomedical Research and Innovation, Kobe, and 2.5 years as a Lecturer at the Center for Quality Assurance in R&D, Kyoto Prefectural University of Medicine. She has managed or supported many R&D projects initiated by academia, including product seeds management, clinical trial protocol development, not only for drugs but also for medical devices and regenerative medicines. She specializes in clinical pharmacology and project management. She is also in charge of research ethics and governance within her university.

Naoki, Tsutsumi, Keiko Inoue (See Part II)

Noriko Iwaya, Degree in English Literature is the mother of a son with a rare disease. She graduated from a junior college and joined a domestic bank, where she was in charge of deposits, disbursements, foreign exchange, and human resource development. In 2006, she headed the administrative department of the Japanese Red Cross Health Care Center. In 2017, she switched to the position of course secretary at a university medical school. At the same time, she established an intractable disease support group, Intractable Disease Support Familia Yamaguchi, which transcends the boundaries of diseases, and continues to plan and operate an intractable disease café. In 2019, she began learning basic knowledge about medicine and drug development after a case in which her son did not participate in an international joint clinical trial. In 2020, she established Association for Congenital Thrombotic Thrombocytopenic Purpura and became its Executive Director (now retired). She has accumulated a wide range of knowledge and practice in the promotion of Patient and Public Involvement, and serves as a mock patient at a pharmacy school and as a member of an Ethics Review Committee. She is also involved in educational activities to promote understanding of e-Sports, the path her son has chosen. A member of the secretariat of Japanese Institute for Public Engagement (Ji4pe). Advisor of Yamaguchi e-Sports Association.

Kyoko Imamura (See Part II)

Introduction to Ethical Innovation for Global Health: Pandemic, Democracy, and Ethics in Research

Chieko Kurihara, Dirceu Greco, and Ames Dhai

Abstract This volume captures the recent evolution of ethics in research involving humans with the experience of COVID-19 pandemic and provides future directions to achieve alternative drug development strategies for equitable global health. It describes the historical context, illustrates the processes in Global South and Asian paradigms to achieve democracy after World War II, how the framework of health research and development could contribute to ensuring fundamental human right to health. The construction is organized into three themed parts: case examples from Global South and Asian regions; historical and international perspectives of principles of ethics in research; and alternative frameworks of clinical development and innovation to achieve global health for all, particularly, those who most need the benefit of scientific achievements.

Keywords Pandemic · Democracy · Right to health · Ethics in research · Global health

1 Purpose of This Publication and Its Background

The purpose of this publication is to clarify an emerging framework of ethics in research and strategies of innovation for global health, in the world where we have experienced COVID-19 pandemic and must prepare for future threats that are prone to happen repeatedly. The issues are not limited to the pandemic situations, but our

C. Kurihara (✉)
Kanagawa Dental University, Yokosuka, Kanagawa, Japan
e-mail: chieko.kurihara@nifty.ne.jp

D. Greco
Department of Internal Medicine, School of Medicine, Federal University of Minas Gerais, Belo Horizonte, Minas Gerais, Brazil

A. Dhai
School of Clinical Medicine, University of the Witwatersrand, Johannesburg, Gauteng, South Africa

C. Kurihara et al. (eds.), *Ethical Innovation for Global Health*,
https://doi.org/10.1007/978-981-99-6163-4_1

goal is to achieve global health for all, especially ensuring the rights of the most vulnerable. The premise of this framework and strategies must be reassurance of globally equal, fundamental human rights [1–3], including the right to health [4].

Most authors of this volume participated in a webinar [5] held 6 months after the first COVID-19 vaccine was developed, where they seriously discussed distributive justice as the one essential hand of successful scientific endeavors. A planning decision for this volume was made 6 months later, and most authors contributed during the ongoing pandemic through 2022, with the last manuscript completed considering the information in May 2023. While this unprecedented pandemic is on its way to being brought under control in high-income countries and even with the World Health Organization (WHO) declaring on May 5, 2023 the end of COVID-19 as a public health emergency [6], there are people across the planet who are, and remain deprived of the benefits of science. Each author described their perceptive awareness of this situation. For the conquest of any disease, not limited to cases of pandemics, research involving humans for scientific development is essential. Research and development programs must be carried out ethically and democratically and aimed at global equality being assured for all. This publication outlines a roadmap in this direction to reform, strengthen and improve ethics in research and health development strategies.

Since the international agreement on inalienable human rights was declared shortly after World War II [1–3], a substantial number of policy documents on ethics in research and on drug development methodologies have been produced, being initially led by Western countries. Simultaneously, people in low- and middle-income countries (LMICs) and/or non-Western regions have been fighting for the establishment of democracy, for decolonization, and for the termination of military regimens. These countries/communities have also faced exploitation of vulnerable populations, with case examples of drug development by the global pharmaceutical industry, with products developed made available only or primarily to wealthy populations.

We saw, in the COVID-19 pandemic, unacceptable inequality of resource allocation that sharpened the acute deterioration of social determinants of health [7, 8]. The syndemic approach [9, 10] reveals important biological and social interactions that affect prognosis, treatment, and health policies. It focuses on the biosocial complex, characterized by biological and social interactions between conditions and states, which increase a person's susceptibility to harm or worsen their health outcomes. This notion was first used for the AIDS pandemic, but it can be applied to the COVID-19 pandemic [11, 12].

Furthermore, the situation revealed the causes of this pandemic to be an increasing disharmonious human way of living with the expansion of human interactions with nature in a globalized industrial society with continuous damage to the ecosystem [13]. Many warned that climate change would facilitate the outbreak of a next pandemic [14, 15]. Globally, ethicists and public health professionals have articulated, put in effort and advocated for both the necessary conceptual and active reform. In the field of ethics in research and drug development strategies, we must emphasize the following notions proclaimed as voices from LMICs:

1.1 No One Is Safe Until Everyone Is Safe [16, 17]

This statement is crystal clear and was made in the context of COVID-19 vaccines, meaning that it will be impossible to control the pandemic unless the vaccines already authorized were equitably distributed to all countries, with emphasis on distribution to the most vulnerable populations/communities. To this end, the COVAX Initiative (GAVI, WHO, CEPI) [18] was a good start as it involved around 190 countries, with the participation of high-income countries (HICs) contributing to access for LMICs. However, COVAX needed to be properly financed to actually reach its objectives, which were modest, as it initially proposed at least 20% of the population of each participating country was immunized [19]. Subsequently the global target was reviewed by WHO to 40% total population coverage by the end of 2021, 70% by mid-2022, just from COVAX, but neither was achieved, especially in LMICs [20]. Thus, other initiatives, such as patent waiving and technology transfer to facilitate production and deployment of affordable vaccines, were required.

Challenges This was a wake-up call to high-income countries (HICs) for them to understand that they could not achieve their health without ensuring the health of those who are worst off. The warning is indeed a steppingstone to ensure that the right to health is universal, with special attention to the most vulnerable populations. Even more important is that of learning to realize that ensuring the health of the most disadvantaged and respecting their dignity are values in themselves, regardless of whether or not these directly affect the privileged populations.

1.2 COVID-19 Vaccines as a Global Public Good [21, 22]

This is a notion associated with the proposed waiver of relevant provisions of the Trade-Related Aspects of Intellectual Property Rights (TRIPS) for COVID-19 related products emanating from South Africa and India [23] and recognized and acknowledged by many other countries, United Nations Educational, Scientific and Cultural Organization (UNESCO) and other international authoritative organizations.

Challenges Immediate suspension of patents is not a single solution and must include, as proposed in October 2020 by South Africa and India, initiation of domestic production of necessary medical products and rapid scaling up of manufacturing capacity. We have seen some initiatives to make local production a reality, so that COVID-19 vaccines and other new technologies are accessible for resource-constrained countries [24]. However, the decision by the WTO was limited to the COVID-19 vaccine [25]. Of note, the WTO did not accept the original proposal for waiver of a broader range of products, "*in relation to prevention, containment or treatment of COVID-19.*" Considering the criticisms [26], the WTO decided to continue discussion on whether or not the waiver could cover COVID-19 diagnostics and therapeutics [27].

In addition, although the waiver of the patent is considered by some as "*treatment of symptom,*" [28] it is fundamentally a start in the sense of recognizing that "*health is not an economic commodity but truly a right for all*" [29].

These notions, which resonate prominently throughout this publication, include and expand the landscape of ethics in research and are in line with the United Nations **Sustainable Development Goals (SDGs)** [30].

2 Structure of This Volume

To achieve the afore-mentioned objectives, this book is organized into three thematic parts:

2.1 Part 1 Relevant Constructions from Global South and/or Asian Paradigms: Brazil, South Africa, Taiwan, South Korea, and Japan

This part focuses on the process in Global South and/or Asian countries in their struggle to achieve democracy after World War II. It covers how the framework of ethics in research was established in those regions and discusses policies implemented to protect research participants from exploitation during new drug development and the ethical requirement of access to treatment for all who need it. A robust framework of ethics in research and concurrent challenges experienced during the pandemic is discussed.

Brazil has achieved an egalitarian, inclusive, accessible to all, universal health system SUS (Sistema Único de Saúde—Brazilian National Health System): It is highly unlikely that the Brazilian response would have advanced in the manner it did without SUS, which was genuinely a major factor in changing Brazilian public health, with their roots established in 1986 at the eighth National Health Conference (NHS). This NHS proposed the bases for the "Health" section of the 1988 Brazilian Constitution, defining Health as "everyone's right and a duty of the State." In 1990 Law 8.080 founded SUS, based on the principles of Universality, Equality, and Integrality in healthcare for everyone, with societal control and user participation playing a decisive part [31]. This was achieved just a few years after the reestablishment of democracy in 1985, after 21 years of a military dictatorship. The Brazilian SUS was fundamental to confront, the AIDS pandemic, providing diagnosis, care, and treatment for all who need it. Conversely, policies during the COVID-19 pandemic put in place by a neoliberal, right-wing government can be considered antithetical, as it denied the severity of the pandemic, negated scientific evidence, and postponed the deployment of the much-need vaccines [32]. Moreover, comprehensive and protective research ethics directives were established in 1996 [33], with exemplary human rights protection for the research participants, ensuring post-trial access and forbidding placebo use when an effective comparator exists.

South Africa abolished apartheid in1994 and implemented in its Constitution in 1996 [34], the UN International Covenant on Human Rights [3] to prohibit experiment without informed consent. The country, being most affected by COVID-19 on the continent pursued the quest for a globally equitable allocation of COVID-19 vaccines. The first technology transfer hub for mRNA vaccines was established in South Africa to scale up production for vaccination on the African continent. It also provides training and licensures for manufacturers in LMICs. Strong partnerships have been developed with the WHO, COVAX initiative, African Centers for Disease Control and Prevention (African CDC) as well as Medicines Patents Pool. To overcome the global situation deemed as "vaccine apartheid," South Africa and India made a proposal for a TRIPS Waiver for COVID-19 related products [23]. Despite some prominent achievements, Africa has also experienced other infections and diseases, which emerged, along with numerous challenges such as lack of well-established public health systems, vaccine hesitancy, and political corruption.

South Korea and Taiwan broke free from Japanese colonialism and established democracies in the 1980s and 1990s, respectively, after overcoming autocratic regimens that lasted while the world experienced the Cold War era [35]. Taiwan established the Human Research Act and the Biobank Act to protect the rights of indigenous people [36] and the country provides additional protection to indigenous research participants in its legal framework. Taiwan's rapid COVID-19 response demonstrates its historical track record, including experiences with the SARS outbreak, other tropical diseases, and the threat of military invasion from China.

South Korea expanded its Bioethics and Safety Act in response to the scandal of scientific misconduct in embryonic stem cell research [37, 38]. The first full and in-depth whistleblower testimony of this scandal is published in this volume, while the author's first attempt of disclosing his experience was reported by Nature [39] and Science [40] in 2014. It reveals the powerful force of politics and honor-seeking greed at the expense of distorting science. The tough difficult struggle of resistance by one individual later expanded to include solidarity of the community of scientists. It led to the victory in the Supreme Court's decision to uphold belief in scientific truth in this particular case, although similar misconducts continue unabated.

Japan reveals a negative example in failing to establish the foundation of human rights of research participants and insufficient reflection of past war crimes of human experimentation. This has led to the government's COVID-19 response characterized as "therapeutic misconception" (misinterpreting "research" as being a "treatment") to promote off-label use of drug, as well as "vaccine nationalism." Japan panic purchased vaccines more than seven times of the entire Japanese population, despite not participating in international placebo-controlled clinical trials to evaluate these vaccines.

2.2 Part 2 Historical and International Perspectives on the Development of Ethical Principles in Research Involving Humans

Part 2 shows how international agreements on human rights and ethics in research were established with the war crimes condemnation of human experiments during World War II by Germany, called Nuremberg Code [41]. Human experiments were also performed by the military of the Japanese Empire [42].

Other examples include The Belmont Report [43], issued in the USA, which is recognized internationally; The World Medical Association's Declaration of Helsinki [44] and the Council for International Organizations of Medical Sciences (CIOMS) International Ethical Guidelines for Health-Related Research [45], and its 2021 publication, Clinical Research in resource-limited settings [46], along with achievements and challenges of these organizations during the COVID-19 pandemic.

Most notably, the HIV/AIDS pandemic that emerged in the 1980s facilitated powerful patient and civil society involvement in human rights advocacy. In the USA, the controversy over access to investigational new drugs (IND) led to a consolidated network of clinical trials and written agency policies by the United States Food and Drug Administration (FDA) to provide unapproved drugs outside the clinical trial projects [47]. After the development of safe and effective therapy for HIV/AIDS in the 1990s, enduring debates about ethics in research emerged, especially on the ethics of placebo-controlled trials and post-trial access, with conflicting values of utilitarianism versus deontology, or otherwise, pluralism or universalism. When there is an established intervention, under what conditions can a placebo-controlled trial be ethically justified? How can post-trial access be provided not only for trial participants, but also for everyone who needs it? These important issues were reflected upon in modifications of international policy documents such as the WMA DoH [44], CIOMS Guidelines [45], and International Council for Harmonisation (ICH)'s guidelines [48].

The topics of placebo use, post-trial access, as well as access to unproven or proven products and resource allocation have been revisited in the COVID-19 pandemic [5]. It should be noted that the WMA has initiated a new revision process for the 2013 version of the DoH. This is planned to be finalized in 2024. Although the publication of this volume has not seen the conclusion of the DoH revision, some of the authors have been involved in the debates or facilitating the discussion on these topics. Notably, a group, the majority of whom are patients and/or the public, contributed to this part with their views on the necessary revision of the DoH.

2.3 Part 3 Alternative Frameworks for Innovation and Drug Development Strategies

Part 3 presents ethical frameworks and innovation alternatives for global health. For many years, marketing-oriented "seeding" trials [49] by global drug companies and impact factor competition in academic research institutes have been dominant. The COVID-19 pandemic has brought to light the limitations of this "closed innovation." Emergency health situations, such as HIV/AIDS and COVID-19, have promoted "open innovation" and "open science" [50], along with patent waiver exemption policies and technology transfer. HIV/AIDS and unequal access to needed treatment underpinned the approval in 2001 of the World Trade Organization (WTO) Doha Declaration on the TRIPS agreement and public health [51]. It was agreed that the TRIPS agreement cannot prevent any member countries from taking measures to protect public health, in particular to promote access to medicines for all. This important intellectual property issue has recently been revisited, starting with South Africa and India's proposal of the TRIPS Waiver. While this achievement was limited to COVID-19 vaccines [25], it suggested a future direction for our global society to pursue to facilitate the protection of public health globally.

Alternative strategies for successful development of vaccines and therapeutic agents for COVID-19 with adequate evaluation of efficacy and safety have been seen as a result of collaboration between private–public organizations and civil society. Such a model was already achieved by non-government organizations such as Medicines Sans Frontieres (MSF) and the Drugs for Neglected Diseases *Initiative* (DND*i*). An innovative model of product development partnerships (PDPs), among institutions from around the world that share a common objective of product development was formed to achieve access to health for the most vulnerable neglected populations. In addition, emerging patient and public involvement (PPI) strategies are also discussed. These collaborative efforts among relevant stakeholders re-enact the notion of the primacy of fair distribution of the results of scientific innovation to those who are most in need of the benefits.

This and other parts also engage with the latest scientific knowledge and methodologies to achieve possible solutions, such as big data collection, storage, analysis, and democratic use of health databases and biobanks, and artificial intelligence (AI) research. The occurrence of the COVID-19 pandemic, in an environment with advanced information technologies promoted rapid disclosures of a substantial number of research results, including preprints, although often prematurely evaluated. It included disclosure and sharing of individual raw data. Such rapid electronic communication produced beneficial outcomes but on the other hand contributed substantially to an "infodemic," that is, the spread of too much information, including false or misleading information, both in digital and physical environments, during the infectious disease outbreak [52]. In addition to the so-called evidence-based policy (based on empirical research results), our society has accelerated "data-driven research" and "data-driven policy making" (based on data being generated) in the emergency situation. In this environment, we need to reconstruct and expand our ethical framework and innovation strategies to protect study participants at large while at the same time, ensuring the availability of scientific results in the global public health system.

3 Historical Analysis on Pandemics and Ethics in Research (Table 1)

Our experience with COVID-19 and other pandemics is analyzed and discussed in each chapter. COVID-19 pandemic is an addition to the recurring infectious threats occurring in our history, and it is inevitable that we will face newly emerging and re-emerging serious health threats [13]. Devastating plagues in the ancient and middle ages caused a significant decline in population. Both peaceful commerce and wartime conflicts have contributed to spreading infections and changing the social fabric. Pandemics have repeatedly witnessed the spread of disinformation and increased discrimination. The "Black Death" in medieval Europe (which resulted in 50 million deaths) in addition caused massacres of Jews, irrationally blaming them for poisoning the wells [55].

Over the past century, several events have killed tens of millions of people, such as the 1918 influenza pandemic (1918, 50 million) and HIV/AIDS (1989, 37 million) (Table 1) [13]. Currently, SARS-CoV 2, which was first reported in December 2019, has been responsible for nearly seven million cumulative deaths (as of the end of May 2023) [56]. It should be noted that the HIV/AIDS pandemic, even with its devastating impact, has revolutionized the innovative global health approach to drug development, the social and cultural behavioral change for prevention and boosted civil society participation in the fight against discrimination and stigmatization [57]. Unfortunately, despite the successful development of new prevention tools, and diagnostic and therapeutic products, their access continues to remain unequally distributed [57]. To neutralize this disparity, in 1996 the Joint United Nations Programme on HIV/AIDS (UNAIDS) was created, co-sponsored by six organizations of the United Nations family: UNICEF, UNDP, UNFPA, UNESCO, WHO, and the World Bank [58]. In addition, the Global Fund for AIDS, Tuberculosis, and Malaria, an international partnership aimed at accelerating the end of these pandemics was created. To tackle other diseases, called neglected diseases, the WHO in 2021 published a roadmap that sets global targets and milestones to prevent, control, eradicate, or eliminate 20 neglected tropical diseases (NTDs), as well as cross-cutting targets aligned with the UN Sustainable Development Goals [30]. These NTDs "*impose a devastating human, social and economic burden on more than 1 billion people worldwide,*" predominantly in tropical and subtropical regions among the most vulnerable marginalized populations [59]. This NTD concept must remind the world what our society has neglected. More in-depth analysis is needed: many times what is neglected is not some types of diseases, but rather our global society has neglected some specific populations or individuals, which are out of the scope

Table 1 The history of pandemic and evolution of ethics in research and key events for product development strategies

Notable infectious diseases (the texts in italic represent the number of deaths)	Notable events to consider human rights and infectious diseases (regional, related to this volume, and global)	Emerging research ethics/bioethics policies, related to this volume
430 BCE Plague at Athens *~100,000* First identified trans-regional pandemic, killed *1/4 of people in Athens* [53] 541 Justinian plagues *30–50 million, killed half of world population* 1340s "Black Death" *~50 million* Killed *1/4 of world, nearly 2/3 of western Eurasian population* [54] 1494 Syphilis *>50,000* c. 1500 Tuberculosis *High millions*	431–04 BCE Peloponnesian War, pandemic caused Athens' defeat by Sparta 1688 England, Bill of Rights 1776 US Declaration of Independence 1789 French Declaration of Human Rights 1791 US Bill of Rights 1822 Brazil, independence from Portugal	Hippocratic Oath (Hippocrates: 460–370 BC) "do no harm" (non-maleficence), privacy protection and other physician's professional codes
1918–1919 "Spanish" influenza *~50 million*	1914–1918 First World War, expanded transmission of "Spanish" flu 1939–1945 Second World War Human experimentations by Nazi Germany and Japanese Imperial military	

(continued)

Table 1 (continued)

	1946 WHO Constitution to establish Right to Health 1946 Japanese democratic constitution under US occupation 1948 UN Declaration of Human Rights 1966 UN International Covenant of Human Rights	1947 Nuremberg Code 1947 Establishment of WMA 1962 US Kefauver-Harris amendment: Informed consent, well-controlled trials 1964 WMA Declaration of Helsinki (first adopted) 1974 US National Research Act: Informed consent and research ethics committee 1975 WMA DoH: Research ethics committee
1976–2020 Ebola *15,258* 1981—HIV/AIDS *~37 million*	1985 Brazil, end of military dictatorship (1964–), establishes Universal Public Health system 1987 South Korea, Declaration of Democratization ending military dictatorship (1981–1988), although first constitution was in 1948 1995 WTO TRIPS agreement 1996 South Africa Constitution to abolish apartheid, implementing UN International Covenant 1996 Taiwan, democratization by direct president election	1979 Four/three principles by Beauchamp & Childres/Belmont Report in US 1982 CIOMS first ethical guidelines 1983 US first written policy on "treatment IND" through controversy on access to investigational HIV/AIDS drug 1996 WMA DoH: First placebo clause 1996 ICH-GCP 1997 Council of Europe: Human Rights and Biomedicine Convention 2000 WMA DoH: Expand the scope to study with data, biological materials; most stringent placebo and post-trial access clauses; publication ethics clauses, causing debates to weaken the placebo and access clauses in 2002, 04, 08 and current 2013 versions. 2001 EU Clinical Trial Directive, replaced with Regulations in 2014
2002 SARS *813* 2009 H1N1 "swine flu" *284,000*	2001 WTO TRIPS Doha Declaration 2002 Global Fund for AIDS, Tuberculosis, and Malaria was established 2012 WHO's first roadmap for 20 neglected tropical diseases (NTDs)	2004 ICMJE statement for registration of clinical trials in public database 2004 Korean Bioethics and Safety Act, expanded in 2012 responding to stem cell scandal 2005 UNESCO Declaration of Bioethics 2010 Taiwan Biobank Act 2011 Taiwan Human Research Act
2019—COVID-19 *Seven million~*	2020 TRIPS Waiver proposal from South Africa and India 2021 UNESCO statement on COVID-19 vaccine to be a global public good	To protect indigenous people's rights 2014 WHO policy for use of investigational drug, responding Ebola outbreak 2022 WHO policy for emergency use of unproven intervention

This table was prepared by authors, adding information from the authors' analysis to the information of notable infectious diseases (leftmost column) obtained from a table in Morens et al. [13]

of the world seeking economic growth instead of the guarantee assurance of equal human rights.

The experience with Ebola, SARS, MARS, H1N1, Zika has provided tools for understanding the interactions among scientific and public health measures together with social and value systems. Recently, we have faced the initially unpredictable features of the threat of the COVID-19 pandemic. Now, effective vaccines have been developed at unprecedented speed, with successful use of novel technologies, and through a massive collaborative effort among governments, industry, academic researchers, and the general public. However, there was unacceptable inequality in the distribution of these vaccines globally [60]. The global calamity brought by the COVID-19 pandemic, coupled with the crises associated with it (social, political, economic and sanitary) must be used as a steppingstone for the necessary changes, not only socioeconomic but also cultural, to achieve a more just world founded on solidary and cooperation. Lessons learnt can help prepare for future pandemics [61], as well as equitable assurance of health for all.

4 Expected Audience and how to Use this Book

This volume captures recent developments in ethics in research and innovative strategies for establishing justice in global health. The foundation of international norms of human rights and human dignity is already established, but implementation has yet to take place, which seems not to be easy because of the many impediments to justice as fairness in achieving quality health for all. It requires the conscious and effective commitment of each individual and effective participation of governments and international institutions. This volume provides an informative resource for academics, researchers, governments, the global pharmaceutical industry, civil society, and other relevant stakeholders. Readers will be able to revisit, reconstruct, and achieve targeted results, together with the reconsideration of the definition of product development priorities and the expected resolution of conflicting arguments, in all stages of the formulation and implementation of institutional, national, and international policies.

We envisage that this book will also be useful for those involved in discussing emerging frameworks for ethics in research and innovative strategies in pandemic situations, as well as for preparedness for the future pandemics. It may assist with contributing to activities where results of innovative products are delivered to those who most need implementation of the results of research and development. In addition, superimposing the history of the struggle for democracy in Global South and Asian countries after World War II will provide perspectives toward reshaping the current political and socioeconomic systems as well as their foundations, such that current inequalities can be meaningfully confronted.

5 Conclusion and Opening to the Following Chapters

We have seen policy statements emerging, building on already established norms, but being continually challenged or sometimes misused. Globally, in history, pandemics or other public health crises have sharpened conflicting values on the protection of individual rights versus public interests. Our vision is that they should be pursued simultaneously. Westernized ethics articulated the alternatives of deontological norms and utilitarianism. This dualistic theory of pointing to individual versus public as conflicting values has been wrongly implemented prioritizing the interest of the richest with detriment to the rights of vulnerable individuals [62].

Our attempt is also to present ethical and human rights considerations in current clinical trials and any other types of studies involving humans, not allowing exploitation of study participants, ensuring access to research interventions found to be effective and their implementation to all who need it globally, in line with individual and public health needs. We understand that this can be achieved through "emancipation" of the most oppressed and marginalized and vulnerable to seek their rights [63].

Therefore, to overcome the depicted inequalities, people living in resource-limited and in disaster settings have the right to participate effectively in ensuring the protection and promotion of ethics and human rights not only in research settings, but also in the implementation of new innovative methodologies, alternative product development and strategies for equitable access to global health. All of this is aimed at achieving adequate, universal health care, which includes but is not limited to equitable global access to developed medicines and other equipment needed to address health challenges. Moreover, this must not be limited to pandemic situations.

Conflict of Interest There is no conflict of interest to be disclosed related to the content of this manuscript.

References

1. United Nations. The universal declaration of human rights. https://www.un.org/en/about-us/universal-declaration-of-human-rights. Adopted 10 Dec 1948
2. United Nations. International covenant on economic, social and cultural rights. Adopted by General Assembly resolution 2200A (XXI) of 16 Dec 1966, entry into force 3 Jan 1976. https://www.ohchr.org/en/instruments-mechanisms/instruments/international-covenant-economic-social-and-cultural-rights.
3. United Nations. International covenant on civil and political rights. Adopted by General Assembly resolution 2200A (XXI) of 16 Dec 1966, entry into force 23 Mar 1976. https://www.ohchr.org/en/instruments-mechanisms/instruments/international-covenant-civil-and-political-rights.
4. World Health Organization. Constitution of the World Health Organization. 22 Jul 1946. https://www.who.int/about/governance/constitution.
5. COVID-19 and bioethics part 3: pandemic and research ethics-democracy, placebo and post-trial access. Clin Eval 2021;49(Sup 38). http://cont.o.oo7.jp/49sup38/49sup38contents_e.html.
6. World Health Organization. Statement on the fifteenth meeting of the IHR (2005) Emergency committee on the COVID-19 pandemic. 5 May 2023. https://www.who.int/news/item/05-05-2023-statement-on-the-fifteenth-meeting-of-the-international-health-regulations-(2005)-emergency-committee-regarding-the-coronavirus-disease-(covid-19)-pandemic. Accessed 31 May 2023.
7. Abrams EM, Szefler SJ. COVID-19 and the impact of social determinants of health. Lancet Respir Med. 2020;8(7):659–61. https://doi.org/10.1016/S2213-2600(20)30234-4.

8. Abedin M, Wahab A, Rahman FN, Omi FR, Shareen S, Rakhshanda S, Islam L, Mayaboti CA, Saha UK, Faruque F, Fletcher LM, Mashreky S. Impact and variability of social determinants of health on the transmission and outcomes of COVID-19 across the world: a systematic review protocol. BMJ Open. 2021;11(12):e053481. https://doi.org/10.1136/bmjopen-2021-053481. PMID: 34853106; PMCID: PMC8637315.
9. Singer M. A dose of drugs, a touch of violence, a case of AIDS: conceptualizing the SAVA syndemic. Free Inquiry in Creative Sociology. 2000:28;1:13–24.
10. Singer M, Bulled N, Ostrach B, Mendenhall E. Syndemics and the biosocial conception of health. Lancet. 2017;389(10072):941–50.
11. Singer M, Clair S. Syndemics and public health: reconceptualizing disease in bio-social context. Med Anthropol Q. 2003;17(4):423–41.
12. Horton R. Offline: COVID-19 is not a pandemic. Lancet. 2020;396:874. https://www.thelancet.com/action/showPdf?pii=S0140-6736%2820%2932000-6.
13. Morens DM, Fauci AS. Emerging pandemic diseases: how we got to COVID-19. Cell. 2020;182(5):1077–92. https://doi.org/10.1016/j.cell.2020.08.021. Epub 2020 Aug 15. Erratum in: Cell. 2020 Oct 29;183(3):837. PMID: 32846157; PMCID: PMC7428724.
14. Zang SM, Benjenk I, Breakey S, Pusey-Reid E, Nicholas PK. The intersection of climate change with the era of COVID-19. Public Health Nurs. 2021;38(2):321–35. https://doi.org/10.1111/phn.12866. Epub 2021 Jan 31. PMID: 33521994; PMCID: PMC8014081.
15. Carlson CJ, Albery GF, Merow C, Trisos CH, Zipfel CM, Eskew EA, Olival KJ, Ross N, Bansal S. Climate change increases cross-species viral transmission risk. Nature. 2022;607:555. https://doi.org/10.1038/s41586-022-04788-w.
16. Gavi The vaccine Alliance. No one is safe until everyone is safe. https://www.gavi.org/vaccineswork/no-one-safe-until-everyone-safe. Accessed 31 May 2023.
17. United Nations. No one is safe, until everyone is. https://www.un.org/en/desa/%E2%80%9Cno-one-safe-until-everyone-%E2%80%9D#:~:text=Deputy %20Secretary%2DGeneral%20 Amina%20Mohammed,safe%20until%20everyone%20is%20safe.%E2%80%9D. Accessed 31 May 2023.
18. World Health Organization. COVAX: Working for global equitable access to COVID-19 vaccines. https://www.who.int/initiatives/act-accelerator/covax. Accessed 31 May 2023.
19. World Health Organization. Allocation logic and algorithm to support allocation of vaccines secured through the COVAX Facility: Explainer based on commonly asked questions. 15 Feb 2021. https://www.who.int/publications/m/item/allocation-logic-and-algorithm-to-support-allocation-of-vaccines-secured-through-the-covax-facility. Accessed 31 May 2023.
20. World Health Organization. Achieving 70% COVID-19 immunization coverage by mid-2022. 23 Dec 2021. https://www.who.int/news/item/23-12-2021-achieving-70-covid-19-immunization-coverage-by-mid-2022#cms. Accessed 31 May 2023.
21. Joint Statement by the UNESCO International Bioethics Committee (IBC) and the UNESCO World Commission on the Ethics of Scientific Knowledge and Technology (COMEST). UNESCO's ethics commissions' call for global vaccines equity and solidarity joint statement. SHS/BIO/IBC-COMEST/COVID-19 vaccines. Paris. 24 Feb 2021.
22. Joint Statement of UNESCO Ethics Commission ensuring equal access for all to vaccines and therapeutic development to confront COVID-19; joint statement of the UNESCO International Bioethics Committee (IBC) and UNESCO World Commission for the Ethics of Science and Technology (COMEST). SHS/IBC-COMEST/COVID-19 Vaccines IP Paris. 21 Sep 2021. https://unesdoc.unesco.org/ark:/48223/pf0000379042.
23. Council for Trade-Related Aspects of Intellectual Property Rights. Waiver from certain provisions of the TRIPS Agreement for the prevention, containment and treatment of COVID-19: Communication from India and South Africa. 2 Oct 2020; IP/C/W/669. https://docs.wto.org/dol2fe/Pages/SS/directdoc.aspx?filename=q:/IP/C/W669.pdf&Open=True.
24. Usher AD. South Africa and India push for COVID-19 patents ban. Lancet. 2020;396(10265):1790–1. https://doi.org/10.1016/S0140-6736(20)32581-2. PMID: 33278923; PMCID: PMC7831528.

25. World Trade Organization, Ministerial Conference Twelfth Session Geneva, 12–15 June 2022. Ministerial decision on the TRIPS agreement. 22 June 2022. WT/MIN(22)/30 WT/L/1141 https://docs.wto.org/dol2fe/Pages/SS/directdoc.aspx?filename=q:/WT/MIN22/30.pdf&Open=True. Adopted on 17 June 2022.
26. World Trade Organization. Decision text on extension of the 17 June 2022 ministerial decision to COVID-19 therapeutics and diagnostics: communication from the delegations of Argentina, Bangladesh, the Plurinational state of Bolivia, Egypt, India, Indonesia, Pakistan, South Africa and the Bolivarian Republic of Venezuela on behalf of the co-sponsors of the IP/C/W/669/REV.1 proposal. 6 Dec 2022; IP/C/W/694. https://docs.wto.org/dol2fe/Pages/SS/directdoc.aspx?filename=q:/IP/C/W694.pdf&Open=True.
27. World Trade Organization, Council for trade-related aspects of intellectual property rights. Paragraph 8 of the ministerial decision on the TRIPS agreement Adopted on 17 June 2022: Report to the General Council. 16 Dec 2022; IP/C/95. https://docs.wto.org/dol2fe/Pages/SS/directdoc.aspx?filename=q:/IP/C/95.pdf&Open=True.
28. Aloudat T. Médecins Sans Frontières (MSF) access campaign for equitable access in the world. Clin Eval. 2021;49(Suppl XXXVIII):72–7. http://cont.o.oo7.jp/49sup38/p61-109.pdf.
29. Kurihara C, Saio T, Imamura K, Greco D. Editorial: pandemic and research ethics – democracy, placebo, and post-trial access. Clin Eval. 2021;49(Suppl XXXVIII):4–7. http://cont.o.oo7.jp/49sup38/49sup38editorialeng.pdf.
30. United Nations. Resolution adopted by the General Assembly on 25 September 2015 - Transforming our world: the 2030 Agenda for sustainable development. 21 Oct 2015. https://www.un.org/en/development/desa/population/migration/generalassembly/docs/globalcompact/A_RES_70_1_E.pdf.
31. The Commonwealth Fund. Tikkanen R, Osborn R, Mossialos E, et al. International health care system profiles – Brazil. 5 Jun 2020. https://www.commonwealthfund.org/international-health-policy-center/countries/brazil. Accessed 31 May 2023.
32. Galvão-Castro B, Grassi MFR, Castilho EA, Greco DB. HIV/Aids and COVID-19 in Brazil: in four decades, two antithetical approaches to face serious pandemics. Mem Inst Oswaldo Cruz. 2021;116:e210071. https://doi.org/10.1590/0074-02760210071. PMID: 34190877; PMCID: PMC8238441.
33. Brazil-Guidelines and standards regulating research involving human beings – Resolution 466/2012, which succeeded Resolution 196/96.
34. South African Government. The constitution of the Republic of South Africa. As adopted on 8 May 1996 and amended on 11 Oct 1996 by the Constitutional Assembly. https://www.gov.za/documents/constitution/constitution-republic-south-africa-1996-1.
35. Wakabayashi M. Democratization of the Taiwanese and Korean political regimes: a comparative study. Dev Econ. 2007;35:422–39. https://www.ide.go.jp/library/English/Publish/Periodicals/De/pdf/97_04_05.pdf.
36. Chan I. Challenges in regulating biomedical research: the human subjects research act and the human biobank management act in Taiwan. Clin Eval. 2014;42(2):459–66.
37. Hwang WS, et al. Evidence of a pluripotent human embryonic stem cell line derived from a cloned blastocyst. Science. 2004;303:1669–74. Retracted in: Kennedy D. Editorial retraction. Science. 2006; 311(5759):335. Epub 2006 Jan 12.
38. Hwang WS, et al. Patient-specific embryonic stem cells derived from human SCNT blastocysts. Science. 2005;308:1777–83. Retracted in: Kennedy D. Editorial retraction. Science. 2006; 311(5759): 335. Epub 2006 Jan 12.
39. Cyranoski D. Whistle-blower breaks his silence. Nature. 2014;505(7485):593–4. https://doi.org/10.1038/505593a. PMID: 24476864.
40. Benderly BL. Paying a price for truth. Science. 2014. https://doi.org/10.1126/science.caredit.a1400028.
41. International Military Tribunal. Trials of war criminals before the Nuremberg Military Tribunals under Control Council law no. 10 Nuremberg, October 1946–April 1949, vol. 2. p. 181–2. (The Nuremberg Code.)

42. The Khabarovsk trial: the Soviet riposte to the Tokyo tribunal. In: Sellars K, editor. Trials for international crimes in Asia. Cambridge University Press; 2016. p. 121–44. https://ihl-databases.icrc.org/applic/ihl/ihl.nsf/Comment.xsp?action=openDocument&documentId=3DEA78B5A19414AFC1258585004344BD#73; https://www.cambridge.org/core/books/abs/trials-for-international-crimes-in-asia/khabarovsk-trial/C89A371724C32B3CBACA7CF0BEF15A6B.
43. The National Commission for the Protection of Human Subjects of Biomedical and Behavioral Research. The Belmont Report: Ethical principles and guidelines for the protection of human subjects of research. 18 Apr 1979. https://www.hhs.gov/ohrp/regulations-and-policy/belmont-report/read-the-belmont-report/index.html.
44. World Medical Association. WMA declaration of Helsinki: ethical principles for medical research involving human subjects. Adopted by the 18th WMA General Assembly, Helsinki, Finland, June 1964, and last amended by the 64th WMA General Assembly, Fortaleza, Brazil. 2013. https://www.wma.net/policies-post/wma-declaration-of-helsinki-ethical-principles-for-medical-research-involving-human-subjects/.
45. Council for International Organizations of Medical Sciences. International ethical guidelines for health-related research involving humans. 2016. https://cioms.ch/publications/product/international-ethical-guidelines-for-health-related-research-involving-humans/.
46. Council for International Organizations of Medical Sciences. Clinical research in resource-limited settings. 2021. https://cioms.ch/publications/product/clinical-research-in-low-resource-settings/.
47. Institute of Medicine (US) roundtable for the development of drugs and vaccines against AIDS, Nichols E. Expanding access to investigational therapies for HIV infection and AIDS: March 12–13, 1990 conference summary. Washington (DC): National Academies Press (US); 1991. PMID: 25121238.
48. International Conference on Harmonisation of Technical Requirements for Registration of Pharmaceuticals for Human Use. ICH Harmonised Tripartite Guideline: Choice of control group and related issues in clinical trials E10. 20 Jul 2000. https://database.ich.org/sites/default/files/E10_Guideline.pdf.
49. Sox HC, Rennie D. Seeding trials: just say "no". Ann Intern Med. 2008;149(4):279–80. https://doi.org/10.7326/0003-4819-149-4-200808190-00012. Erratum in: Ann Intern Med 2008 Sep 16;149(6):439. PMID: 18711161.
50. United Nations Educational, Scientific and Cultural Organization. UNESCO recommendation on open Science. November 2021. https://en.unesco.org/science-sustainable-future/open-science/recommendation.
51. World Trade Organization. Doha declaration on the TRIPS agreement and public health. Adopted by the WTO ministerial conference of 2001 in Doha on Nov14, 2001. https://www.wto.org/english/thewto_e/minist_e/min01_e/mindecl_trips_e.htm.
52. World Health Organization. WHO competency framework: building a response workforce to manage infodemics. 15 Sep 2021. https://www.who.int/publications/i/item/9789240035287. Accessed 31 May 2023.
53. Littman RJ. The plague of Athens: epidemiology and paleopathology. Mt Sinai J Med. 2009;76(5):456–67. https://doi.org/10.1002/msj.20137. PMID: 19787658.
54. Spyrou MA, Musralina L, Gnecchi Ruscone GA, Kocher A, Borbone PG, Khartanovich VI, Buzhilova A, Djansugurova L, Bos KI, Kühnert D, Haak W, Slavin P, Krause J. The source of the Black Death in fourteenth-century central Eurasia. Nature. 2022;606(7915):718–24. https://doi.org/10.1038/s41586-022-04800-3. Epub 2022 Jun 15. PMID: 35705810; PMCID: PMC9217749.
55. Glatter KA, Finkelman P. History of the plague: an ancient pandemic for the age of COVID-19. Am J Med. 2021;134(2):176–81. https://doi.org/10.1016/j.amjmed.2020.08.019. Epub 2020 Sep 24. PMID: 32979306; PMCID: PMC7513766.
56. World Health Organization. WHO coronavirus (COVID-19) dashboard. https://covid19.who.int/. Accessed 31 May 2023.

57. Brandt AM. How AIDS invented global health. N Engl J Med. 2013;368(23):2149–52. https://doi.org/10.1056/NEJMp1305297. PMID: 23738542.
58. UNAIDS. Facts about UNAIDS: an overview. https://data.unaids.org/publications/irc-pub03/una96-2_en.pdf. Accessed 31 May 2023.
59. World Health Organization. Ending the neglect to attain the Sustainable Development Goals: A road map for neglected tropical diseases 2021–2030. 2021. https://www.who.int/teams/control-of-neglected-tropical-diseases/ending-ntds-together-towards-2030.
60. The Independent Panel for Pandemic Preparedness and Response. Johnson Sirleaf HEE, Helen Clark RH, et al. Losing time: end this pandemic and secure the future. 2021. https://theindependentpanel.org/documents/. Accessed 31 May 2023.
61. World Health Organization. COVID-19 strategic preparedness and response plan: 1 Feb 2021 to 31 Jan 2022. 2021 License: CC BY-NC-SA 3.0 IGO. https://apps.who.int/iris/handle/10665/340072.
62. Reid L. Triage of critical care resources in COVID-19: a stronger role for justice. J Med Ethics. 2020;46(8):526–30. https://doi.org/10.1136/medethics-2020-106320. Epub 2020 Jun 16. PMID: 32546657; PMCID: PMC7316108.
63. Freire P. Pedagogy of the oppressed (Portuguese: Pedagogia do Oprimido). 1968.

Part I
Relevant Constructions from Global South and/or Asian Paradigms: Brazil, South Africa, Taiwan, South Korea, and Japan

Democracy Restoration in Brazil, *the Constitutional Guarantee of Health as a Right for All, Giving Rise to a Universal Health System* (SUS) and of a National Research Ethics Commission

Dirceu Greco and Bernardo Galvão-Castro

Abstract The creation of a universal health system (SUS), a national research ethics commission (CONEP), and the role of emancipation, societal participation, and democracy in these achievements and the fight against AIDS and COVID-19 in Brazil are described.

In the 1970s a Sanitary Reform movement aimed at strengthening public health policies, involving all relevant stakeholders contributed to the end of the 21-year military dictatorship (1985). The notion of health as a right led to the creation of SUS in 1990 based on four principles: Universality; Comprehensiveness; Equity; Decentralization of actions to the states/municipalities and effective community participation in the planning, and execution of public health services. Health as a right and as a duty of the State was included in the 1988 Federal Constitution. In 1996, CONEP was created to ethically supervise all research projects involving human subjects and the Brazilian Research ethics guidelines guarantee post-trial access to developed products and limit the use of placebo to situations where there are no effective comparators.

Restoration of democracy was fundamental for the creation of SUS and its role in the confrontation of AIDS (1985 onward) is an example to many countries, as all persons living with HIV are guaranteed, by law, access to prevention, diagnosis, and treatment without any out-of-pocket contribution.

In contrast, 40 years later (2020), the response to COVID during a far-right government was antithetical with many flaws, including an unacceptable death rate.

To face worldwide disparities which are fertile ground for infectious diseases, actions to address the social determinants of health (SDH) must be included in all discussions, guidelines, and covenants on emerging/re-emerging infectious diseases.

D. Greco (✉)
School of Medicine, Federal University of Minas Gerais, Belo Horizonte, Brazil

B. Galvão-Castro
Escola Bahiana de Medicina e Saúde Pública, Salvador, Brazil

C. Kurihara et al. (eds.), *Ethical Innovation for Global Health*,
https://doi.org/10.1007/978-981-99-6163-4_2

Keywords Democracy · Brazil · Universal health system · Health as a right · Research ethics

1 Health Emergencies in Brazil

In the last three centuries, Brazil, similar to many other countries, has been plagued by several epidemics/pandemics such as yellow fever, bubonic plague, smallpox, Spanish flu, AIDS, Zika, and Sars-CoV-2/COVID-19. Between 1850 and 1889, the country experienced a major outbreak of Yellow fever, whose etiologic agent is a Flavivirus transmitted by the Aedes aegypti mosquito. In 1899, the bubonic plague, caused by Yersinia pestis, arrived in Brazil through the port of Santos, with devastating consequences in several cities around the country. "Spanish flu," caused by the H1N1 influenza virus, reached Brazil in 1918, killing around 35,000 people. Smallpox, caused by the variola virus, hit the city of Rio de Janeiro in 1904, leaving 3500 people dead in that year alone.

After the first case of AIDS in Brazil was officially reported in 1982, human immunodeficiency virus (HIV) spread rapidly throughout the country, and from the 1980s until December 2019, AIDS caused 349,784 deaths; currently, 694,000 individuals are undergoing antiretroviral therapy (ARV) in the country.

The first Brazilian case of SARS-CoV2 infection was reported in São Paulo on February 25, 2020. Since then, there have been 34.096.935 confirmed cases of COVID-19 with 680.786 deaths, reported to WHO (15 August 2022) [1].

To confront these health emergencies, several sectors, including health, economic, and political, have either/both positively or/and negatively impacted outcomes. In most cases, despite government intervention usually being proposed, action was only actually taken after the spread of infectious agents, resulting in exacerbated numbers of cases and deaths. In some situations, the government took decisions without involving relevant stakeholders. One emblematic example is related to the smallpox outbreak in the city of Rio de Janeiro in 1904. In October that year, the federal government passed a Smallpox Mandatory Vaccination law, which was not accepted by the population, triggering what became known as "the Vaccine Revolt" a month later [2].

At the beginning of the twentieth century, Rio de Janeiro, the capital of Brazil, suffered from serious inadequacies in infrastructure: it lacked adequate water and sewage systems, had irregular garbage collection, and much of the population was living in overcrowded housing conditions. Although this environment was conducive to the spread of yellow fever, smallpox, and the bubonic plague, the government of Rio de Janeiro launched radical urban reform; however, this action was taken not necessarily with the aim of improving sanitation/living conditions, but rather due to serious economic consequences arising from the difficulty of establishing trade with other countries. With the objective of sanitizing the city, the eminent scientist Oswaldo Cruz led the execution of these public works [3].

Articles in the press not only criticized the government's actions, but also disseminated fake news about possible risks of the vaccine. A large part of Brazilian society also aligned with the opposition due to a perceived violation of individual rights. These events culminated in the "Vaccine Revolt," resulting in 30 deaths, 110 injured persons, and over a thousand arrests; half of the detainees were banished to the State of Acre in northern Brazil. In the end, these crises were eventually overcome, the spread of disease was controlled, and smallpox vaccination finally became accepted.

In other cases, while scientists recommended control measures based on scientific evidence, government authorities complied with or often did not accept the recommendations, arguing, among other reasons, that this would have negative economic impacts. Two striking examples, one of inadequate response and the other of adequate response, were related to the last two pandemics that overwhelmed the country in the last four decades, AIDS and COVID-19, respectively, which will be detailed later.

2 Geopolitical Scenario

The Federative Republic of Brazil is the largest country in Latin America, covering an area of 8.5 million square kilometers, divided into five geopolitical regions: the North, Northeast, West/Central, the Southeast, and South. Currently, the country's approximately 210 million inhabitants identify themselves as brown (46.8%) or black (9.4%), white (42.7%), and either of Asian descent or indigenous (1.1%) [4]. Brazil's political system involves three levels of government autonomy: the federal government, 26 states and a federal district, and 5563 municipalities. Despite Brazil being the 12th largest economy in the world and the largest in South America, with a 2020 Gross Domestic Product (GDP) of $1.44 trillion, the population of this continental-sized, multi-ethnic country suffers from high levels of socioeconomic and health inequality [5].

3 The Creation of Brazilian Unified Health System (SUS)

In 1937, President-elect Getúlio Vargas in a coup d'état established a quasi-totalitarian constitution and the legislature was shut down. His deposition in 1945 restored democratic rules and in 1946 a new constitution was adopted. In 1964, another coup d'état gave rise to a military dictatorship which lasted until 1985. Although social protection in Brazil was expanded during this period, government reforms were mostly aimed at private health care and eventually were extended, through social security, to rural workers. However, protections were fragmented and unequal before 1988 (when the new Federal Constitution was promulgated), with the public health system only serving those who made social security contributions. Thus, public health care services were accessible exclusively for formally contracted workers through Retirement and Pension Institutes (*Institutos de Aposentadoria e Pensão- IAP*), delivered according to each individual's original job category (e.g., civil servant, bank employee, mariner, commercial, industrial or railroad worker). By contrast, those with informal employment or the unemployed were forced to depend on charity or pay out-of-pocket for private health care.

Brazilian health sector reform began to take shape in the mid-1970s amid movements promoting democracy, gaining ample societal support. A Sanitary Reform movement was established as part of the resistance to the dictatorship and aimed at the strengthening of public policies and construction of the bases for a Social Welfare State [6]. It involved civil society, trade unions, health-related/social professionals and organizations, universities, students, and academia among others, who together considered health as a social and political issue, not only a biological question dependent exclusively on medical care. The pressure of these movements culminated in 1985 with the end of the 21-year military [7].

Soon after the restoration of democracy [8] was promulgated (1988), defining health as a universal right and a responsibility of the State, giving rise to SUS, which became indoctrinated into law in 1990.

SUS was established as part of an ample National Health Sector reform, a hallmark in the history of public health in Brazil, based on four fundamental principles: Universality (access to all); Comprehensiveness (comprising prevention, care, and treatment); Equity (the health of all citizens is equally guaranteed by the State); Decentralization of actions from the Central government to the states and on to municipalities; Societal participation (effective participation of the community in the planning, control, and execution of public health services). The right to health was clearly specified in 1988 Federal Constitution article 196: "Health is a right of all and a duty of the State and shall be guaranteed by means of social and economic policies aimed at reducing the risk of illness and other hazards, and at universal and equal access to actions and services for the promotion, protection and recovery of health," and article 198: "Health actions and public services integrate a regionalized and hierarchical network and constitute a single system, organized according to the following directives: I – decentralization, with a single management in each sphere of government; II – full service, priority being given to preventive activities, without prejudice to assistance services; III – participation of the community." The creation of SUS was fundamental to the establishment of the exemplary confrontation of the AIDS Pandemic [9].

Over the course of the years confronting AIDS in Brazil, it is worth addressing some of its achievements and the role of SUS.

3.1 Combining Forces with a Common Goal

As soon as the first cases of AIDS were reported in the early 1980s several factors played important roles in the insertion of public health policy in the government's agenda, such as mobilization of the academic community, favorable political conditions, and, above all, the participation of civil society. In the confrontation of AIDS, the combined national response involving the State, Civil Society, and Academia, was based on human rights and the principles enshrined in SUS. Metaphorically speaking, the Brazilian response resembled something opposite of a plane crash, in which small events add up to finally bring the airplane down. Fortunately, the stakeholders overcame difficulties in working together and united forces to confront this important health issue, putting pressure on the federal government to ensure adequate action and thus share responsibility for the success of the Brazilian HIV/AIDS response.

In 1985–1986, before the creation of SUS, with the participation of all the combined forces the national AIDS program was established, leading to the formation of inter-institutional state commissions as well as a national policy to confront AIDS while emphasizing the respect and protection of human rights. Also in 1986, the AIDS program implemented the National AIDS Commission (CNAIDS), involving the participation of civil society, academia, and government, to discuss and recommend actions for the continuous enhancement of program policies.

However, it was the creation of SUS in 1990, based on principles of equality and integrality in health care, that ensured the expansion of the AIDS program, allowing for a comprehensive response to the pandemic.

The role of SUS was reinforced by the sanctioning of a seminal law (no. 9313) in 1996, under pressure from organized civil society, which transformed universal access to antiretroviral (ARV) therapy into a right through SUS. This move went against common sense, wrote the World Bank declaring that low- and middle-income countries (LMIC) should concentrate their efforts on prevention [10], since the complexity of treatment schemes would hinder adherence, thereby increasing the risk of resistant strains. This stance was clearly incorrect, as between 1996 and 2002 Brazilian investment to curb the AIDS pandemic reached approximately US$ 1.6 billion. In addition to the invaluable social impact of reducing mortality, morbidity, hospitalizations, and early retirements [11], an estimated savings of approximately US$2 billion was achieved. The result of this bold action received international recognition, and the prevalence of resistant viruses in Brazil has remained at or even lower than levels in high-income countries (HIC).

3.2 Participation of Civil Society

The participation of the citizenry, together with NGOs promoting human rights for lesbian, gay, bisexual, transgender, queer and intersex (LGBTQI) persons, and sex workers, was fundamental in the establishment and maintenance of the Brazilian policy response to HIV/AIDS. This involvement led to the emancipation of citizens to exert power over public health policy. One example of this is the participation of people living with HIV/AIDS (PLWHA) in demanding their rights, both nationally and internationally, making their voices heard by Health Councils, government institutions, regulatory agencies, and the World Health Organization (WHO-UNAIDS). At the same time, the participation of people exposed to other endemic diseases, such as schistosomiasis, malaria, or Chagas Disease, was and continues to be less significant in this respect, despite millions being afflicted. Emancipation is a necessary step to involve all stakeholders and is referenced here in accordance with the Brazilian educator Paulo Freire's [12] publications on education for freedom, meaning that liberation and autonomy are inherent elements of citizenship, rights, and the fight against inequality. Freire adds that emancipation will not happen by chance, nor by concession, but will rather be a conquest carried out by the human praxis, which calls for uninterrupted struggle.

3.3 The Universities/Health Secretariats/Professionals/ Academia

The participation of public health institutions and *lato* sensu, academia, was and continues to be an important factor in the response to the AIDS pandemic. It has allowed for the establishment of quality health services, bolstered by technical and academic support for specific policies based on scientific evidence, involving diagnosis, care, virologic and immunologic assessment, as well as research.

4 Scientific, Economic, Social, and Ethical Impacts of the AIDS Pandemic

4.1 In Science

With the restoration of democracy, great advances were made in the fields of science, technology, and public health in Brazil [13]. Scientific efforts aimed at understanding retroviral infection, together with the knowledge gained from the AIDS pandemic since the first reported cases in 1981, were essential to the identification of HIV as the etiological agent of AIDS [14], making it possible to implement serological diagnostic techniques and control bloodborne transmission. However, at the time, diagnosis was only readily available in HIC, due to the high cost of patent-protected techniques. Upon receiving donated HIV cell lines from the US National Institutes of Health in July 1985, the Oswaldo Cruz Foundation (FIOCRUZ), a leading Brazilian public health and research institution, was able to quickly implement an immunofluorescence serological assay to screen for HIV in blood banks and at public laboratories throughout Brazil [15].

In 1987, for the first time in Latin America, Brazilian researchers successfully isolated HIV [16]. The researchers' expertise, the infrastructure in place and collaboration with Pan American Health Organization (PAHO)/UNAIDS allowed for the rapid transfer of technology and the implementation of a Brazilian network to systematically monitor HIV polymorphism [17, 18], as well as the Brazilian network for HIV Drug Resistance Surveillance [19].

Fifteen years later (1996), the positive results from combined antiretroviral therapy brought hope to PLWHA; however, not all shared in this scientific victory. For many years this therapeutic option was restricted to HICs, highlighting the urgent need to ensure the right of access to scientific and technological progress for everyone who needs it. In this scenario, the public access to antiretroviral provided by SUS was a counterpoint.

4.2 Ethics, Research, and Access

The AIDS pandemic amplified the need for effective international collaboration and solidarity, with the WHO/UNAIDS playing an unquestionable role. AIDS-related NGOs created worldwide established much-needed advocacy for individuals affected by AIDS. Other non-governmental institutions, such as the Global Fund to Fight Malaria, Tuberculosis, and AIDS and UNITAID, as well as private foundations, e.g., Clinton Health Access Initiative (CHAI), Bill and Melinda Gates Foundation, took aim at financing the confrontation against the AIDS pandemic. While these organizations indeed provided substantial financial resources, this effort also reduced the pressure on governments to actively participate in the public health response. Moreover, these funders had discretion over setting the agenda, with associated political and ideological implications. An example of this was the reduction or interruption of external funding, which led to difficulties in maintaining adequate treatment and care for PLWHA in Africa [20].

4.2.1 The National Health Council (CNS) and the Role of the Brazilian Research Ethics Commission (CONEP) in Protecting Research Participants

The CNS, created in 1937, was adapted to the 1988 Federal Constitution and was promulgated by Law 8142/1990 as part of SUS [21], is a deliberative and permanent committee linked to the Brazilian Ministry of Health. Its mission is to participate in the formulation and control the execution of National Health Policy on a federal level, as well as systematically monitor public health policies. It is comprised of 48 members and their respective substitutes, elected to represent public health system users, workers, SUS managers, and health service providers.

The CNS created the National Research Ethics Commission (CONEP), to be responsible for the ethical supervision of all research projects involving human subjects in Brazil, which are submitted to more than 800 regional Research Ethics Committees (CEP) distributed throughout the country.

This CEP-CONEP System was established by Resolution CNS 196/96, succeeded by Resolution 466/2012, which has since 1996 set guidelines and regulatory norms for all research involving human subjects in Brazil. Brazil objected to the proposed changes to the ethical requirements related to placebo and post-trial access to the 2008 version of the Declaration of Helsinki. In the previous (2000) DoH placebo use was restricted to situations where no effective control was available and ensured that research participants received post-trial access to developed products. Both items were looser in the 2008 version. In a sovereign and bold decision, CNS approved Resolution 404/2008 [22], which guaranteed that, at the end of a clinical trial, all participants must be ensured access to the best proven prophylactic, therapeutic, and diagnostic options identified by their respective clinical trial. This was subsequently expanded and incorporated into Resolution 466/2012 [23]. As detailed in items III.1 "Ethics of research involves: a) respect for the research participant's dignity and autonomy, recognizing his/her vulnerability, ensuring his/her will to contribute and remain, or not, in the research study, by means of express, free and informed consent;" and III.3 "All research using experimental methodologies in the biomedical area, involving human beings… b) when using placebo, such use shall be fully justified as to its non-maleficence and methodology requirements, where the benefits, risks, difficulties and effectiveness of a new therapeutic method shall be tested, comparing it to the best current prophylactic, diagnostic and therapeutic methods. No use of placebo or any other treatment is permitted in studies where there are no proven methods of prophylaxis, diagnosis or treatment; … and d) guarantee to all participants, at the end of the study and for an unlimited time, free access to the best prophylactic, diagnostic and therapeutic methods with proven efficacy; d.1) access will also be guaranteed in the interval between the end of the individual's participation and the end of the study, in which case this guarantee could be granted by means of extension studies, according to a duly justified analysis from the doctor assisting the research participant." These norms may serve as leverage to extend access beyond controlled research environments, reaching the sphere of public health for all who may benefit from the fruits of research.

Property Rights (TRIPS) agreement be made more flexible, but the World Trade Organization (WTO), more importantly, should opt to suspend patents to foster production and the availability of generics [31]. A relevant step to this end was South Africa and India's proposal in October 2020 for immediate suspension of patents, including copyrights, industrial design, and other undisclosed information, for COVID-19 vaccines and other new technologies. This comprehensive proposal would enable many middle-income countries that have well-equipped laboratories and well-prepared researchers to immediately begin producing needed products. However, after 20 months of deliberations (17 June 2022), the WTO did not approve the scope of the patent waiver as proposed, as the approved resolution only waived limited tools for production and does not apply to all countries. It is notable that this inadequate decision occurred during a pandemic that has taken more than 15 million lives, many of which could have been saved had vaccines been equitably distributed [32]. The WTO's decision sets a negative precedent against fair access, not only for future pandemics, but also in the context of many other illnesses that disproportionately affect the poor [31]. It is further disappointing that despite many politicians making commitments and expressing words of solidarity, HICs have squandered an opportunity to try to mitigate unacceptable inequities regarding health access.

7 Perspectives

Pandemics illustrate the urgent need for a universal and inclusive public health system to provide effective prevention and care for all, and the existence of SUS and its role in Brazil serves as a prime example. The same is true regarding having in place mechanisms to protect research participants, avoiding double-standards, independently of their country of origin or economic conditions.

Procedural steps: Most of what is needed to respond to emerging and re-emerging diseases ethically and scientifically is already known and available. This includes, but it is not limited to:

- An established international framework to identify and respond to outbreaks, such as International Health Regulations -IHR (2005). The purpose and scope of the IHR (2005) are "to prevent, protect against, control and provide a public health response to the international spread of disease in ways that are commensurate with and restricted to public health risks, and which avoid unnecessary interference with international traffic and trade" [33].
- Though infectious disease emergencies may affect both developing and developed countries, the likelihood of their emergence and the potential to spread unchecked usually impact more vulnerable countries and communities more severely. All health, medical, and humanitarian communities must abide by existing ethical and research guidance and work toward full compliance with the IHR framework, which is binding for all WHO members.

- The role of regional and international institutions. The delay, until August 2014, in declaring a Public Health Emergency of International Concern (PHEIC) for the 2014–2015 Ebola epidemic in West Africa constitutes an extreme example of institutional failure. In this case, the timing is conspicuous, as it was not until Ebola cases appeared in the USA and Europe that the WHO Director-General declared Ebola a PHEIC on 8 August 2014, in spite of confirmed, intense transmission occurring for months. Only then did the UN Security Council declare the Ebola outbreak in West Africa a threat to international peace and security in its first meeting in history addressing a public health crisis [34].
- Collaborate with the Global Virome Project, a cooperative scientific initiative to improve the capacity to discover potentially new viral threats to human populations. The Project estimates that between 631,000 and 827,000 unknown viruses could be zoonotic and have the potential to infect humans [35].

  – Substantive obstacles hamper the effective application of measures to protect and promote human rights:

- A critical review of the several sets of established ethical guidelines, directed at both "normal" situations and health emergency responses, is needed. A solution to this tangled web of differing guidelines could be the establishment of a single, comprehensive internationally approved set of ethical guidelines that ensures the protection of research participants everywhere, as well as providing equal access to developed products, preventing double-standards both in research and in public health [36]. This effort could be led by the WHO and UNESCO, using the Universal Declaration of Bioethics and Human Rights as a basis.
- The results of research aimed at developing diagnostic tools, drugs, and vaccines must be transformed into true access for all via public health. To accomplish this, robust international collaboration, adequate financing, inclusiveness, and solidarity are essential. Accordingly, it is worth quoting the opening remarks of WHO Director-General Tedros Adhanom Ghebreyesus on the ethics of egalitarian access to COVID-19 vaccines at the 148th Session of the WHO Executive Board (18 January 2021) WHO Director-General's [37]: "I need to be blunt: the world is on the brink of a catastrophic moral failure – and the price of this failure will be paid with lives and livelihoods in the world's poorest countries. Even as they speak the language of equitable access, some countries and companies continue to prioritize bilateral deals, going around COVAX [38], driving up prices and attempting to jump to the front of the queue… This is wrong. Forty-four bilateral deals were signed last year, and at least 12 have already been signed this year. The situation is compounded by the fact that most manufacturers have prioritized regulatory approval in rich countries where the profits are highest, rather than submitting full dossiers to WHO."
- It is noteworthy that more than a year after these remarks, this situation continues to be unacceptable. The urgent need to mitigate inequity is illustrated by the fact that, as of May 2022 only 16% of people in low-income countries have received a single vaccine dose compared to 80% in high-income countries [39].

- Placing limits on the influence of pharmaceutical companies and other commercial interests and establishing mechanisms to mount effective responses to health emergencies to curb unethical practices, such as vaccine hoarding, which may impair effective political decision-making.
- Efforts must be made to promote transparency and ensure access to the timely dissemination of unbiased, clear, culturally adapted and relevant scientific based information to counteract the negative effects of infodemia, an infodemic is an overabundance of information, which includes deliberate dissemination of wrong information to undermine the public health response. Mis- and disinformation can be harmful to people's physical and mental health; increase stigmatization; threaten precious health gains; and lead to poor observance of public health measures, thus reducing their effectiveness and endangering countries' ability to stop the pandemic [40].
- Avoid or limit other conflicts of interest which may delay, or needlessly accelerate, the declaration of a public health emergency, as attributing the status of epidemic can be politically and economically sensitive, leading to ripple effects on international trade (as with COVID-19 in China) as well as negatively impacting sectors such as travel and tourism [34].
- Tackle global disparities by effectively adopting the UN 17 Sustainable Development Goals [41] as a priority. The UN describes it as "an urgent call for action by all countries—developed and developing - in a global partnership. They recognize that ending poverty and other deprivations must go together with strategies that improve health and education, reduce inequality, and spur economic growth—all while tackling climate change and working to preserve our oceans and forests." Meanwhile, include financial support measures for the most socially vulnerable during quarantine periods, and expand discussions on Universal Basic Income. According to Haggh, B, Rohregger, "UBI policies, that is, the first steps toward a UBI, are increasingly perceived as one set of measures that may insulate subsistence guarantees from increased economic pressure; increase the impact of other welfare policies, such as education and health; as well as re-incentivize employment and savings" [42].
- Include the strengthening or establishment of adequate and affordable public health systems everywhere, accessible to all [43].

8 Lessons Learned

8.1 The Impact of the Restoration of Democracy in Brazil

An exemplary case for democracy is the importance of its restoration in issuance of new Federal constitution in 1988 which included health as right of all and a duty of the State and the following creation of a national public health system, accessible to all, without any out-of-pocket payment and funded exclusively by taxpayers.

8.2 In Research

International collaboration in AIDS research has facilitated a re-examination of legal and, particularly, ethical aspects, and has stimulated the implementation of measures to protect the human rights of vulnerable populations. Unfortunately, the urgency in controlling AIDS and, more recently, COVID-19, added to increasing incidence of other illnesses (e.g., hepatitis, malaria, dengue, cholera, SARS-CoV-2), has prompted some to argue for easing ethical requirements in research, especially in developing countries.

With the perspective of ensuring justice (and to counteract these arguments):

- Establish universally respected ethical guidelines, with the UNESCO Universal Declaration of Bioethics and Human Rights [44] serving as a reference.
- Ensure that research project participants have post-trial access to drugs, vaccines, interventions, and prevention strategies shown to be safe and effective, which should eventually be extended to universal public health access. To achieve the latter and to ensure that health products cannot be treated as economical commodities, a discussion of the scope of intellectual property rights embedded in the TRIPS agreement is critical, especially, but not exclusively in the context of health emergencies [45].

8.3 In Public Health

The restoration of democracy, with right to health imbedded in the new constitution and the creation of SUS were fundamental in dealing with the AIDS epidemic in Brazil. And as detailed it prompted clear political decision-making and the establishment of strong partnerships among several stakeholders, including government institutions, civil society, universities, researchers, and health professionals. This partnership facilitated a prominent position for Brazil in various international fora, especially WHO/UNAIDS, in the promotion and defense of human rights, opposition to discrimination and in favor expanded access to antiretrovirals worldwide. Brazil's provision of universal care and ARV access without out-of-pocket participation would not have happened without the existence of its universal health system (SUS), unequivocally showing that a middle-income country can, even in the face of many inequalities, treat people equally regardless of race, sex, sexual orientation, or economic status. And all this sets an example for the world, not exclusively to other LMIC.

Thus, to effectively combat COVID-19, other pandemics and endemic illnesses, the lessons learned from the worldwide response to the AIDS pandemic, which included research, public health practice, societal participation and universal access to adequate health care, can be considered a global health model to be followed [46–48].

9 Conclusions

To truly embrace the lessons learned from the AIDS epidemic, the participation of health, medical, governmental, and international institutions/communities must move beyond research, providing care and emergency assistance, extending even past the goal of making the highest attainable standards of physical and mental health available to all. Imperative actions, in concert with the participation of civil society, must also confront isolationism and xenophobia, promote international solidarity and cooperation, adequately finance scientific research and quality public health services, address anti-science and anti-vaccine rhetoric. and prioritize egalitarian access to technological progress. And take steps to ensure that the 17 UN SDGs will be urgently achieved.

The global community must further support the open science movement and insist on meaningful action to tackle climate change, an important driver of disease outbreaks [49].

The Brazilian experience with confronting AIDS should be used as a good example showing that the restoration of democracy was crucial for the recognition of health as a right and also for the creation of a true universal health system.

And last but not least, an urgent decision on the non-patentability of products for COVID-19 is critical, together with a thorough revision of the TRIPS agreement regarding health products, whose access must be considered a human right, and not a commercial commodity. If this is achieved, it will be easier to tackle emergency health situations and also other illnesses that inflict more aggressively vulnerable populations.

Thus, all must truly work together to tackle the obscene disparities among countries and, in and within, their populations. Mechanisms to mitigate, or, more optimistically, eliminate these prevailing disparities must be included in each and every discussion, guidance paper, and covenant regarding emerging or re-emerging infectious diseases. Only with such involvement it will be possible to address the social determinants of health (SDH) as they have a crucial influence on health inequities, which impact the unjust and avoidable differences in health status seen within and between countries. "In countries at all levels of income, health and illness follow a social gradient: the lower the socioeconomic position, the worse the health" [50]. And they facilitate the establishment and spread of the current syndemic [51, 52]. Tackling the SDH will contribute to timely and adequately confront other future pandemics that are certain to come. In the 2020 WHO International Independent Panel for Pandemic Preparedness and Response the summary of the recommendations states: "To prepare the world for the future so that the next disease outbreak does not become a pandemic, the panel calls for a series of crucial reforms that will address gaps in high-level coordinated leadership globally and nationally, funding, access to what must become global goods, and WHO's independence, focus, and authority" [53]. Of note is the recent World Health Organization's (WHO) declaration in July 2022 that the monkeypox outbreak spreading globally is a public health emergency of international concern (PHEIC) [54].

Acknowledgments The authors would like to thank Andris K. Walter for critical analysis, English language revision, and manuscript copyediting assistance.

Conflict of Interest The authors have no competing interests to declare that are relevant to the content of this chapter.

References

1. WHO Coronavirus (COVID-19) Dashboard [Internet]. Disponível em: https://covid19.who.int. Accessed 15 Aug 2022.
2. Hochman G. Vacinação, varíola e uma cultura da imunização no Brasil. Ciênc saúde coletiva [Internet]. 2011;16(2):375–86. Available in: http://www.scielo.br/scielo.php?script=sci_arttext&pid=S1413-81232011000200002&lng=pt&tlng=pt. Accessed 8 Augs 2022.
3. Stepan N. Beginnings of Brazilian science: Oswaldo Cruz, medical research and policy, 1890–1920. New York: Science History Publications; 1976. 225 p.
4. Azevedo ALM dos S. IBGE - Educa | Jovens [Internet]. IBGE Educa Jovens. Available in: https://educa.ibge.gov.br/jovens/conheca-o-brasil/populacao/18319-cor-ou-raca.html. Accessed 8 Aug 2022.
5. The inequality epidemic in Brazil [Internet]. Oxfam Brasil. Available in https://www.oxfam.org.br/justica-social-e-economica/forum-economico-de-davos/the-inequality-epidemic-in-brazil/. Accessed 8 Aug 2022.
6. Fleury S. Brazilian sanitary reform: dilemmas between the instituing and the institutionalized. Cien Saude Colet. 2009;14(3):743. https://doi.org/10.1590/S1413-81232009000300010.
7. Paim J, Travassos C, Almeida C, Bahia L, Macinko J. The Brazilian health system: history, advances, and challenges. The Lancet [Internet]. 2011;377(9779):1778–97. Available in: https://linkinghub.elsevier.com/retrieve/pii/S0140673611600548. Accessed 8 Aug 2022.
8. Vajda I, Zimbres P, de Souza VT. Constitution of the federative Republic of Brazil. Brasilia: Documentation and Information Center, Publishing Coordination, Chamber of Deputies; 2010.
9. Greco DB. Trinta anos de enfrentamento à epidemia da Aids no Brasil, 1985–2015. Cienc saúde coletiva [Internet]. 2016;21(5):1553–64. Available in: http://www.scielo.br/scielo.php?script=sci_arttext&pid=S1413-81232016000501553&lng=pt&tlng=pt. Accessed 8 Aug 2022.
10. Galvão J. Access to antiretroviral drugs in Brazil. The Lancet [Internet]. 2002;360(9348): 1862–5. available in: https://linkinghub.elsevier.com/retrieve/pii/S0140673602117752. Accessed 8 Aug 2022.
11. Teixeira PR, Vitória MA, Barcarolo J. Antiretroviral treatment in resource-poor settings: the Brazilian experience. AIDS [Internet]. 2004;18(Suppl 3):S5–7. Available in: http://gateway.ovid.com/ovidweb.cgi?T=JS&PAGE=crossref&AN=00002030-200406003-00002. Accessed 8 Aug 2022.
12. Freire P. Pedagogy of the oppressed. 30th anniversary ed. New York: Continuum; 2000. 183 p.
13. Goldenberg S, Rodrigues ML, Savino W. Brazilian science: towards extinction? Mem Inst Oswaldo Cruz. 2022;116:e210357.
14. Barré-Sinoussi F, Chermann JC, Rey F, Nugeyre MT, Chamaret S, Gruest J, et al. Isolation of a T-lymphotropic retrovirus from a patient at risk for acquired immune deficiency syndrome (AIDS). Science. 1983;220(4599):868–71.
15. Castro BG, Fernandez JC, de Castilho EA, Pereira HG, Pereira MS, Cruz FO. Human immunodeficiency virus infection in Brazil. JAMA. 1987;257(19):2592–3.
16. Galvão-Castro B, Ivo-dos-Santos J, Couto-Fernandez JC, Bongertz V, Chequer-Bou-Habib D, Sion FS, et al. Isolation and antigenic characterization of human immunodeficiency virus (HIV) in Brazil. Mem Inst Oswaldo Cruz [Internet]. 1987;82(4):453–6. available in: http://

www.scielo.br/scielo.php?script=sci_arttext&pid=S0074-02761987000400001&lng=en&tlng=en. Accessed 8 Aug 2022.

17. Galvão-Castro B, Couto-Fernandez J, Mello M, Linhares-de-Carvalho M, Castello-Branco L, Bongertz V, et al. A nationwide effort to systematically monitor HIV-1 diversity in Brazil: preliminary results. Mem Inst Oswaldo Cruz [Internet]. 1996;91(3):335–8. Available in: http://www.scielo.br/scielo.php?script=sci_arttext&pid=S0074-02761996000300014&lng=en&tlng=en. Accessed 8 Aug 2022.
18. Bongertz V, Bou-Habib DC, Brígido LFM, Caseiro M, Chequer PJN, Couto-Fernandez JC, et al. HIV-1 diversity in Brazil: genetic, biologic, and immunologic characterization of HIV-1 strains in three potential HIV vaccine evaluation sites: JAIDS J Acquir Immune Deficien Syndrom [Internet]. 2000;23(2):184–93. Available in: http://journals.lww.com/00126334-200002010-00011. Accessed 8 Aug 2022.
19. Arruda MB, Boullosa LT, Cardoso CC, da Costa CM, Alves CR, de Lima ST, et al. Brazilian network for HIV drug resistance surveillance (HIV-BresNet): a survey of treatment-naive individuals. J Int AIDS Soc. 2018;21(3):e25032.
20. Katz IT, Bassett IV, Wright AA. PEPFAR in transition—implications for HIV Care in South Africa. N Engl J Med [Internet]. 2013 [cited 8 Aug 2022];369(15):1385–7. Available in: http://www.nejm.org/doi/10.1056/NEJMp1310982.
21. Conselho Nacional de Saúde. LATINO [Internet]. Disponível em: https://latinno.net/en/case/3097/. Accessed 8 Aug 2022.
22. Conselho Nacional de Saúde [Internet]. Available in: https://conselho.saude.gov.br/resolucoes/reso_08.htm. Accessed 8 Aug 2022.
23. Resolution No. 466, OF 12 Dec 2012 [Internet]. Available in: https://conselho.saude.gov.br/resolucoes/2012/466_english.pdf.
24. Galvão-Castro B, Grassi MFR, de Castilho EA, Greco DB. HIV/Aids and COVID-19 in Brazil: in four decades, two antithetical approaches to face serious pandemics. Mem Inst Oswaldo Cruz. 2021;116:e210071.
25. Hallal PC. SOS Brazil: science under attack The Lancet [Internet]. 2021;397(10272):373–4. Available in: https://linkinghub.elsevier.com/retrieve/pii/S0140673621001410. Accessed 8 Aug 2022.
26. Lasco G. Medical populism and the COVID-19 pandemic. Global Public Health [Internet]. 2020;15(10):1417–29. Available in: https://www.tandfonline.com/doi/full/10.1080/17441692.2020.1807581. Accessed 8 Aug 2022.
27. de Figueiredo AM, de Figueiredo DCMM, Gomes LB, Massuda A, Gil-García E, de Toledo Vianna RP, et al. Social determinants of health and COVID-19 infection in Brazil: an analysis of the pandemic. Rev Bras Enferm [Internet]. 2020;73(suppl 2):e20200673. Available in: http://www.scielo.br/scielo.php?script=sci_arttext&pid=S0034-71672020001400158&tlng=en. Accessed 8 Aug 2022.
28. WTO DOHA Declaration 2001. Available at https://www.wto.org/english/res_e/booksp_e/ddec_e.pdf. Accessed 15 Aug 2022.
29. Greco DB, Simão M. Brazilian policy of universal access to AIDS treatment: sustainability challenges and perspectives. AIDS. 2007;21(Suppl 4):S37–45.
30. Usher AD. South Africa and India push for COVID-19 patents ban. The Lancet [Internet]. 2020;396(10265):1790–1. Available in: https://linkinghub.elsevier.com/retrieve/pii/S0140673620325812. Accessed 8 Aug 2022.
31. TRIPS Council welcomes MC12 TRIPS waiver decision, discusses possible extension [Internet]. Available in: https://www.wto.org/english/news_e/news22_e/trip_08jul22_e.htm. Accessed 8 Aug 2022.
32. MSF [Internet]. Médecins Sans Frontières (MSF) International. Disponível em: https://www.msf.org/lack-real-ip-waiver-covid-19-tools-disappointing-failure-people, https://reliefweb.int/report/world/inability-agree-real-pandemic-intellectual-property-waiver-wto-devastating-global-failure-people-world-over. Accessed 8 Aug 2022.

33. International Health Regulations. 3rd ed [Internet]. 2005. Available in: https://www.who.int/publications-detail-redirect/9789241580496. Accessed 8 Aug 2022.
34. Rull M, Kickbusch I, Lauer H. Policy debate | international responses to global epidemics: Ebola and beyond. poldev [Internet]. 2015;6(2). Available in: http://journals.openedition.org/poldev/2178. Accessed 8 Aug 2022.
35. Jonas O, Seifman R. Do we need a Global Virome project? Lancet Global Health [Internet]. 2019;7(10):e1314–6. Available in: https://linkinghub.elsevier.com/retrieve/pii/S2214109X19303353.
36. Nicholson T, Admay C, Shakow A, Keshavjee S. Double standards in global health: medicine, human rights law and multidrug-resistant TB treatment policy. Health Hum Rights. 2016;18(1):85–102. PMID: 27781001; PMCID: PMC5070682.
37. WHO Director-General's opening remarks at 148th session of the Executive Board [Internet]. Available in: https://www.who.int/director-general/speeches/detail/who-director-general-s-opening-remarks-at-148th-session-of-the-executive-board. Accessed 8 Aug 2022.
38. COVAX [Internet]. Available in: https://www.who.int/initiatives/act-accelerator/covax. Accessed 8 Aug 2022.
39. COVAX calls for urgent action to close vaccine equity gap [Internet]. Available in: https://www.who.int/news/item/20-05-2022-covax-calls-for-urgent-action-to-close-vaccine-equity-gap. Accessed 8 Aug 2022.
40. Managing the COVID-19 infodemic: promoting healthy behaviours and mitigating the harm from misinformation and disinformation [Internet]. Available in: https://www.who.int/news/item/23-09-2020-managing-the-covid-19-infodemic-promoting-healthy-behaviours-and-mitigating-the-harm-from-misinformation-and-disinformation. Accessed 8 Aug 2022.
41. Home - United Nations Sustainable Development [Internet]. Available in: https://www.un.org/sustainabledevelopment/. Accessed 8 Aug 2022.
42. UBI policies and their potential for addressing health inequities. Louise Haagh [Internet]. Available in: https://louisehaagh.net/ubi-policies-and-their-potential-for-addressing-health-inequities/. Accessed 8 Aug 2022.
43. Future C on a GHRF for the, National Academy of Medicine S. Strengthening public health as the foundation of the health system and first line of defense [Internet]. The neglected dimension of global security: a framework to counter infectious disease crises. National Academies Press (US); 2016. National Academies Press (US): https://www.ncbi.nlm.nih.gov/books/NBK368392/. Accessed 8 Aug 2022.
44. https://plus.google.com/+UNESCO. UNESCO and the universal declaration on human rights [Internet]. UNESCO. 2018. National Academies Press (US): https://en.unesco.org/udhr. Accessed 8 Aug 2022.
45. Greco DB. Ethical limits to placebo use and access to Covid-19 vaccines as a human right. IJME [Internet]. 2021;06(02):106–12. National Academies Press (US): https://ijme.in/articles/ethical-limits-to-placebo-use-and-access-to-covid-19-vaccines-as-a-human-right/. Accessed 8 Aug 2022.
46. Piot P, Quinn TC. Response to the AIDS pandemic—a global health model. N Engl J Med [Internet]. 2013;368(23):2210–8. National Academies Press (US): http://www.nejm.org/doi/10.1056/NEJMra1201533. Accessed 8 Aug 2022.
47. Brandt AM. How AIDS invented global health. N Engl J Med [Internet]. 2013;368(23):2149–52. National Academies Press (US): http://www.nejm.org/doi/10.1056/NEJMp1305297. Accessed 8 Aug 2022.
48. How AIDS changed everything—MDG6: 15 years, 15 lessons of hope from the AIDS response [Internet]. National Academies Press (US). Available in https://www.unaids.org/en/resources/documents/2015/MDG6_15years-15lessonsfromtheAIDSresponse. Accessed 8 Aug 2022.
49. Watts N, Amann M, Arnell N, Ayeb-Karlsson S, Belesova K, Boykoff M, et al. The 2019 report of The Lancet Countdown on health and climate change: ensuring that the health of a child born today is not defined by a changing climate. The Lancet [Internet]. 2019;394(10211):1836–78.

Available in : https://linkinghub.elsevier.com/retrieve/pii/S0140673619325966. Accessed 8 Aug 2022.

50. Social determinants of health [Internet]. Available in: https://www.who.int/health-topics/social-determinants-of-health. Accessed 8 Aug 2022.
51. Singer M, Clair S. Syndemics and public health: reconceptualizing disease in bio-social context. Med Anthropol Q dezembro de. 2003;17(4):423–41.
52. Horton R. Offline: COVID-19 is not a pandemic. The Lancet [Internet]. 2020;396(10255):874. Available in: https://linkinghub.elsevier.com/retrieve/pii/S0140673620320006. Accessed 8 Aug 2022.
53. Sirleaf EJ, Clark H. Report of the independent panel for pandemic preparedness and response: making COVID-19 the last pandemic. The Lancet [Internet]. 2021;398(10295):101–3. Available in: https://linkinghub.elsevier.com/retrieve/pii/S0140673621010953. Accessed 8 Aug 2022.
54. Monkeypox declared a global health emergency by the World Health Organization [Internet]. UN News. 2022. Available in: https://news.un.org/en/story/2022/07/1123152. Accessed 8 Aug 2022.

Equitable Access to COVID-19 Vaccines, Vaccine Research, and Vaccine Apartheid on the African Continent: Challenges and Recommendations

Ames Dhai

Abstract Globally no country was prepared for the devastation caused by the COVID-19 pandemic. The most brutal economic, social, and health impacts have been experienced among the most vulnerable. With the focus being on COVID-19, other infections and diseases emerged in Africa. Despite the evidence that COVID-19 vaccination will result in bringing the virus and its variants under control, numerous challenges have impeded wide-scale vaccination. These include vaccine apartheid (inequity), a lack of established and well financed public health systems, poor to no infrastructure for the development, supply, and administration of vaccines to its populations, lack of confidence by individuals, vaccine hesitancy, and corruption. Africa's reliance on foreign suppliers for vaccines impacts public health security. Research during the pandemic further exposed Africa's problems, which range from moral imperialism in international research to struggling research ethics committees. The African Union and the African Centers for Disease Control and Prevention have called for a New Public Health Order to safeguard the health and economic security of the continent, with a key pillar being that of expanding the local manufacture of vaccines, diagnostics, and therapeutics. Interventions must be considered a public good. Ethical leadership and governance are ethically essential.

Keywords Vulnerable · Equity · Vaccine apartheid · Public good · Corruption · Vaccine hesitancy · Moral imperialism

A. Dhai (✉)
School of Clinical Medicine, University of the Witwatersrand, Johannesburg, Gauteng, South Africa
e-mail: Ames.Dhai@wits.ac.za

C. Kurihara et al. (eds.), *Ethical Innovation for Global Health*,
https://doi.org/10.1007/978-981-99-6163-4_3

1 Introduction

Similar to other regions in the world, the African continent has been affected considerably by the COVID-19 pandemic with health, economic, and social impacts being most dire and unprecedented on individuals, families, and communities. However, the most brutal economic, social, and health impacts of the pandemic have been on the most vulnerable worldwide.

Globally no country was prepared for the devastation caused by the pandemic. Moreover, even countries with the best healthcare systems in the world did not have the public health capabilities to cope with the pressures brought on by the virus. Africa entered the pandemic era with already overburdened and in some cases failed health and other social systems [1]. As the pandemic progressed and with all attention being focused on managing it, other viral infections and diseases re-emerged on the continent. For example, Ebola re-surfaced in the Democratic Republic of the Congo, Lassa fever in Guinea and Liberia, Rift Valley Fever in Kenya, and Marburg Virus Disease in Nigeria, Sierra Leone, and Republic of Guinea [2–6]. Hence, in the absence of an effective, affordable, and easily available treatment regime, the value of vaccinating populations widely against SARS-CoV-2 must be underscored. It is well established that COVID-19 vaccination will result in bringing the virus and its variants under control. Unfortunately, due to numerous challenges the world has grappled with, despite confronting the virus with available vaccines, several variants of concern emerged causing a number of related waves in various regions, including Africa. Vaccine inequity, also known as vaccine apartheid is a substantive cause of the development of variants of concern and the World Health Organization has clearly articulated that for as long as this global injustice persists, the greater the potential for emergence of variants that could evade vaccines [7]. Other reasons for lack of access to COVID-19 vaccines in Africa include a lack of established and well financed public health systems [8], poor to no infrastructure for the development, supply and administration of vaccines to its populations, lack of confidence by individuals, communities, and populations in the vaccine (vaccine hesitancy) and corruption re-emerging as a parallel plague at a number of levels. Africa has also been subject to controversy in the context of COVID-19 vaccine research ranging from moral imperialism to the public health imperative to test novel COVID-19 candidate vaccines on the African population in background settings of the already struggling research ethics review processes being further pressurized.

The objectives of this chapter are to:

- describe the current context with regard to vaccine development and manufacture on the continent,
- discuss the African Union and Africa CDC approach to ensuring access to COVID-19 vaccines on the continent,
- analyze the notion of COVID-19 vaccines as a global public good and discuss the reasons for vaccine apartheid,
- examine impediments to access including corruption, vaccine hesitancy, and poor uptake of COVID-19 vaccines, and

- explore the ethical issues associated with COVID-19 vaccine research in Africa.

2 The Current Context with Regard to Vaccine Development and Manufacture on the Continent

Effective COVID-19 vaccines will limit person-to-person transmission and disease severity. Various vaccine development platforms have been advanced against the virus. These include the live attenuated virus, viral vectors, inactivated virus, subunit vaccines, recombinant DNA and protein vaccines [8]. Despite the initial delay in the pandemic reaching Africa, it rapidly picked up speed with many confirmed cases and deaths. For a significant period, Africa received only relatively small doses of vaccines. Frequently, the supplies were unpredicted and there was little warning on when the vaccines would arrive and what type of vaccines they would be [9]. Therefore the need for developing local manufacturing capacity and increasing that capacity where already present, together with ensuring procuring necessary raw materials and available infrastructure for transport and administration of the vaccines became evident very early in the vaccination process during the pandemic.

Challenges to vaccine manufacturing in Africa include weak investments for vaccine manufacturing by African governments; weak regulatory capacity for vaccine research, development, and production; low interest in vaccine production in Africa by global vaccine stakeholders; uncertainties in the demands for vaccines made in Africa by African countries [10]; manufacturers ceasing production as a result of inability to compete with imported vaccines; dependence on global supply chains, which at times are hampered by intellectual property issues, trade barriers, monopolized supply and export bans; dependability on external suppliers posing severe problems to Africa's health resilience; and a preference by African countries and their governments for ready-made vaccines that they can import [9]. Therefore the focus has been on securing deliveries rather than manufacturing. It is easier to secure funding or pool resources from foreign donors to facilitate payment and procurement as compared to the cost of vaccine development, preclinical testing, and clinical trials. While significant investments and grants, backed by the state are made available to produce vaccines in wealthier countries, African manufacturers face the high costs of vaccine production with little or no funding support from the states and a lack of political commitment in this context. This will have bearing on problems of sustainability of any vaccine manufacturing initiative undertaken. In addition to insufficient funds to medical scientists, research, and development, African technological innovations have been undermined to create markets for the Western-led products including vaccines [9]. Africa's reliance on foreign suppliers for vaccines impacts public health security, as evidenced during the COVID-19 pandemic. This means that Africa could remain last-in-line and face significant procurement challenges in future. These many longstanding barriers faced by African research centers and biotechnology facilities in accessing several important steps of the value chain, such as preclinical research and good manufacturing practice (GMP) batch manufacturing for clinical trials mean that it is currently not possible to move a vaccine concept from research through to clinical trials entirely on the African continent. Furthermore, projects cannot be advanced and commercialized because of the lack of required partners, sponsors, investors, technological transfer, and know-how. While expertise does exist in Africa, it is spread across the continent, with limited connections. Moreover, because of a paucity of open and well paid posts, graduates of training programs frequently leave their countries for opportunities in higher-income regions [11].

For decades, there have been huge concerns at the lack of vaccine production in Africa and prior to the pandemic, 99% of Africa's vaccines were manufactured outside the continent. Currently, only.1% of the global supply of vaccines is produced in Africa. Accordingly, millions of lives had been endangered and social and economic progress on the continent impeded. This is a market failure which needs to be addressed urgently. Seven out of ten vaccines used in Africa are donated to the continent by the Global Vaccine Alliance (Gavi). Most are for childhood immunization programs and are manufactured in India or by multinational vaccine manufacturers in North America or Japan [12]. It can be stated that these donations perpetuate dependence and add to the impediments to development of vaccines and other interventions against diseases in Africa. Moreover, donations could result in the perception among donees that the challenges resulting from patent law are no longer an issue that needs to be confronted. The COVID-19 pandemic underscored how fatal this dependency on imported vaccines could be. In addition, Africa cannot rely on fellow states in the Global South. This is well illustrated by India halting vaccine supplies to the continent early in 2021, at the height of the delta variant outbreak when only 1.5% of the African population had received a vaccine dose at that time [12].

Fortunately influential role players on the continent like the Presidents of Rwanda and South Africa and African private sector business executives have seen the need for and support the establishment of an African pharmaceutical industry to supply Africa's needs albeit building such an entity could take longer than two decades and cost billions of dollars. Moreover, a foundation to provide financial and strategic support for the development of the pharmaceutical industry and the consolidation of regional vaccination programs in Africa has been established by the African Export-Import Bank and the African Development Bank. Currently, about 33% of African countries pay for their vaccine needs. The Partnership for African Vaccine Manufacturing (PAVM) projects that the value of the total African market could reach between $3billion and $17biliion by 2040 [12].

Development and/or expansion of national vaccine industries commenced in several countries from 2020. Egypt, South Africa, and Senegal have each partnered with private sector manufacturers to expand volume capacity. Ghana has reconfigured part of its pharmaceutical industry to make vaccines. Rwanda has commenced on work to manufacture mRNA vaccines from scratch. These countries have moved forward, signed major agreements, and continue raising finance. Nevertheless, there is a need for coordination at a continental or even regional level to assist with ensuring long-term sustainability and equitable access to vaccines [11, 13].

Even though only 1% of vaccines used in Africa are manufactured on the continent, there are more than 30 new vaccine manufacturing initiatives underway, with momentum gathering to make the expansion possible. The first technology transfer hub for messenger RNA COVID-19 vaccines was established in South Africa in the second half of 2021 to scale up production and access to doses across the continent. The hub also provides training on mRNA technologies for manufacturers from low- and middle-income countries (LMICs) and supplies them with licenses so they can move forward with manufacturing. The World Health Organization (WHO) has taken the lead in establishing the hub in partnership with partners in the international COVAX initiative, Biovac, a South African company which serves as a vaccine developer, Afrigen Biologics and Vaccines, a South African company which serves as a manufacturer, universities, and the Africa Centres for Disease Control and Prevention (Africa CDC). The WHO has received assistance from the Medicines Patents Pool to negotiate with technical partners and support governance of the hub [14]. In February 2022, South African scientists at this hub reproduced Moderna's COVID-19 vaccine, achieving a major milestone for the hub and the continent [15].

3 The African Union and Africa CDC Approach to Ensuring Access to COVID-19 Vaccines on the Continent [11]

To address the many bottlenecks described above, The African Union (AU) and the African Centers for Disease Control and Prevention (Africa CDC) have called for a New Public Health Order, the aim of which is to safeguard the health and economic security of the continent, with a key pillar being that of expanding the local manufacture of vaccines, diagnostics, and therapeutics. In April 2021, at a gathering of African leaders, the Partnerships for African Vaccine Manufacturing (PAVM) was established by the African Union (AU) to deliver on the goal of enabling the African vaccine manufacturing industry to develop, produce, and supply over 60 percent of the total vaccine doses for Africa by 2040. The interim goals are 10 percent by 2025 and 30 percent by 2030. They do acknowledge that the vaccine ecosystem is a dynamic one and the initiative is quite ambitious, as 1.5 billion doses of vaccines per year will have to be produced by 2040. The need for this massive and complex undertaking is evident as the COVID-19 pandemic made it clear that the supply of outbreak vaccines remains uneven and unreliable, even though supply of routine vaccines has been stable. While vaccine manufacturing in Africa is currently nascent, this will have to change with increasing demand. Self-financing countries make up 30% of the African vaccine market, an amount of $419 million. It has been projected that the share of self-procuring countries will grow in the next decade with countries moving away from Global Alliance for Vaccines and Immunization (Gavi) support. 90 percent of total vaccine production volumes are currently for Gavi supported countries where the spend is only about one-third of the price per dose compared to self-financing countries.

For the vision of the New Public Health Order to be realized, an integrated ecosystem approach will be necessary. Investment will be required in all steps of the vaccine manufacturing supply chain. This includes research and development (R&D), drug substance, and fill and finish. R&D will need to increase to include preclinical and clinical trials. Supporting industries will need to grow as well for the provision of raw materials, active ingredients, inactive ingredients, and consumables like vials, sterile bottles, syringes, and rubber stoppers. Key benefits of developing an enabling environment include reducing production costs and increasing sustainability of vaccine production on the continent, thereby promoting self-reliance and health security.

In order to scale up vaccine manufacturing a Framework for Action (FFA) has been developed by the AU and Africa CDC, which has identified eight enablers and has commenced work on the following programs to take forward these enablers:

- *Program 1: Creating an African Vaccines Procurement Pooling Mechanism*

- There is emerging consensus that a pooled-procurement mechanism can be used to procure routine vaccines for all AU countries by 2040. This will allow for achieving sustainable and reliable volumes with economies of scale. Currently, there is an existing entity, the African COVID-19 Vaccine Acquisition Task Team (AVATT). The AVATT is to be expanded to include products beyond the COVID-19 vaccines and countries not supported by or transitioning from Gavi could be incentivized to procure vaccines through AVATT. In addition, expertise from organizations such as Africa Medical Supplies Platform (AMSP), Afreximbank, and UNICEF will be drawn upon.
- *Program 2: Establishing a vaccine manufacturing Deal Preparation Facility and supporting fundraising for ecosystem enablers*
- Sustained investment into manufacturing capacities and the broader enabling ecosystem will be necessary for Africa's vaccine industry to develop. The program works to overcome several long-standing challenges that are an impediment to the vaccine manufacturing industry, which include limited numbers of truly bankable projects and an environment that is not conducive to investment. Machinery and equipment need to be imported and setting up facilities is exorbitantly costly with return on investment for vaccine research being lower than other biopharmaceutical opportunities. This has to be viewed in light of the long-term economic growth to the continent from the vaccine industry which includes the potential for exporting vaccines internationally. Establishing a vaccine manufacturing deal preparation and financing facility with a one-stop-shop approach would be critical to confronting these challenges and to create a meaningful pipeline of viable projects of different risk profiles. Fundraising initiatives through the PAVM to support the enablers would complement such a facility.
- *Program 3: Strengthening National Regulatory Agencies and Regional Centers of Regulatory Excellence to build vaccine regulatory excellence*

- It is critical that well-functioning National Regulatory Authorities (NRAs) are developed. NRAs are to be strengthened so that African manufacturers are able to achieve World Health Organization prequalification (WHO PQ), which is a prerequisite for vaccine manufacturers to export products. African NRAs will need to achieve at least WHO maturity level three (ML3) status. Given that most African NRAs are still at ML1 or ML2, PAVM will work through the African Medicines Regulatory Agency and the Africa Medicines Regulatory Harmonization Program to strengthen the regulatory system for vaccines and develop a harmonized African ecosystem.
- *Program 4: Supporting the transfer of vaccine technologies and intellectual property through a TT and IP Enablement Unit*
- Local manufacturing will require an estimated minimum of 23 technology transfers in the next 20 years. Through this program a vaccine technology transfer and intellectual property (IP) enablement unit will be established to facilitate the transfer of technologies and IP by addressing the many barriers so that successful execution of technology transfers on the continent is accelerated. The progress in technology development during the COVID-19 pandemic will be further enhanced. Knowledge gaps will be addressed by transferring the technologies and know-how from experienced manufacturers to local manufacturers.
- *Program 5: Creating Regional Capability and Capacity Centers to support talent and critical skills development*
- The total vaccine workforce will need to quadruple to approximately 12,500 full-time employees (FTEs). Hence, about 10,500 new FTEs will need to be trained. This is based on 9,500 new jobs and 10 percent brain drain. Through this program, Capability and Capacity Centers (CCCs) will be created to address identified gaps in expertise and associated challenges. Partnerships between research institutions, manufacturing companies, and educational institutions will be fostered in order to build a much stronger bridge between education and full-time work.
- *Program 6: Putting in place Vaccine Research and Development Centers and a Research and Development Coordinating Platform*
- Through this program a system of regional R&D centers will be established connected by a single continental R&D coordinating platform. The function of the R&D centers will be that of consolidating infrastructure, assets, and expertise in each region. The R&D coordinating platform will work on setting up continental R&D priorities and collating funding for use at the R&D centers. The platform coordinates collaboration opportunities with established international or continental organizations.
- *Program 7: Undertaking Advocacy for enabling trade policies for vaccines*
- This program will address the problem of overall trade integration being low across the different regions of the continent. Trade bans during the pandemic period have also underscored the risk of individual country protectionism during outbreaks. Trade policy within regions and member states will be enabled through a number of projects to be facilitated through the program.
- *Program 8: Ensuring an effective Continental Strategy for delivery and oversight*

- Several parallel programs and supporting initiatives and pilots are planned, including the consolidation and oversight of ongoing vaccine manufacturing projects, the creation and maintenance of investment relationships, the streamlining of existing and proposed capability and capacity centers, and advocacy for supportive policies across the ecosystem.

This plan by the AU and Africa CDC is laudable albeit ambitious. It will require positive political will and shared responsibilities from within individual countries on the continent to see it executed successfully. Currently, the pandemic of corruption in Africa poses a significant impediment to achieving the envisaged New Public Health Order. What is also required to successfully implement this plan are ethical African leadership and governance.

4 COVID-19 Vaccines: A Global Public Good VS Vaccine Apartheid

The COVID-19 pandemic exposed the excessive self-egotism and greed of many of the globe's rich countries who, in this unprecedented crisis entirely disregarded the global nature of the problem [16]. By March 2021, globally, 559 million doses had been administered, with 10% of the economies, accounting for 77% of the total number vaccinated [17]. This situation made apparent the depth of unfairness globally and that the right to health for all was once again being denied. As with other crises, exacerbations of pre-existing inequalities across the world became patently evident, with the most vulnerable being affected the most. At that stage, even the United Nations strongly stipulated that vaccine equity affirms human rights and that vaccine nationalism denies it, stating also that vaccines must be a global public good, accessible and affordable to all [18].

At this juncture, it would be prudent to have an elucidation of what a global public good is and why it should apply to the COVID-19 vaccines. Generally it is understood as a good whose impacts are equitably spread across the globe without causing division [19]. No price can be placed on the benefits of these goods and hence the principle of exclusion cannot be applied to them. When one individual uses these goods, their availability to others cannot be allowed to be reduced [20]. The good and its benefits are not marketable and therefore must be available at negligible or zero cost to all in the global village [19]. Thus, two criteria determine a public good: non-rival in consumption and being non-excludable. Non-rivalry entails that the consumption by one person must not interfere with the goods being available to others equally [20]. For goods to be non-excludable, suppliers cannot deny them to those who are unable to pay its market price. In the arena of public goods and costs, climate change is often quoted as an example. Climate change is usually caused by unsustainable practices of the richest countries but impacts all regions and in particular the poorest [19]. Because of the transboundary implications of public goods, international cooperation and action is ethically imperative.

Reflecting on what we witnessed with COVID-19 vaccines, it becomes clear that for rich economies, the notion of global public good did not form part of their lexicon. What we saw instead was morally unconscionable vaccine apartheid. The term apartheid is derived from the South African Afrikaans language and means "the state of being apart" or "separateness." It illustrates the situation of systematic institutionalized racial discrimination that was present in Southwest Africa and South Africa until their transition to democracy [21]. It is an apt description of the "separateness" from vaccine access and programs for LMICs. From late 2020, while many high-income countries (HICs) reduced severe outcomes against COVID-19 with vaccination, LMICs globally and in particular in the poorest countries suffered mortality and severe morbidity due to vaccine inequity [22]. Early in November 2020, HICs were exposed to being well into a "shopping spree" for COVID-19 vaccines. A global assessment of purchasing agreements for the vaccine revealed that HICs and a few middle-income countries with manufacturing capacity had already purchased almost 3.8 billion doses, with options for another 5 billion. The USA already had agreements to buy up enough doses for 230% of its population, and with time could control about 1.8 billion doses, i.e., a quarter of the world's supply [23]. This COVID-19 vaccine shopping and hoarding meant that the populations in many of these countries could be vaccinated several times over while billions of people in poorer countries would have to wait until much longer. This was an unambiguous repeat of the H1N1 influenza pandemic in 2009, when nearly all the vaccines that were available were bought up by HICs [24]. It was only after these countries received adequate supplies that LMICs were allowed access [25], with the vaccines arriving in most of these countries when the pandemic was over. The WHO, in its concern of a repeat of vaccine nationalism spearheaded a coalition, the COVID-19 Vaccine Global Access (COVAX) facility with the Global Alliance for Vaccines and Immunization (Gavi), and the Coalition for Epidemic Preparedness Innovations (CEPI). This was an ambitious effort to create equitable access to effective vaccines globally. It tried to create a global risk-sharing mechanism for pooled procurement and equitable distribution of the vaccines when registered for use [26]. However, several signatories to the facility, including Canada, the UK, and the European Union, by negotiating bilateral deals directly with industry through advanced market commitments, undermined COVAX's goal to counter vaccine nationalism, and hence paid lip-service to its principles of global equitable access and fairness [24].

Vaccine apartheid led to those who had the means and power to procure vaccines to hurriedly clinch the "me first" deals. The poorest and most vulnerable in the world were left behind. By mid-January 2021, more than 39 million doses were administered in 49 higher-income countries, but almost no vaccines had been administered in low-income countries. The Director General of the WHO stated that there had been a "catastrophic moral failure" in the sharing of COVID-19 vaccines [27]. Given this approach by rich countries, it is not surprising that harms to public health and the global economy are perpetuated. Already, at that stage, it was shown that twice as many deaths could have been averted, and it could cost the global economy up to USD 1.2 trillion in GDP if vaccine inequity continued and LMICs were not granted equitable access to this public good [27]. In the meantime, the virus continued disrupting global supply chains and economies in the world as LMICs struggled to procure vaccines and protect their populations. HICs seemed to be slow in understanding that poor vaccine acquisition and rapidly replicating viruses were the perfect ingredients for the development of mutations and variants, and that global herd immunity would not be attainable if vaccine supply to LMICs lagged behind. Moreover, viruses ignore national boundaries which remained porous to the virus during the pandemic [24].

An additional complexity to the notion of COVID-19 being a public health good was the "business as usual" implementation of intellectual property rights (IPR) by vaccine manufacturers. Several HICs robustly opposed the application lodged by South Africa and India to the World Trade Organization (WTO) for a temporary waiver of IPR for COVID-19 vaccines during the pandemic [28]. A waiver, together with technology transfer and building of infrastructure would assist in ensuring fair and equitable access to the much-needed public good. Given that manufacturers were resistant to sharing, and that they were supported by HICs, the non-excludable criterion for the vaccine being a public good was impeded. They did not even consider the many calls made by the UN and WHO for global equitable access to the vaccines. Moreover, they seemed oblivious to the UNESCO's Statement on Global Vaccine Equity and Solidarity which refers to the COVID-19 vaccines as a "Global common good" (article3). The Statement firmly rejected vaccine nationalism as a "predatory rush," raised ethical concerns on the regulation of patenting and ownership rights and stressed that responses to the pandemic needed to be built on equality, justice, and solidarity. For true equity in the global access to the vaccines, a shared understanding of health as a global common good without territorial limits and new global legal instruments for economic and political treaties were required. The Statement further highlighted that the Agreement on Trade Related Aspects of Intellectual Property Rights (TRIPS) and the agreements of the WTO were not designed to manage situations such as pandemics [29].

Despite the challenges outlined above, i.e., the global risks of vaccine nationalism, the erosion of global public health principles, and calls being made by several international bodies and scholars [30], currently no international law makes legal provision for the implementation of interventions, even in a pandemic, as a global public good. International legal reform, which takes into consideration international solidarity, fairness, and multilateral support, is key as we prepare for future pandemics. Moreover, it must include a thorough review of the TRIPS agreement and an immediate ban on any of the many TRIPS+ bilateral agreements.

5 Impediments to Access Including Corruption, Vaccine Hesitancy, and Poor Uptake of COVID-19 Vaccines

Poor infrastructure for the development, supply, and administration of vaccines to populations in Africa coupled with the response to this by the African Union and Africa CDC has been described in detail above. The impact of vaccine apartheid has been discussed in the previous section. Most countries in Africa were already struggling with poor health infrastructure before the onset of the COVID-19 pandemic. Negatively politicizing the vaccine in Tanzania also led to obstructions in access. The immediate response of the late president was to downplay the severity and seriousness of the COVID-19 pandemic including doubting the vaccine [31]. Early in 2021, the health minister announced that there were no plans in place to accept the vaccines [32]. Fortunately, with the swearing in of the new president, there was a change in approach and COVID-19 vaccines were accepted into the country [31]. There is little doubt, however, that the seeds for vaccine hesitancy had already been implanted in the country. In this section, the effect of corruption and vaccine hesitancy is considered.

6 Corruption

Corruption, which represents an abuse of trust and intentional violation of duty is described as the "use, misuse, or abuse" of public office or resources for private gain may be actual or potential, financial, or even political [33]. Corruption impacts negatively on population health outcomes, in particular for the poor and disadvantaged. The devastating effects of the COVID-19 pandemic were worsened by widespread and pervasive corruption on the African continent, some of which were from funds from multilateral organizations. Several donor agencies supported African governments with various financial packages. Together, the International Monetary Fund and World Bank together provided about $57-billion to help Africa [34].

Procurement of personal protective equipment, essential goods and services, and vaccines was fast-tracked. Given that most of these items were procured under a certificate of emergency, public scrutiny was evaded and accountability measures to monitor the use of these funds were relaxed. Corruption related to COVID-19, mainly in procurement, was seen across the continent. A 2021 audit in Cameroon revealed the misuse of about $333-million meant for the pandemic response in 2020. In Malawi some government officials colluded with the private sector and squandered $1.3 million of COVID-19 funds through procurement and allowance irregularities. In Kenya, about $400 million which was meant to buy medical equipment was allegedly stolen by the Kenya Medical Supplies Authority. In addition, the Kenyan Ethics and Anti-Corruption Commission revealed further irregular expenditures of about $71.96 million. The Kenyan Auditor-General's report on COVID-19 expenditure revealed that over $69-million of COVID-19 funds had been misused.

The Federal Ministry of Health in Nigeria allegedly bought 1,808 face masks for $96,000. Uganda saw four top officials being arrested for allegedly overpricing COVID-19 food relief items, leading to a loss of $528,000. The health minister in Zimbabwe was dismissed reportedly for illegally awarding a multimillion-dollar contract which resulted in inflating the cost of medical equipment. In Ghana, the health minister purchased the Russian Sputnik V vaccine at a unit price of $19 instead of the $10 factory price and in South Africa the health minister was placed on leave while irregular contracts of about $10 million were being investigated. He subsequently resigned. In addition, country-wide, there was public anger over the suspected inflation of government contracts for the purchase of $900-million worth of medical supplies [34].

South Africa was the most affected country on the continent with COVID-19, given the large number of infections and deaths as compared to the rest of Africa. It would seem that South Africa was the country most steeped in COVID-19 corruption as well. As early as April 2020, the National Treasury relaxed its procurement regulations in order to make it easy for government entities to procure required materials for emergency response. This lowering of the threshold for accountability provided fertile ground for perpetuation of corruption in an already corrupt country and opened the floodgates of corruption by both political leaders and the private business community. Billions of South African Rands were made available by the government, attracting corrupt individuals and groups in the political and business sectors to syphon the money. Just some examples are: medical equipment and goods were supplied by companies not registered nor licensed to supply them; prices were inflated by some companies registered to supply medical equipment; politicians pushed for tenders to be awarded to relatives and friends; and relief goods for the poor were stolen to use as political capital for campaigning, or for distribution to families and friends [35].

Civil society response to this corruption was phenomenal and a number of different activities were taken to expose the corrupt. Investigations were undertaken by established organizations on combating corruption, and reports were produced for the public and the government. A number of civil society organizations combined to petition the government to act on the exposed corruption. Equally active was the media in its investigation and making public of the corruption. Media also urged government to redress these ills [35]. Hence, civil society and media were instrumental in highlighting the issues and compelling government to act. The South African government had no option but to implement consequence action in order to hold the perpetrators accountable. Such corruption during a pandemic is morally offensive. South Africa's Bill of Rights of its Constitution is one of the most progressive globally. Government officials take an oath when they are sworn in to uphold and protect the Constitution. There was an unashamed disregard of their oaths when members within the government structures were involved in the COVID-19 corrupt activities.

Over and above the civil society reaction and media response, what is required in Africa and probably other parts of the world are ethical leadership and ethical governance processes. Implementation of these processes is ethically imperative. With trust in governments in Africa already being low at the onset of the COVID-19 pandemic, taking advantage of the relaxation of accountability processes because of the emergency to further their corrupt actions clearly deepened the distrust even further, such that individuals and communities were suspicious of and resisted uptake of preventative measures instituted by the state to curb spread of the virus.

7 Vaccine Hesitancy

At the time of writing this chapter, the COVID-19 pandemic is in its third year and is not yet over. In Africa, with only 11% of its adult population fully vaccinated, there is risk of COVID-19 resurgence in several states because of low testing and low vaccination rates and poor adherence to public health measures. Moreover, mass gatherings continue. Many health systems are struggling or depleted. Care for other diseases and infections must also be provided for [2]. While much has been said about lack of equity, once vaccines are available for use, achieving immunity in communities is largely dependent on their willing uptake by populations. Any initiative toward allocating vaccines fairly would fail if people were unwilling to receive them. It has been stated that vaccine hesitancy, i.e., the reluctance to receive vaccines is one of the top ten threats to global health [2].

The much anticipated COVID-19 vaccines managed to curb the disease substantially. However, currently, the pandemic is still with us despite decline in disease severity, hospitalizations, and deaths. Moreover, vaccine hesitancy and refusal continue globally. Vaccine hesitancy is a complicated phenomenon. Instrumental factors propelling vaccine hesitancy are context-specific and include concerns about safety and efficacy, lower education, mistrust in science, health authorities, governments, and misinformation. Vaccine acceptance could also be curbed because of the limited efficacy of the current COVID-19 vaccines in preventing infection against new circulating variants [34]. In Africa, additional factors for vaccine hesitancy being high among certain population groups were bad experiences during previous frequent unethical medical experiments and poor messaging. Given that access to technology on the continent improved during this pandemic, the use of social media increased. Concerns have been raised with regard to the spread of misinformation across different social networks proliferated through the anti-vaccination movement. Where social media was used to spread healthy messages by healthcare workers, compliance with public health strategies improved [2]. Socio-demographic factors also played a role in vaccine hesitancy. For example, vaccine hesitancy was found to be significantly higher in males in Nigeria, not having a university degree in South Africa, and those earning more than the median income in Ghana [36].

As we continue our journey through COVID-19 and prepare for the next pandemic, it would be prudent for us to include programs to counter resistance to public health measures, including early, honest, culturally appropriate, and focused messaging by our governments and health authorities. This should not only be through mainstream media but also via use of technologies including social media. Policies to counter spread of misinformation could be considered. Most importantly though, is that governments need to work very hard at regaining trust, because vaccine hesitancy is high where there is mistrust of governments. For the trust deficit to be eradicated, it is requisite that governments exercise political will to tackle corruption head-on.

8 Ethical Issues Associated with COVID-19 Vaccine Research in Africa

Globally, there are many challenges to conducting research during a pandemic. Some of these challenges are specific to LMICs. A pandemic in itself puts pressure on the need for research, in particular where treatment options are limited and mortality and morbidity are high. While expedience is necessary, it does not justify unethical practice. Local contexts need to be understood, voices from the community and other relevant stakeholders heard and included, dignity respected, and rights upheld [37].

Early in the pandemic (April 2020), two French doctors participated in a television debate where they discussed whether COVID-19 vaccine trials should be done in Africa. They suggested that the trials should be done on the continent considering there were no masks, no treatments, no resuscitation, and hence the populations would be highly exposed. Considering that this would not be acceptable in HICs, these two doctors proposed a perpetuation of the double standard in research [38]. Despite these two doctors being forced to publicly apologize, it became patently clear that ethics dumping from HICS continues in Africa despite international and local norms, standards, and guidelines. What is also underscored is that moral imperialism and colonialist thinking in some multinational clinical trials persist and perhaps researchers from HICs see no need for ethical guidelines to apply in regions like Africa where research that will not be allowed in their countries can be outsourced to the continent [38].

It is therefore not surprising that communities in Africa are suspicious of researchers from HICs and will be reluctant to participate in multinational clinical trials. Hence, building trust among these communities is essential. It is critical that local communities are consulted and engaged with early on during the planning of research, and in particular pandemic vaccine research [37]. Because of the pressure for research to be carried out speedily, balancing the benefits vs the harms to individuals and communities will be requisite. In addition, respecting autonomy and dignity must be ensured albeit language and cultural barriers. This entails that research participants will need to be properly informed and understand the risks and benefits before being enrolled in studies.

A further challenge is the capacity of research Ethics Committees (RECs) to review and approve the research in a timely manner. African governments and institutions have not adequately invested in setting up RECs. Poor infrastructure leading to delays and inefficiencies in the administrative processes is commonplace. However, during a pandemic, rapid review and approval of novel approaches are necessary to avoid delays in research. RECs on the continent did struggle to respond to the needs of the COVID-19 pandemic. REC over-reach and paternalism was also found to be a problem. For example, in one of the sites in South Africa, after unblinding in the Chadox/AstraZenica vaccine study, participants under 30 years were not allowed by the REC from receiving the investigational product, contrary to the regulatory and other REC approvals [37].

Because of the unconscionable vaccine nationalism as discussed above, going forward, it would be essential that the principle of proportionality is implemented in research. The clinical trials were conducted globally and included LMICs. Hence, both HICs and LMICs contributed to the development of the vaccines. Justice as in fairness makes it an ethical requirement that all countries contributing to the research are given a fair chance to access these interventions if proven efficacious. The principle of proportionality could facilitate the application of fairness by making it necessary for countries to be given the opportunity to purchase the vaccines in line with the proportion of research participants enrolled and their contribution to the studies. This could mean that Africa would not be last in the queue to purchase vaccines, as has been the case until now [37].

9 Conclusion

This chapter has briefly discussed some pertinent aspects of the COVID-19 pandemic and the vaccine in Africa. As the COVID-19 pandemic evolves, many key issues have arisen necessitating critical reflection at global, regional, and country levels. National pandemic preparedness has been low on the priority list of most African countries where failing or failed healthcare infrastructure was further exposed by the current pandemic. Equitable access to health care and in particular to interventions developed for management of pandemics as a global public good needs to be considered a priority with regard to international policy making. Ethical leadership and governance and the eradication of corruption are necessary, especially if the populations are to take public health preventative measures seriously. African governments must be trustworthy and must ensure accurate, reliable messaging for citizens to transition from vaccine hesitancy to vaccine confidence. Ethical imperialism and ethics dumping in Africa and other LMICs has to be safeguarded against, even at the height of a pandemic. It is hoped that the initiatives emanating from the African Union and Africa CDC are successfully implemented and not left as paper to gather dust on the shelves of the respective organizations.

Conflict of Interest There is no conflict of interest to be disclosed related to the content of this manuscript.

References

1. Dhai A, Ballot D, Veller M. Introduction to pandemics in healthcare: principles, processes and practice. In: Dhai A, Ballot D, Veller M, editors. Pandemics in healthcare: principles, processes and practice. 1st ed. Cape Town: Juta and Company (Pty) Ltd; 2021. p. 1–10.
2. Osuagwu, et al. The impact of information sources on COVID-19 vaccine hesitancy and resistance in sub-Saharan Africa. BMC Public Health. 2023;23:38. https://doi.org/10.1186/s12889-022-14972-2.
3. Riaz MMA, Ahmad U, Mohan A, Dos Santos Costa AC, Khan H, Babar MS, et al. Global impact of vaccine nationalism during COVID-19 pandemic. Trop Med Health. 2021;49(1):101.
4. Uwishema O, Okereke M, Onyeaka H, Hasan MM, Donatus D, Martin Z, et al. Threats and outbreaks of cholera in Africa amidst COVID-19 pandemic: a double burden on Africa's health systems. Trop Med Health. 2021;49(1):1–5.
5. Okonji OC, Okonji EF, Mohanan P, Babar MS, Saleem A, Khawaja UA, et al. Marburg virus disease outbreak amidst COVID-19 in the Republic of Guinea: a point of contention for the fragile health system? Clin Epidemiol Glob Health. 2022;13:100920.
6. Hasan MM, Mohanan P, Emmanuella N, dos Santos Costa AC, Ahmad S, Mousavi SH, et al. Challenges of HIV amidst COVID-19 in Africa: can we conquer them? J Med Virol. 2022;94(1):11–3.
7. WHO Director-General's opening remarks at the 71st session of the Regional Committee for Europe. 2021. Accessed 16 Jan 2023.
8. Kentikelenis A, King L, Martin McKee M, Stuckler D. The International Monetary Fund and the Ebola outbreak. The Lancet. 2015;3:e69–70. https://doi.org/10.1016/S2214-109X(14)70377-8.
9. Lamptey E, Senkyire EK, Benita DA, Boakye EO. COVID-19 vaccines development in Africa: a review of current situation and existing challenges of vaccine production. Clin Exp Vaccine Res. 2022;11:82–8. https://doi.org/10.7774/cevr.2022.11.1.82.
10. Nweneka CV, Disu T. The future of vaccine manufacturing in Africa. Brookings. 2022. Foresight Africa 2022 (brookings.edu). Accessed 16 Jan 2023.
11. African Union, Africa CDC. Partnerships for African Vaccine Manufacturing (PAVM) framework for action. https://africacdc.org/download/partnerships-for-african-vaccine-manufacturing-pavm-framework-for-action/. Accessed 16 Jan 2023.
12. Erondu N. Now is the moment to launch an African vaccine industry. Chatham House. International Affairs Think Tank. 2022. https://www.chathamhouse.org/publications/the-world-today/2022-08/now-moment-launch-african-vaccine-industry. Accessed 16 Jan 2023.
13. Berkley S. Ramping up Africa's vaccine manufacturing capability id good for everyone. Here's why. World Economic Forum Annual Meeting. 2022. Why ramping up African vaccine manufacturing benefits all. World Economic Forum (weforum.org). https://www.weforum.org/agenda/2022/10/ramping-up-africa-s-vaccine-manufacturing-capability-is-good-for-everyone-heres-why/. Accessed 16 Jan 2023.
14. Jerving S. New tech hub aims to boost access to mRNA COVID-19 vaccines in Africa. Devex. 2021 https://www.devex.com/news/new-tech-hub-aims-to-boost-access-to-mrna-covid-19-vaccines-in-africa-100193. Accessed 16 Jan 2023.
15. Editorial. Africa is bringing vaccine manufacturing home. Nature. 2022;602:184. https://doi.org/10.1038/d41586-022-00335-9. Accessed 16 Jan 2023.
16. Dhai A. Access to COVID-19 Vaccines as a global public good: a coordinated global response based on equality, justice and solidarity is key. S A fr J Bioethics Law. 2021;14(1):2–3. https://doi.org/10.7196/SAJBL.2021.v14i1.768.
17. Email from Dr Tedros Adhanom Ghebreyesus: communications@who.int. 1 Apr 2021.
18. UN Security Council demands COVID-19 vaccine ceasefires; WHO pushes for more actions to speed up inoculations. 2021. https://news.un.org/en/story/2021/02/1085942. Accessed 1 Mar 2023.

19. van den Berg RD. Evaluation of the funding of global public goods note for the OECD/DAC evaluation network. 2015. https://www.oecd.org/dac/evaluation/Evaluation-of-Global-Public-Goods-Evalnet-note.pdf. Accessed 01 Mar 2023.
20. Chowdappa VA. Public goods vs private goods and merit goods. 2016. https://www.slideshare.net/CHOWDAPPAVA/publicprivate-and-merit-goods. Accessed 1 Mar 2023.
21. Lanziotti VS, Bulut Y, Buonsenso D, et al. Vaccine apartheid: this is not the way to end the pandemic. J Paediatr Child Health. 2022;58:228–31.
22. Prasad S, Shahid A, Lorraine E, et al. Vaccine apartheid: the separation of the world's poorest and most vulnerable and the birth of Omicron. Ther Adv Vaccines Immunother. 2022;10:1–5. https://doi.org/10.1177/25151355221107975.
23. Launch and Scale Speedometer. Africa: new study shows rich country shopping spree for COVID-19 vaccines could mean fewer vaccinations for billions in low-income countries. All Africa. 2020. https://allafrica.com/stories/202011020103.html. Accessed 29 Nov 2022.
24. Dhai A. COVID-19 vaccines: equitable access, vaccine hesitancy and the dilemma of emergency use approvals. S Afr J Bioeth Law. 2020;13(2):77–8.
25. Usdin S. COVAX created to try to avoid global bidding frenzy for COVID-19 vaccines. Biocentury. 2020:1–3.
26. Global alliance for vaccines and immunisation. COVAX. Gavi. 2020. https://www.gavi.org/covax-facility. Accessed 29 Nov 2022.
27. Vaccine nationalism, global public good and the poor. 2021. http://www.icccad.net/dailystar-articles/vaccine-nationalism-global-public-good-and-the-poor/. Accessed 1 Dec 2022.
28. Quin C. Rich vs poor (again) at WTO. 2021. https://foreignpolicy.com/2021/03/10/wto-intellectual-propert-waiver-india-south-africa/. Accessed 1 Dec 2022.
29. UNESCO's Ethics commissions' call for global vaccine equity and solidarity. https://unesdoc.unesco.org/ark:/48223/pf0000375608. Accessed 1 Dec 2022.
30. Greco D, Rueda E. A critical commentary on the arguments against waiving vaccine patents. Commentaire critique sur les arguments contre la renonciation aux brevets sur les vaccins. IBC Newsletter. 2021;2:16–20. https://unesdoc.unesco.org/ark:/48223/pf0000378406.
31. Hamisi NM, Dai B, Ibrahim M. Global Health Security amid COVID-19: Tanzanian government's response to the COVID-19 pandemic. BMC Public Health. 2023). https://bmcpublichealth.biomedcentral.com/articles/10.1186/s12889-023-14991-7. Accessed 23 Mar 2023.;23:205. https://doi.org/10.1186/s12889-023-14991-7.
32. Makoni M. Tanzania refuses COVID-19 vaccines. The Lancet. World Report. 2021;566. https://www.thelancet.com/journals/lancet/article/PIIS0140-6736(21)00362-7/fulltext Accessed 22 Mar 2021.
33. Naher N, Hoque R, Hassan MS, Balabanova D, Adams AM, Ahmed SM. The influence of corruption and governance in the delivery of frontline health care services in the public sector: a scoping review of current and future prospects in low and middle-income countries of south and south-east Asia. BMC Public Health. 2020;20:880. https://doi.org/10.1186/s12889-020-08975-0.
34. Aikens ER. Institute for Security Studies. Corruption in Africa deepens the wounds of COVID-19. https://issafrica.org/iss-today/corruption-in-africa-deepens-the-wounds-of-covid-19. Accessed 25 Mar 2023.
35. Aseulime RA, Ndimande-Hlongwa N. Civil society, COVID-19 pandemic and state corruption implications for nation building in South Africa. J Nat Build Policy Stud (JoNPS). 2022;163–81. ISSN: 2516-3124 (Print) ISSN: 2516-3132 (Online).
36. Lazarus JV, Wyka K, White TM. A survey of COVID-19 vaccine acceptance across 23 countries in 2022. Nature Medicine. 2023;29:366–75. https://doi.org/10.1038/s41591-022-02185-4.
37. Dhai A, Gray G, Mokhachane M, Mdhluli M. Health research, ethics and pandemics: challenges and recommendations. In: Dhai A, Ballot D, Veller M, editors. Pandemics in healthcare: principles, processes and practice. 1st ed. Cape Town: Juta and Company (Pty) Ltd; 2021. p. 75–93.
38. Hellman F, Williams-Jones B, Garrafa V. COVID-19 and moral imperialism in multinational clinical research. Arch Med Res. 2020;51:572–3.

Response to COVID-19 Pandemic and Ethical Innovations in Taiwan

Ian Chen, Yvonne Chen, and Daniel Fu-Chang Tsai

Abstract Taiwan has established itself in recent decades as a democracy that values and protects human rights. By extension, the rights and well-being of research subjects are clearly defined through legislation. Two notable cases detailed here are the Biobank Act and the law concerning research involving indigenous people. During COVID-19, however, citizens were quick to surrender their privacy and freedom in support of intrusive public health measures that aimed to combat the pandemic. This research investigates the reasons behind this swift shift in general public's tolerance toward intrusion of privacy and limitation of freedom. The results show that Taiwan's public health history, social and geopolitical background, and geographical and demographic characteristics all played a role in successful COVID-19 policy compliance. Additional factors contributing to the country's effective management of COVID-19 were innovative contact tracing design and proactive communication strategies.

Keywords COVID-19 · Pandemic · Democracy · Biobank · Indigenous people

1 Introduction

Having transitioned from an autocratic regime to a multi-party democracy in the late 1980s, Taiwan has established itself as one of the most democratic countries in Asia in the past three decades. Legal infrastructures have been developed to

I. Chen
Human Research Protection Center, National Taiwan University Hospital, Taipei, Taiwan

Y. Chen
School of Business, Shenandoah University, Winchester, VA, USA

D. F.-C. Tsai (✉)
Graduate Institute and Department of Medical Education and Bioethics, National Taiwan University College of Medicine, Taipei, Taiwan
e-mail: fctsai@ntu.edu.tw

C. Kurihara et al. (eds.), *Ethical Innovation for Global Health*,
https://doi.org/10.1007/978-981-99-6163-4_4

facilitate governance and protect rights for all. In the same vein, a strong legal and ethical framework is present to safeguard the rights and well-being of research participants. Two examples to be discussed here are the Biobank Act and the legislation concerning research involving Taiwanese indigenous peoples.

Despite the individual rights widely celebrated by its citizens, in the face of COVID-19 pandemic during 2020–2022, many were surprised by how easily the Taiwanese people traded their much-cherished individual liberty and privacy for public health enforcement. Citizens, by and large, willingly gave up their individual freedom to protect the health of communities and the safety of their country.

This chapter will investigate the two seemingly contradictory scenarios involving medical research and public health. At issue is what factors and/or context explains Taiwan's pragmatic approach to individual right and privacy.

2 Country with the Most Biobanks per Capita

After World War II, Taiwan became a single-party state. Human right violations were pervasive. As the country gradually became a true democracy, new legislation was enacted to place an emphasis on human right protection. The Biobank Act of 2010 is a prominent example.

The Act dictates that the collection of biospecimen and personal information for biomedical research, especially in genetics, must meet a set of stringent requirements. For instance, the potential research participant must be made aware of the data collection and storage in detail. The biobank is to be certified by the government to ensure that its policies and procedures of operation and governance are meeting the requirements. The certification is valid initially for 3 years and is subsequently subject to renewal every 3 years [1].

The Biobank Act thus imposes on the scientific community that having a certified biobank is the only lawful way to collect human specimen for future use. In compliance with the new law, all major hospitals and research institutions started creating their own biobanks. Taiwan currently has 35 certified biobanks in operation under government oversight, which is one of the countries in the world with the highest density per capita in biobanks. However, most of the biobanks are small-scale banks affiliated with hospitals. The usage of biospecimen has been limited, not justified by the substantial amount of resource invested in its creation. To address the issue of insufficient user demand, the government created a platform listing the information on inventories of each bank. Researchers can then use this platform to scan for biospecimen suitable for their pursuits. This information platform could be considered as a remedy for the Biobank Act, which inevitably restricts researcher's access to human biospecimen for the sake of protecting its donors.

3 Research Involving Indigenous People

Taiwan has a population of 23 million, and 2.3 percent of it are identified as indigenous peoples of Austronesians [2]. However, the genetic study reveals that more than half of the Taiwanese has Austronesian-related ancestry [3]. Among Taiwanese aborigines, there are 16 different ethnic groups, and each group has its own culture and language. For centuries, these indigenous peoples were subject to repression and discrimination.

To protect the right and well-being of Taiwanese indigenous peoples, both individually and collectively as an ethnic group, a new legislation in 2011 requires that all research involving indigenous peoples shall be subject to additional review and approval by the corresponding representatives of indigenous peoples. After the new law was introduced, researchers were reluctant to pursue indigenous studies until the regulation of the review process was streamlined and finally promulgated in 2017. In short, an investigator needs to first submit their research protocol to the Council of Indigenous Peoples, and the Council then forward it to the relevant review committee, which includes either the tribal meeting, the indigenous township meeting, or the central review committee of the Council, depending on the extent and the area of indigenous peoples involved in the study. Due to this review requirement, investigators usually communicate with the tribal groups proactively when designing a research study.

A provision of the Indigenous Peoples Basic Law of 2005 also states protection of the indigenous peoples in scientific research. Anyone engaging in academic research in the indigenous reserve, tribe, or their adjoin-land shall consult with and obtain permission from the indigenous peoples or tribes. In 2020 an incident drew the public attention to this law—a rocket launch site was built on indigenous land without first consulting with the indigenous community [4]. The site, next to a beach in southeastern Taiwan, was planned to be used to launch a domestically made commercial rocket for research. The launch facility was eventually abandoned after strong and vocal protest from the local community.

4 Response to COVID-19 Epidemic

While human rights and individual freedom are revered in the democratic Taiwan, its citizens' responses to the public measures taken during COVID-19 pandemic, from 2020 to 2022, showcase a balance between individual rights and public health. It was illuminating, and perhaps puzzling to some, that most Taiwanese quickly embraced strict quarantine measures, trading their liberty and rights for the pandemic containment.

When the COVID-19 virus first started to spread from China, Taiwan was expected to be hard hit as the two neighboring countries are tightly integrated economically and socially. However, Taiwan managed to maintain one of the lowest case rates in the world for the first 2 years of the pandemic. Case in point, there was not a single local case for more than 200 days in 2020 [5]. The Taiwanese strategies for COVID-19 containment generally included wearing facial mask and mandating hands hygiene in all public areas, banning international travel, shutting down certain businesses (such as pubs, gyms, and night clubs), enforcing contact tracing, testing high-risk populations, and quarantining the infected and their close contacts. Individuals were required to register their cell phone numbers when entering stores, restaurants, and various community facilities. Although these measures were lawfully authorized by the government, they severely compromised individual liberty and privacy.

Many of these measures were implemented swiftly at the beginning of the pandemic. For instance, at the midnight of December 31, 2019, Dr. Yi-Chun Lo, Deputy Director-General of the Taiwan Center of Disease Control (CDC), noted an intense discussion on a popular website, the Bulletin Board System, about the transmission of an unknown type of pneumonia in Wuhan, China, similar to the severe acute respiratory syndrome (SARS) of 2002. Deputy Lo also noted the alarming internet discussions among health professionals of Wuhan. Taiwan CDC immediately alerted the World Health Organization (WHO) of the developing endemic, requested WHO for further information and instituted quarantine measures for all flight arrivals from Wuhan [6]. Two weeks later representatives of Taiwan CDC went to Wuhan to evaluate the situation. The Taiwan government soon designated the Wuhan pneumonia as a novel coronavirus infection and mandated comprehensive monitoring. One more week later, by late January 2020, Taiwan identified its first COVID case through an onboard screening on a plane arrived from China [7].

How was it possible for Taiwan to act swiftly and decisively at the onset of COVID-19? The answer was lessons learned. The devastating aftermath two decades ago from SARS has prepared Taiwan for its subsequent public health crisis, including the COVID-19 pandemic.

4.1 Severe Acute Respiratory Syndrome Outbreak in 2003

Taiwan experienced the severe acute respiratory syndrome outbreak in 2003. Similar to COVID-19, SARS was a coronavirus infection imported from China, but with a significantly higher rate of mortality. The mortality rate of SARS was observed to exceed 20% in Taiwan. At the onset of SARS, China concealed the information, and the WHO did not provide any assistance to Taiwan. This lack of information and external support caused a widespread panic in the country. Moreover, the authority's response to SARS was slow to form, haphazard in implementation, and not effectively communicate to the public.

Learning from its past mistakes, Taiwan identified the weaknesses in its infrastructure for crisis management and proceeded to reform the public health system [8]. To name a few, the major legislation of communicable disease control was overhauled to equip the designated government agency with the authority to issue mandatory measures during public health crisis, the post-graduate training for medical residents and other health professionals expands its curriculum to include infection control concerning epidemics, and the command system for an epidemic was reconfigured and streamlined between the central and local government. The CDC also established a routine surveillance procedure and system for early detection and risk assessment of any outbreak [9].

The COVID-19 pandemic, occurring 18 years later, put into test this surveillance system and its corresponding contingency measures. It has worked as expected for Taiwan this time. Lessons learned!

4.2 Constant Threat of China Invasion

In addition to the SARS epidemic, there are additional factors contributing to Taiwan's perspective on collective safety and security during the pandemic. The pandemic was akin to a war, specifically a biowarfare, in the mind of Taiwanese, threating the survival of its people. Facing an existential crisis, Taiwanese, just like citizens of other countries, will go to battle to fight for their survival. Adding to this urgency is the fear that a weakened society from rising mortality and a collapse of its economy might bring about an invasion from China.

For 80 years, the neighboring China has relentlessly threated to invade and annex Taiwan with force. In addition to flexing its military muscle to harass Taiwan, China has isolated Taiwan from the world arena and made numerous attempts to undermine the island nation's democracy. In view of the constant threats, Taiwanese lives with a strong sense of insecurity, but also urgency. Had the pandemic brought about panic and chaos, it might present China with an opportunity to invade. During the SARS outbreak, critically and unfortunately, Taiwan learned that China would not be transparent in data collection and sharing. Despite the disease was originated from the country, China did not help Taiwan to combat the epidemic by providing crucial data. Drawing from the past experiences and political reality, it is prudent of Taiwan to closely monitor the public health situations in China to timely detect cross-border disease transmission.

The COVID-19 pandemic quickly invoked a wartime mentality among the public. Many demanded the authority to issue stringent measures that could effectively contain the virus transmission. There was also a high degree of consensus to penalize and publicly shame those who were in violation of the quarantine policy or spreading rumors and fake news.

4.3 A Large Population on a Small Island

Taiwan is one of the most densely populated countries in the world, with 23.2 million inhabitants on an area of 35,808 square kilometers (13,826 sq. mi). High population density is very conducive for virus transmission. The country is also an island surrounded by oceans. There is no place to escape if the pandemic progresses. Therefore, the sensible decision for its people would be to make the necessary sacrifice for the safety and security of the homeland. A popular slogan circulated during the pandemic, "One island, one life," resonates with many in Taiwan.

4.4 History of Combating Infectious Diseases

The country was known to be vulnerable to infectious diseases in the past, thanks to its tropical marine climate and geographical location. Over a century ago, when the Qing Dynasty ceded Taiwan to the Emperor of Japan, the Japanese troops inflicted severe casualties in Taiwan, many people were killed by cholera and malaria [10]. The fear of infectious diseases is an enduring memory for generations of Taiwanese.

Fortunately, since 1960s a large scale of public health measures has been implemented and Taiwan has either contained or eradicated many serious infectious diseases. People began to realize that infectious diseases could be defeated by science and public health measures. As healthcare professionals are much respected and trusted in the Taiwanese society, many came out to inform the public about the science and advocate for the policy during the COVID-19 pandemic. This collective effort in the community made consolidating public support for the COVID-19 policy easier.

5 Innovation for the Response to COVID-19 Pandemic

When implementing public health measures to prevent disease transmission, there is a fine line between the common good and the individual liberty and privacy. In many countries, such is the case of the USA, tensions were high between the public health authorities and citizens during the pandemic. Many took part in large-scale protests on mask wearing and quarantine. To balance the societal interest with the individual rights, Taiwan adopted several innovative strategies outlined below.

5.1 *Strategy for Communicating Effectively*

Policy implementation cannot be successful without getting buy-in first from its stakeholders, and effective communication is the key. Policy makers in Taiwan believe that effective crisis communication should be guided by the following tenets: transparency, accuracy, timeliness, empowerment, and an emphasis on due process. Based on these principles, a communication matrix of meeting venue, purpose, medium, frequency, and audience was developed.

Since COVID-19 was designated as a novel epidemic, the Taiwan's Central Epidemic Command Center began to hold daily press conference, including in weekends. The Command Center used these sessions to brief the public data from domestic and international sources and explain the responsive measures taken by the government and the cooperative actions expected of its citizens. The questions from the press and the public were discussed and answered in an open forum. False claims were thoroughly disputed and corrected, and information clarified. To convey to citizens the urgency of evolving situations, the press conference was jointly held by the Minister of Health, the head of CDC, the Deputy Minister of Internal Affair, the lead expert on infectious diseases, and the Minister of Economic Affairs. The esteemed "cast" of the press conference represents the top authorities and leading experts in charge of the pandemic, and it demonstrated to the public the government's commitment to get the job done.

Taking note from the press conferences, the media channels then disseminated this information on a real time basis to the wider public. The mass audience then provided their feedback or questions on the message areas of news channels or on talk show websites. The following day, the public could expect responses from the Command Center addressing the concerns raised. This process of frequent and timely information exchange allows the public to be readily informed of developing situations and be equipped with tools to assess their individual risk and make sound decisions.

In hindsight, this innovative way to exchange information was effective as it educates and empowers the audience. Not only did the Epidemic Command Center garner widespread support from the public, but many citizens began to monitor compliance and encourage others to take the anti-epidemic measures seriously. Case in point is that the public now became an active and empowered ally of the healthcare authorities in fighting the disease, taking destiny into their own hands, instead of merely being told to give up their right and follow the leader.

There is also an innovation in terms of an emphasis on *due process*. Due process was thoughtfully taken into consideration in the written form of communications, such as an announcement of quarantine which details why a person is isolated, for how long, at which facility, what resources will be provided for, and where to file a grievance. Due process was also incorporated into daily press conferences to ease the public anxiety and fear, to explain why certain action is taken or not taken for the time being, and to provide a reasonable expectation according to the situation.

This respectful and constructive two-way communication style could also ease the tension between the government and citizens.

5.2 Strategy for Balancing Individual Rights with Public Health

The fight against coronavirus is multi-front, which requires a dynamic strategy. The infectious rate and mortality rate of a virus may change, the virus may evolve to new variants, and the pandemic may last for years. Therefore, prevention and quarantine measures should be subject to continuous review and frequent revisions. One of most important goals for regular adjustments is to strike a balance between the need to protect individual rights and the objectives to promote population health.

For example, staying in a quarantine facility for weeks could be mentally taxing. Some measures have been adopted to ease the difficulty of confinement. For example, one can choose from a variety of quarantine hotels, or to stay at a free government-sponsored quarantine facility. The quarantine facility usually provides a private suite for one person with access to internet and cable television. Meals are provided, but residents can order food delivery, or have their family delivered meals. The government quarantine facilities were managed by public hospitals, and healthcare consultations were provided for residents. When a resident needs healthcare, the facility will find a suitable hospital for treatment and coordinate patient transportation. The staff at the facility also are trained to provide mental health support [11].

Another example is to design a system which does contact tracing of the infected with minimal infringement of privacy. The tracing system is a key to identify and contain possible transmission before an infected person is diagnosed. A mass contact tracing system was established in Taiwan which requires individual to scan a quick response (QR) code when visiting a store or other venue. The information will then be texted to a database for future use. The process takes only 3 s and is free of charge. However, the tracking is a serious infringement of privacy and there was concern of the misuse of personal information. When designing such system, Taiwan's Minister of Digital Affairs, Audrey Tang, resorted to the "G0v" for solutions. G0v, pronounced as "gov zero," is a crowd sourced development process by a group of digital-savvy workers, including designers, programmers, and activists [12]. The final solution was a system that collects minimal amount of information, including only cell phone numbers, code of the venue and time. The data will be stored only at the telecom operators and will be deleted after 28 days. The data will only be used for epidemic investigation when needed. Moreover, the system allows an individual to inquire whether his data was ever accessed during the 28-day period and by whom [13]. The tracing system operated for a year and was promptly deactivated after the pandemic policy goal changed from virus containment to coexisting, which rendered contact tracing unnecessary.

The conscientious design of G0v tracing system and its careful handling of personal data serve as a prominent example of government balancing the public health with protecting individual privacy. Gaining public trust and confidence, authorities in Taiwan were able to convince citizens that they are not snooping on them under the pretense of public health.

5.3 *Strategy for Prioritizing Resource Utilization*

Public health crisis usually leads to, or exacerbates, resource shortages, which makes setting priorities for resource allocation critical for managing public health. Taiwan, at the various stages of the pandemic, had experienced shortage of PPEs (personal protective equipment), alcohol-based disinfectant, vaccines, antigen screening kits, and antiviral medicines. Clear guidelines were set for the allocation and distribution of limited PPE supply. For example, at the early stage of the pandemic, facemasks were reserved for the health professionals and patients to prioritize saving lives while safeguarding the lives of healthcare workers [14]. The government subsequently banned all export of facemasks and expropriated all production capacity to meet the domestic demand for PPEs. After securing and boosting production of PPEs, a large portion of the production was rationed to every individual and distributed via the National Health Insurance system (NHI), a compulsory social insurance in Taiwan. The government subsidized masks were sold at designated vendors and could only be purchased using an NHI card [15]. Issues such as price gouging and excessive demand were consequently avoided. Following this PPE production and distribution plan, the masks were made affordable to all, and the problem of shortage was successfully addressed.

6 Conclusion

The case study of Taiwan presents two vastly different approaches in protecting the rights and safety of its individuals versus the public health. As a democratic society, the rights and well-being of research subjects are protected by stringent policies, whether it is an individual donating biospecimen for future research or an experiment involving a group of indigenous people. However, when the COVID-19 pandemic struck, Taiwanese people were quick to make the compromises needed for public health, social economic, and national security. This effective crisis response could be attributed to Taiwan's history, social and geopolitical background, prior experience in public health crisis including SARS and tropical infectious diseases, the constant military threat from China, and its unique geography of being a densely populated country on a small island not connected to any land masses.

To address the concerns for limiting individual privacy and liberty, Taiwan implemented several provisions and measures. The authority adopted innovative system for contact tracing that designs to minimize invasion of privacy. PPEs were rationed and subsidized to mitigate the issue of access shortage. Simultaneously, the frequent dissemination and active exchange of information kept the public well informed and empowered. This carefully orchestrated public health policy effectively aligned the community's interest with the government's objective for managing a severe public health crisis. The result was a success story in Taiwan's handling of COVID-19.

The purpose of our case study was to investigate Taiwan's pandemic response in hopes of finding best practices helpful for future public health management. We are cognizant of the fact that public health strategy is not a one-size-fit-all playbook, as seen in how part of the U.S. population turned mask wearing into a cultural war during COVID-19 [16] and how constrained the American health authority was by the diverse public opinions toward vaccines [17]. Nevertheless, our research was able to identify a common thread among the best practices, and that is transparency and effective government health risk communication [18]. This key element of the policy helps forge a close partnership between authority and the public, empower its people, and promote citizen advocacy. This baseline communication strategy combined with conscientious and innovative public health measures, taking into account the country's cultural, demographic, political, and educational circumstances, could provide a successful roadmap to managing public health crisis.

Conflict of Interest There is no conflict of interest to be disclosed related to the content of this manuscript.

References

1. Human Biobank Management Act, Art. 4, 16 and 23. https://law.moj.gov.tw/ENG/LawClass/LawAll.aspx?pcode=L0020164. Accessed 13 July 2022.
2. The web site of council of indigenous peoples. https://www.cip.gov.tw/en/index.html. Accessed 13 July 2022.
3. Lo YH, et al. Detecting genetic ancestry and adaptation in the Taiwanese Han people. Mol Biol Evol. 2021;38(10):4149–65.
4. Fang JY. Rocket launch site dispute in Taitung. [Mandarin] UpMedia. https://www.upmedia.mg/news_info.php?Type=24&SerialNo=78160 Accessed 13 July 2022.
5. Hale E. How Taiwan used simple tech to help contain Covid-19. 25 BBC. Feb 2022. https://www.bbc.com/news/business-60461732. Accessed 13 July 2022.
6. Yang S. Evidence of Taiwan health official learning about Wuhan virus human-to-human transmission in Dec. Taiwan News. 16 Apr 2020. https://www.taiwannews.com.tw/en/news/3917429. Accessed 13 July 2022.
7. Watt L. Taiwan says it tried to warn the world about coronavirus. Here's what it really knew and when. TIME. 2020 May 19. https://time.com/5826025/taiwan-who-trump-coronavirus-covid19/. Accessed 13 July 2022.
8. Chen JJ, et al. Evaluate Taiwan's infrastructure of public health and healthcare to embrace a crisis like SARS: review and outlook. [mandarin]. Commun Develop J. 2003;104:12–6.
9. Jian SW, et al. Risk assessment for COVID-19 pandemic in Taiwan. Int J Infect Dis. 2021;104:746–51.

10. Takekoshi Y. Japanese rule in Formosa. London, New York, Bombay and Calcutta: Longmans, Green, & Co; 1907.
11. Ministry of Health and Welfare, The Story of Quarantine Facilities (2022). [mandarin] Taipei. United Daily News Group.
12. How Taiwan used simple tech to help contain Covid-19. 25 BBC. Feb 2022. https://www.bbc.com/news/business-60461732.
13. Taiwan Centers for Disease Control. The data collected from the Contact Tracing System is used in a lawful, appropriate and necessary way. [mandarin] Jun 29, 2021. https://www.cdc.gov.tw/Category/ListContent/EmXemht4IT-IRAPrAnyG9A?uaid=HS0hjvHxAOTCPtNPmDo7Bw. Accessed 13 July 2022.
14. Tsai FC, Chou YC. Combat the Wuhan pneumonia with facemask in Taiwan. Humanit Soc Scie Newslett Quart [mandarin]. 2021;22:4.
15. Su YF, et al. Masks and medical care: two keys to Taiwan's success in preventing COVID-19 spread. Travel Med Infect Dis. 2020;38:101780.
16. Malina D, et al. Mask wars. 2022. N Eng J Med. https://www.nejm.org/doi/10.1056/NEJMp2214296. Accessed 7 Feb 2023.
17. Hu T, et al. Revealing public opinion towards COVID-19 vaccines with twitter data in the United States: spatiotemporal perspective. J Med Internet Res. 2021. https://www.jmir.org/2021/9/e30854. Accessed 8 Feb 2023.
18. Kim DKD, et al. An analysis of Government Communication in the United States during the COVID-19 Pandemic: recommendations for effective government health risk communication. World Medical Health Policy. 2023. https://onlinelibrary.wiley.com/doi/10.1002/wmh3.363. Accessed 8 Feb 2023.

Between Truth and Profit: Scientific Misconduct Case of Human Cloned Embryonic Stem Cell and Revisiting Cases During COVID-19 Pandemic

Young-Joon Ryu

Abstract This article discusses the case of Woo-Suk Hwang, who was convicted in the "Human Cloned Embryonic Stem Cell Science Paper Manipulation Case" (also known as "the Hwang WS Scandal") as described by a whistleblower in the case. Despite being convicted, Hwang has consistently engaged in persistent activities to return to society without showing any remorse, including directly filing criminal charges to remove intrusive informants, for more than two decades. This article is the world's first detailed formal record and academic statement on the case, including its ongoing progress despite it being thought to have ended. It also covers ethical issues related to the COVID-19 pandemic as well as scientific research irregularities committed by parents to help their children gain admission to college that occurred during the period of the pandemic. Despite the considerable social cost to Korean society, this article aims to highlight the common phenomenon in Korea and to promote international cooperation in addressing these shared challenges.

Keywords Research integrity · Scientific misconduct · Hwang Woo-Suk scandal
Embryonic stem cell research · Somatic cell nuclear transfer · Whistleblower

1 Introduction

Before I start, I would like to clarify one thing. I was the one who first informed the world about the fraudulent paper on stem cells from cloned human embryos in 2005. At the time, I was a PhD course graduate student studying stem cells, but after

Y.-J. Ryu (✉)
Department of Medical Ethics and Medical Humanities/Department of Pathology, Kangwon National University College of Medicine, Chun-Cheon, South Korea
e-mail: mindmade@kangwon.ac.kr

C. Kurihara et al. (eds.), *Ethical Innovation for Global Health*,
https://doi.org/10.1007/978-981-99-6163-4_5

the incident I made a new start and returned to the medical community, majoring in pathology and medical humanities. I am currently a university professor studying those same subjects. I was in the master's and doctorate course in veterinary medicine at Seoul National University (hereinafter SNU) from 2002 to 2006, and my advisor was Hwang Woo-Suk (hereinafter Hwang WS) [1].

After learning how Hwang's lies were deep-rooted and organized while working full-time in graduate school, I reported the facts to MBC's television program "PD Note," civic groups, and academia in 2005. He was expelled from SNU after fabrication of scientific papers in 2004 [2] and 2005 [3] was confirmed in 2006 and as a scientist in the past who was expelled from the scientific community, he is listed as a representative case in various papers and textbooks in the world. Despite irrecoverable sacrifices, this incident was seminal in revealing scientific misconduct in Korea resulting in benchmarks and directions for future research. People often think that this was a one-off event in 2005. However, until recently, it was an ongoing case socially and for me, personally. Twelve years after the paper was found to have been fabricated, I was sued directly by Hwang WS for defamation in 2016. The outcome of the trial was entirely in my favor. During the trial, I contemplated where Hwang WS bold behavior had come from. His 2016 lawsuit was proof that he still did not understand the nature of the 2005 case and his own problems. The conclusion of the 2018 trial was a good opportunity to clear up the decades-long dispute. If Korean society and other parts of the international community of science do not realize the essential problem of the misconduct incident, we will see the future appearance of a second Hwang. I hope my experience can help prevent such an unfortunate incident.

2 The Hwang WS Scandal: A Strange Case of Scientific Misconduct in Korea

2.1 Case Outline

The "Hwang WS scandal" refers to the 2005 scientific misconduct involving a paper at SNU in Korea. The official title used in the prosecution's investigation was "Fabrication Case of Papers on Cloned Human Embryonic Stem Cells." Hwang WS, then 53 years old professor at SNU College of Veterinary Medicine, fabricated and ordered and aided the fabrication of the papers "Evidence of a Pluripotent Human Embryonic Stem Cell Line Derived from a Cloned Blastocyst" [2] and "Patient-Specific Embryonic Stem Cells Derived from Human SCNT Blastocysts" [3] published in the journal *Science* in 2004 and 2005, respectively.

Based on these two papers, he received around 37 million dollars in research funding from the Korean central and local government and private companies, becoming a world-famous researcher of human cloning and embryonic stem cells. However, his misconduct as a scholar was revealed to the world not long after the papers were published. On June 1, 2005, I reported to the MBC's investigative program "PD Note" (PD stands for "producer") that Hwang et al.'s papers were fabricated, and the broadcasting company wanted to verify my report. Hwang WS tried to cover up the incident with lies, conciliations, and threats and buried the informant and MBC with his influence over the media, political, and financial power stemming from school, regional, and blood ties. The truth finally came out with the help of a small number of journalists and scientists who made great sacrifices. On January 10, 2006, the SNU Investigation Committee announced that the two papers were fabricated and that Hwang WS had, directly and indirectly, led the fabrication. In addition, it announced that no patient-specific human cloned embryonic stem cells existed. The results of this investigation represented conclusion of this issue in the academic field.

However, numerous people who believed in Hwang WS lies and the media, politicians, and chaebol who defended him used every means to prevent his downfall. Thus, the case turned from a scientific misconduct case into a social and political one and was transferred to prosecutors. On May 12, 2006, after a five-month investigation, the prosecution identified Hwang WS and Sun-Jong Kim (main party of Hwang WS fraudulent scandal) as the culprits behind the paper's fabrication. The prosecution charged Hwang WS with fraud, defrauding government-funded research funds, embezzling privately funded research funds, defrauding of privately funded research funds, and violating bioethics laws [4] Over the next 8 years, Hwang WS was convicted in all three trials, and on February 27, 2014, he was sentenced to one and a half years in prison and 2 years of probation as four charges were eventually upheld by the South Korean Supreme Court, ending the judicial process [5].

Hwang WS scientific deception, confirmed by SNU's announcement and through the trial process, swallowed up Korean society and the international scientific community beyond the huge scientific lie [6]. However, the Korean scientific community proved its ability to self-purify through a few unnamed individuals' sacrifices for the truth. Since then, the case has been included in numerous papers and textbooks at home and abroad as a representative example of a violation of scientific research ethics.

2.2 *Hwang WS First Public Lie in Science* [1]

Hwang WS political and social power revealed by the 2005 scandal was not achieved overnight. If Hwang WS power had resulted from a legitimate effort, it would never have collapsed because of a small crack like my report. People who do not know the truth or want to disregard it still believe in his achievements because of Korea's first cloned milk cow, "Young-Long"; the world's first cloned Korean cow, "Jin-Yi"; the pregnancy of the cloned Baek-Du mountain tiger; and the cloning of a genetically modified BSE-resistant cow. All of these were his lies with no scientific paper, even evidences. When did he start telling such bold lies? To answer this question as an ethicist, I thought about the Hwang WS case, and based on Hwang WS own statements recorded in media reports or books and what he publicly revealed through his legal representative.

[1] This paragraph is based on already published what Hwang Woo-Suk or his legal representative said in the media, booklets, and interviews. All others have been cited.

Through this process, I found understanding Hwang WS social view and habits made easier, through the main characters that influenced him and the detailed situation. In summary, he was born in a poor farmhouse in the countryside called Buyeo, South Chungcheong Province. Since he could not study because of his difficulty in making a living, he left home and moved to Daejeon city with the help of his relatives to go to school. His academic performance or grades were not good until the beginning of third year in high school, but he tried hard to go to SNU. He entered SNU College of Veterinary Medicine in 1972 with great effort. This is very important because it is compulsory for men to join the army in South Korea. Little is known about Hwang's military service. It takes about 3 years to serve in the military at that time, so average college graduation age for Korean men is of 27. However, Hwang WS graduated from SNU at 25, which is a two-year discrepancy from a typical Korean man. It can be inferred that he served 6 months as supplementary military service and left 2 years early using the system at the time. After graduating from college in 1977, Hwang WS began a full-time graduate career with a major in veterinary theriogenology at SNU. The following year in 1978, he married a woman from a wealthy family at age 26 with whom he had two sons. In 1979, he earned a master's degree at age 27 and started a Ph.D. in the same graduate school. Three years later, in February 1982, Hwang WS earned his doctorate. In 1982, Hwang WS applied for a professorship at SNU but failed. He gave up his career, sold his small apartment in Seoul, purchased land in Toe-Chon, Gyeonggi province, and returned to farming in 1983 with his mother. Later, in 1984, he went to study as a visiting researcher at the University of Hokkaido in Japan at the suggestion of the then dean of the college of veterinary medicine. [2] Hwang WS was appointed assistant professor of veterinary theriogenology at SNU in 1986 at age 34.

Two years later, he was diagnosed with liver cancer at age 36 and underwent surgery twice at SNU Hospital in June and August 1988. In the same year, Hwang WS divorced his wife after 10 years of marriage. Nine years later, on April 9, 1997, after many twists and turns, Hwang WS remarried a woman from a powerful family, [3] and had a daughter with her. [4, 5] Concerning Hwang WS life before 1997 (when he first appeared in the media), I learned more through oral statements by his middle

[2] Lee Jang-Rak was not directly mentioned in Hwang Woo-Suk's statements. But Hwang WS testified that the dean at the time recommended it to him, and Lee Jang-Rak's tenure as the dean of the veterinary school was 1981–1985.

[3] Many rumors have circulated about his second marriage, SNU professor Ahn Gyu-Ri, and his relationship with Park Ki-Young, an aide to Cheong Wa Dae (Blue house), but this will not be discussed here.

[4] *Hwang Woo-suk, a World-changing Scientist*, Jaeum and Moeum; *Hwang Woo-suk, the Stubborn Bull*, Blue Bicycle; *Hwang Woo-suk's Story*, A-One Books; *Hwang Woo-suk's Dream*, Dongseo Munhwasa; *Hwang Woo-suk in Comics*, Dong-A Science; *Dr. Hwang Woo-suk's Beautiful Path of Life*, Ire Media; *The Magic*, Purple; *Let's Learn about the Success of Hwang Woo-suk*, Dongseo Munhwasa; *The Boy Hwang Woo-suk*, MK; *Dr. Hwang and Stem Cells*, Hakwonsa; *The Hwang Woo-suk Report*, Nature and Freedom.

[5] Hwang Woo-Suk's direct statement through his legal representative Moon Hyung-sik, Lee Geon-haeng.

school teacher. [6] The teacher, who I met by chance, recalled Hwang as a diligent student. Hwang's appearance in 1990s can also be found in the records of a reporter Lee for monthly magazine Shin Donga [7].

The first time she met Hwang WS was in the fall at a Korean restaurant in Seoul. At that time, the reporter was writing for a KBS science program and the meeting was arranged by KBS reporter Mr. Cho, an alumnus of Daejeon High School.

She said Dr. Hwang introduced himself as an ordinary veterinary school professor and told her that he received his degree in artificial insemination from SNU, but they did not hire him as a professor, so he went to Japan to study. He also talked about his childhood, saying that he grew up very poor and her mother did not care if he was sick, but she valued cows so much that he would make a fuss if they were sick, so he wanted to be a veterinarian.

Meanwhile, after remarrying in 1997, Hwang WS appeared in the media with a "very different look than before." The reason for this was the so-called Brucella vaccine incident that occurred in 1998. At the time, *Brucella* was already vaccinated against in the U.S., Australia, New Zealand, and Canada, but not in Korea; Brucella was also prevalent in cows. To solve this situation, Professor Baek of the Veterinary College of Chungbuk National University brought over an American *Brucella* strain called RB51 and conducted a domestic adaptation experiment. With research funding from the government, he officially inoculated 390,000 cows nationwide through the Central Livestock Infectious Disease Research Institute and the Korea Microbiological Research Institute. However, many vaccinated cows suddenly had miscarriages or gave birth prematurely, and the prosecution launched an investigation to determine the cause. The prosecution discovered that the vaccine was pathogenic, concluded that Brucella was transmitted through it, and arrested Professor Baek, who was eventually kicked out of the university. However, it turned out that the vaccine was not pathogenic; it was contaminated due to storage-related mistakes at the farm. In November 2005, the Supreme Court confirmed this as well as the fact that the Brucella vaccine which Professor Baek had developed was not pathogenic, acquitting him of the prosecution's indictment.

Once acquitted, Professor Baek gave a shocking statement to the media. Hwang WS, an SNU professor at the time, had gone to the prosecution as a witness and essentially stated that the vaccine was a sham, a statement which the prosecution used as decisive grounds for prosecuting Baek.

The following is Hwang WS's statement in the prosecution's statement as revealed by Professor Baek [8]:

> *'I don't want to believe that it's true, but the pathogenic expression of RB51, which was manufactured and vaccinated at Jungang and Hanmi, was partially confirmed...' (Prosecution records, p. 552). 'The Jungang and Hanmi products that were inoculated went into the intestines, the uterus, and the breast, causing the fetus to miscarry... I think that some of the cows vaccinated with RB51 were clearly pathogenic' (p. 556). 'The normal miscarriage and premature birth rate of cows is 5.3%. However, the miscarriage and pre-*

[6] In 2004, the author met Hwang's teacher in person at an event held at SNU. He introduced himself as his homeroom teacher in the second grade of middle school.

mature birth rate of vaccine lot 4 produced by Jungang was 40–54.3%, and the rate for Hanmi was 39.8% as only lot 1 was inoculated. It was a big shock to see that when a normal cow was vaccinated, it was infected with Brucella' (p. 560). 'Some scholars suspect that what was developed domestically may have been pathogens and not vaccines' (p. 572). 'The U.S. RB51 vaccine also caused miscarriages and premature births' (p. 3184).

During his own three trials, Hwang's words were revealed to be false and Baek's claims were acknowledged as true. It was unfortunate that professor Baek was dismissed from the university, but more serious is that the Brucella vaccine is no longer being developed or used in Korea after the incident. This is because people avoid taking risks after seeing individuals being harmed without any evidence, as in this case. More than 18 years later, around 10,000 cows die every year of Brucella, with people being infected as well; thus, the aftermath of Hwang's false statements continues [9].

There is something I do not understand here. If Hwang WS was no vaccine expert and his statement had no explicable scientific basis, how did he exercise so much power over the prosecutor? Hwang's main field was veterinary theriogenology, and he only gained laboratory experience in artificial pregnancy during his 2 years of study abroad. To understand this, we need to see how the name SNU works in Korean society, along with the scientific ignorance of the media and the prosecution. For the past 70 years, SNU has been the most difficult school to gain entrance to in Korea and the peak of high school students' competition in college entrance exams. In short, it has been a symbol of social hierarchy since the university's founding. For several decades, key posts of high-ranking government officials such as ministers, vice-ministers, judges, and prosecutors were occupied with SNU graduates and the school ties created therein turned Korea into a rigid society based on academic elitism. Thus, discrimination still exists in Korea, depending on whether you are from SNU [10]. Second, the prosecutor in charge was ignorant about *Brucella* and its vaccines and was therefore incapable of detecting Hwang's lies that had no scientific basis. Consequently, he may have trusted the title of "SNU Professor Hwang Woo-Suk." Due to these two reasons, the testimony of an SNU professor who neither specialized in the field in question nor could provide a scientific basis for his statements exerted more influence on the prosecution than the true statements by Professor Baek, a vaccine expert. Through this incident, Hwang WS probably came to see that his most substantial asset was his position as a "professor of SNU." A graduate and professor of SNU, Hwang WS saw the logic of power exerted in Korea through this incident and took full advantage of his position for his own success. [7]

Meanwhile, on December 14, 1998, Professor Lee and Kim's team at Kyung-Hee University Medical Center Infertility Clinic announced to the press their development up to the four-cell stage through human somatic cell nuclear transfer [11]. At the time, Hwang was part of a due diligence team dispatched by the Korean Medical Association along with Professor Seo (SNU, Biochemistry), Professor Moon (SNU, Obstetrics and Gynecology), and Dr. Lee (Biotechnology Research

[7] Hwang often said things in his laboratory that showed his awareness of the SNU premium (former researcher's statement).

Institute). The team concluded its investigation by saying that the cloning experiment should be discontinued immediately, and the publication postponed. [8] In a TV interview, even Hwang WS seemed to oppose human embryo cloning, citing the possibility of human cloning. [9] This incident is presumed to be for Hwang WS how socially influential "human somatic cell cloning" was, just as with Dr. Wilmut's Dolly, the cloned sheep announced in 1997. However, while human eggs were essential for experiments, he would have hesitated for a moment because of his limitations of procurement of human oocytes as a veterinarian and waited for an opportunity.

In 1999, Hwang WS completely fabricated the reveal of Korea's first cloned milk cow "Young-Long." Since 1997, Hwang WS had been unsuccessfully trying to clone cows, which were his main expertise. At the time, scientists at the National Livestock Technology Research Institute, Professor Park's lab at Gyeongsang National University's Livestock Department, Konkuk University's Livestock Department, Maria Obstetrics and Gynecology, Kun-Kuk University and ACT KOREA were studying this field competitively. When Hwang's research failed despite his efforts, he heard rumors that a cloned cow became pregnant at the National Livestock Technology Research Institute. Hwang decided that he could not miss such a good opportunity to gain hegemony in the field of cloning, and this was how Hwang WS came to fabricate Young-long, the cloned cow.

What choice did he make without scientific evidence or papers to prove his research presentation? He planned a puppet show, taking advantage of the fact that the Ministry of Science and Technology and the media were careless about verification. What presented Hwang WS as a "great scientist" to Koreans was the idea that he was the fifth person in the world and the first in Korea to clone a cow, and this was only due to large-scale reports by the Korean media. On the morning of February 19, 1999, Hwang WS sent out press material to reporters, telling them to come cover the birth of a cow (which was yet to be born) at a farm in Hwaseong-gun, Gyeonggi province, and ordered the Ministry of Science and Technology to prepare to distribute the press material [12]. Many reporters recorded the scene of the cow's birth through footage and photos and sent out the news as a headline without few evidence or published results, instead simply relying on Hwang's words, "This cow is a cloned cow." [10] Behind this false report, which was presented to the whole nation, was the title "Professor Hwang WS of SNU," Korea's top undergraduate school, and the naive belief that he could not have lied. His bold first puppet shows thus succeeded. Later, when MBC's "PD Note" asked for verification of Young-Long, Hwang WS was unable to submit any genetic evidence that it was a clone and tried to avoid the issue by saying he had "lost the data while moving" and "Since it was already the fifth clone, there was no journal that would accept it" [13]. Moreover, to

[8] Results of the human embryo cloning experiment and Hwang Woo-suk's Interview Video, Korean Medical Association.

[9] Screen shots from MBC's "PD Note."

[10] Refer to broadcast by MBC's "PD Note."

cover up suspicion, Hwang WS even leaked news to the media that cloned cow, Young-Long had given birth to a baby [14]. The scientific community showed great interest in the cloned cow but overlooked the fact that no paper was published on it. Thanks to this, he formed close relationships with high-ranking government officials such as the president through the Minister of Agriculture [15].

Hwang WS had an excellent helper to reporters in playing the media this way. This was Mr. Cho, a KBS reporter who was an alumnus of Hwang's high school. He seems to have naturally introduced Hwang to other reporters. Cho was from the same hometown and same high school and had worked for a long time as a weather reporter at KBS. Cho was well aware of the nature of media and science journalists, so it was natural that he introduced Hwang to several of them [7]. Acting in front of reporters without any evidence would have been impossible without the certainty that the media's capacity for verification was poor. I point out here that Hwang WS took full advantage of his regional and academic background [16]. This is also supported by the fact that several of the "Chungcheong Hometown Association" members were among the 33 members of the so-called National Assembly committee members in support of Hwang Woo-Suk, who were decided at the National Assembly in 2005 [17].

Hwang WS' next public lie in science was about the first cloned Korean cow "Jin-Yi." Hwang WS started to lie more deliberately. Instead of disclosing it to the media immediately after birth, he delayed because he was conscious of the criticism in the Young-Long case that the announcement was made immediately without time for genetic testing. However, the real reason for the delay was something else. A day before the announcement, Hwang WS had attended the National Science and Technology Commission chaired by then President and reported that he was "the first person in the world to clone a Korean cow," and the president named it "Jin-Yi" to "become a cow loved by the people," like Hwang Jin-Yi, "who left behind work that was praised through the years" [18] (Hwang Jin-Yi was a beauty that appears in Korean history). It was another typical lie without evidence or publication, but the media had no reason to doubt Hwang WS, professor of SNU, who had already become famous. After this incident, Hwang WS was selected as a "new intellectual person" by that government, a title given to prominent figures in each field, and was later appointed as a National Science and Technology Commission member. This lie that he had cloned a Korean cow turned an ordinary professor into a "patriotic scientist" and put him in a committee that decided the Korean government's scientific policy.

Hwang's next lie was about the successful cloning of a BSE-resistant cow. In the early 2000s, BSE (bovine spongiform encephalopathy) and human mad cow disease hit the world. Cows not walking properly, collapsing on the news, and humans contracting the same disease through contaminated beef consumption drove the entire nation into fear. Hwang WS did not miss this opportunity. In 2003, Hwang said he had developed a so-called BSE-resistant cow by inserting somatic cells, from which genes that created "prion protein" (which allegedly caused BSE) were removed, into the cow's eggs. He was confident, saying he would send the cow to the Animal Hygiene Laboratory in Tsukuba, Japan for verification. Even those suspicious of his previous achievements began to believe him, saying that he could not have done so if it were not real. However, his confidence did not come from his success. The cow was "discovered by luck," meaning the gene that created the "prion" had naturally mutated as well as no genetic modification had occurred. At a laboratory meeting, Hwang WS described it as a "gift from heaven." Of course, under any test, the cow with naturally modified prion genes looked like a cow whose prion-producing genes were modified by humans. Some strongly deny that prions are the cause of mad cow disease, [11] so the cause of the disease remains unclear, but that could not stop Hwang's claims. This incident impressed the public: Hwang WS, a scientist full of academic confidence, would even send the cow to Japan for verification.

In the process, Hwang WS began to receive another suspicious question from scientists. How can he conduct techniques such as gene editing despite his poor laboratory career? Actually, Hwang's career focused on clinical activities dealing with cows on farms, and his laboratory activities related to artificial pregnancy were limited. Even then, he hired laboratory technicians from junior colleges to carry out the duties and he simply received the results rather than directly participating. To dispel these doubts, he recruited Professor Kang Sung-Geun, who was studying at the University of British Columbia in Canada. (He was from the same high school as junior Professor Lee BC in Hwang's department.) The work in the SNU lab is divided into laboratory and clinical part. Hwang WS' work was done entirely in the clinic, while the laboratory work was done by Sung-Geun Kang. Lee BC's role was in the same clinical field as Hwang, so Hwang no longer needed Lee BC, and Hwang's attitude made him appear to be discriminating against Lee BC. As an example, when Lee BC successfully cloned the dog "Snuppy," Hwang WS took all the related patent rights. After leaving Hwang's team, Lee BC filed a lawsuit over the patent rights, but lost the trial.

Hwang's next public lie in science was the announcement of an ultrasound showing a cloned Mt. Baek-Du tiger fetus. In 1999, he announced a project to clone the Baek-Du mountain tiger [19]. (Mt. Baek-Du is a sacred mountain of Korean people and it is something that could play a role in diplomacy with North Korea.) The media's ensuing behavior was surprising. As if it were certain that a cloned Baek-Du mountain tiger would soon be born [20], numerous reporters wrote their articles without verification, believing Hwang's ungrounded words [21] and including announcements such as "The birth of the first Baek-Du mountain tiger in July" [22]. Hwang WS did collect somatic cells from tigers at Seoul Grand Park Zoo once or twice. Later, he replaced tiger oocyte with cat's because it was difficult to obtain tiger oocytes, and when that did not work, he committed the ridiculous act of transplanting tiger somatic cells into pig eggs for years under the guise of an "experiment." An embryo created by transplanting tiger somatic cells into a pig's egg was transplanted into the pig's womb, along with the pig's cloned embryo. There was a plausible theory called "helper embryo," graduate school student Kim JT, the project team leader at that time, voiced opposition to Hwang WS but was ignored. However, he endured for the sake of his Ph.D. and was able to graduate only after

[11] Statement by Woo HJ, Professor of Veterinary Medicine, SNU.

submitting his dissertation on the topic. I remember his conscience suffering because of Hwang's ridiculous instructions. Cho JG, the leader of the cow team, complied with Hwang's instructions, kept the truth about Young-Long and Jin-Yi a secret, and eventually became a professor at Chungnam University. Another team leader of the porcine(pig) team, Kim DY and Hyun SW, became professors at Gachon University and Chungbuk University, respectively. Also, Dr. Kim, who graduated without disclosing the secret about the "cloned Mt. Baek-Du tiger," became an official at the Korean Intellectual Property Office. However, some graduate students who resisted Hwang's instructions either did not graduate or had to change careers. In Korean universities, the relationship between professors and graduate students has a clear hierarchical and authoritarian structure, so it would not be an exaggeration to call it a master–slave relationship. Such a relationship makes it impossible to voice criticism when something goes wrong, thus causing people to remain silent about research ethics and integrity violations.

2.3 *The "Fabrication Case of Papers on Cloned Human Embryonic Stem Cells" and Its Timeline: 2005–2013*

2.3.1 An Unfettered Race

I was born in Busan, Korea in 1973 and later informed the world of Hwang's violation of research integrity as a whistleblower. After graduating from Kosin University Medical School in 1998, I entered the military and contemplated where to pursue a life as a basic scientist. Around that time, in 1996, Dr. Ian Wilmut of the United Kingdom created the cloned sheep Dolly [23], and in 1998, Dr. James Thomson's team at the University of Wisconsin-Madison developed and announced the first technology to isolate and grow human embryonic stem cell lines [24]. I thought that these two technologies would be great sources for future research in the fields of embryology, drug development, and clinical treatment. After my mandatory military service, I decided to end my career as a clinician after my internship at the Asan Medical Center in Seoul. In 2022, I enrolled in the master's program at the Graduate School of SNU and started working with academic advisor Hwang WS.

At that time, Hwang WS was an accomplished scholar who had received media exposure for about 2–3 years, starting with Young-Long, a clone cow in 1999. He also received significant research funds from the nation. Thus, it seemed obvious that it would be a great study environment. I worked in the laboratory as a full-time graduate student from 6 a.m. to 12 p.m. without taking weekends off, and as a result, I was able to manage the human stem cell team. As a team leader, I had the opportunity to plan and lead research. Professor Cho JJ, who attended Albert Einstein Medical School in New York, and was also recruited as a team advisor, established the fundamental framework for human research. Hwang's limitation of not being able to save one's identity and human eggs as a veterinarian was finally solved when I was able to collaborate with medical schools and hospital. Therefore, it was only natural that Hwang WS showed great trust in me, it was because I was an essential team member playing such a role. Moreover, as the research progressed smoothly, and my ability was consequently proven, Hwang WS and I developed a tighter bond.

However, after entering the graduate school, I became suspicious of Hwang's achievements, which were known to the media. At that time, under the guise of "security," Hwang WS thoroughly separated the human, cow, pig, dog, and tiger cloning teams in the graduate lab, but he could not block secretive communication between individuals. My suspicions were confirmed by information that slowly emerged from some overly suppressed graduate students. For example, I was informed by some leaders or members, who participated in the process of creating the clone "Young-Long," that the contents were also unknown to them. This came as a shock to me. It was confirmed that Hwang's accomplishments about cow-cloning were not only groundless but did not actually happen and were proven to be fabrications based on scientific deception and media manipulation of his position as a professor at SNU, regionalism, and academic ties.

After confirming this fact, when I received my master's degree and started my doctoral degree, my feelings toward Hwang WS were bound to change. I had no choice but to give up my dream of conducting basic research because Hwang, my advisor, was an unreliable person who did not adhere to the essential discipline of a scholar. However, the timing was not right to put this decision into action. Since I was the team leader for human stem cells in the effort to create cloned human embryonic stem cell lines, it was practically impossible for me to quit the research group immediately.

While the chaotic situation continued, I was able to establish one embryonic stem cell line (SNU-NT-1) that I accidentally obtained through the cloning process. However, knowing that it likely occurred through parthenogenesis rather than cloning, I was not surprised that the actual verification process led by professor Kang proved that it was not a clone. For me, stem cells obtained from parthenogenesis were still valuable. However, Hwang WS differed in opinion. If the SNU-NT-1 had undergone parthenogenesis rather than cloning, it would have been meaningless to Hwang's plan; thus, they must have been cloned no matter what. Our different perspectives caused friction between myself and Hwang WS, who was in charge of overseeing the entire thesis writing process; eventually, the issue started to spill over into the outside world. Hwang WS wanted to publish the findings in prestigious science journals such as *Nature* and *Science* and reiterated that it should be stem cells from cloning rather than parthenogenesis. In this case, there was no problem publishing it in Science because it was the first case in the world of stem cells obtained by parthenogenesis. The conflict within me consequently grew intense. Since I vehemently disagreed with his beliefs rather than his scientific reasoning, my inner conflict deepened. Since then, Hwang WS had delegated the paper that I was working on to Professor Kang, and at his instruction, more obvious fabrications were in full swing. However, for me, this situation made it easier to end my participation.

After finishing this paper, I decided to leave the lab and take action. The research, which featured extensive direct manipulation by Hwang WS, was released from the lab and appeared on the cover of the academic journal *Science* in May 2004. To make a living, I had no choice but to quit the path of a basic scientist and return to being a clinician who sees patients in hospitals. However, I still attended graduate school because I was still working on my doctoral degree course work. I chose to major in neurosurgery in the faint hope that I would be able to develop the first planned stem cell therapy for spinal cord injuries and apply it in clinical practice. From January 2005, I started working as a neurosurgery resident at Korea Cancer Center Hospital in Seoul.

Around end of March 2005, a patent attorney, Dr. Kim SW, a graduate school classmate at SNU, came to see me. The purpose of his visit was to revise the new additions to the first patent for human embryo cloning, written by me in 2004. At this meeting, I heard disturbing news. In just a few months since I left, a total of 11 patient-specific stem cells had been created. I could not believe it. The human stem cell team, which was formed when stem cells were first discovered in 2004, consisted of six people, but because 4 members already left, it was technically impossible to clone and cultivate 11 stem cells in such a short period. In short, I thought Hwang WS was telling lies again, and his motive was to win the Nobel Prize. In 2005, Mr. Kim, a member of the National Assembly's Science and ICT Committee, appeared on a broadcast and testified as follows [25]:

> *We know that the Roh Moo-Hyun government's Nobel Prize project worked, and even high-ranking national officials introduced Professor Hwang to the head of the Swedish Nobel Prize committee. High-ranking diplomats and other working-level government teams worked hard at the UN on the debated stem cell research, particularly because of the bioethics issues surrounding embryonic stem cell research, for which the prohibition treaty was adopted not as a treaty, but as a prohibition of embryonic stem cell research. I received a tip from a high-ranking diplomat in office.*

In the 2011 testimony of former member of Parliament Mr. Kim, Sweden's Nobel Prize Committee Director, Dr. Han, a special advisor to the Nobel Foundation, made the following statement to the Korean media [26]:

> *Dr. Hwang Woo-Suk would have won the Nobel Prize in Physiology and Medicine if his research on embryonic stem cells had not been fabricated. I even invited the head of the Swedish stem cell research institute and arranged a week-long joint study with Dr. Hwang. At that time, the head of the Swedish research institute was also deceived. As a result, my credibility within the Nobel Foundation has suffered tremendously. I was the one who actively informed the Nobel Foundation of Dr. Hwang's research accomplishments at the time.*

Hwang WS, who ingratiated and connected with the Kim Dae-Jung government through two false cow clones, was again strongly associated with the Roh Moo-Hyun government, which had continuity with the previous government through lying about human embryo cloning. The government, thirsty for scientific achievements, promised Hwang WS support at the government level, followed by Park, Presidential Secretary for Science and Technology. Was it in exchange for this assistance? She received a gift from Hwang WS to become the author of *Science* research paper in 2004 [27] before she was an aide; she also received 250 million won in research funds from Hwang WS before becoming Presidential Secretary [28]. Park violated the authorship ethics by making no scientific contribution to the 2004 scientific paper before being a presidential aide; this could be interpreted as an act of Hwang's interest in knowing that he was already close to power.

What was more shocking was the second piece of news delivered by a patent attorney. Hwang WS twice intended to conduct experiments on humans, around May and October 2004. The first subject was a 10-year-old boy whom I knew well and had collected somatic cells from at the hospital operating table. It really startled me. I became angry and saying, "Hwang's lies are now beyond honor, money, and other people's lives!" I realized that I could no longer watch Hwang's mindless rush. I sent three verification request messages to Professor Cho JJ of SNU College of Dentistry, who was supervising the human stem cell team at the time, on behalf of Hwang's research team. He received report from me and twice delivered it directly to professor Lee and Kang, and once directly to Hwang WS, but they were all ignored; furthermore, Professor Cho was forced out of Hwang's research team.

The first thing I did was to visit Sun-Jong Kim, who was working at the MizMedi Hospital, to check the facts. I excluded Dr. Yoon HS, the head of a MizMedi research institute, and questioned Kim SS when we were alone on the hospital rooftop.

Is it true that 11 were created?

You know that. That I have no choice but to do it.

Kim's words strongly implied to me that the paper was telling a lie and there seems to be manipulation. Hwang's conducting of a clinical test on a human body with such false cells would endanger human life.

After hearing about the creation of 11 stem cell lines from a patent attorney, I thought that many professors and doctors who had been on Hwang's research team for about 3 months would not merely remain silent in the face of such fraud. However, professors at SNU hospital, including Professor Ahn and Moon, who were MD professors on Hwang's team, were not only silent, but also eager to use the opportunity for their own advancement. Eventually, on May 20, 2005, Hwang WS and his sympathizers announced through the American Association for the Advancement of Science (AAAS) that they had successfully isolated 11 patient-specific stem cells. At this time, Hwang's triangular alliance between the government, press, and conglomerate was complete; even the people joined in, and an unstoppable race began without any criticism or verification.

2.3.2 Ryu's Report and MBC PD Note's Request for Verification

After hearing the news, for about 3 months I made every effort to find a way to verify my suspicions about the lies told by Hwang WS. It was also confirmed at that time that there were few laws or state protections for whistleblowers in Korea, and even if reported to the police, prosecutors, the National Assembly, and media, it caused significant damage. Some media outlets and civic organizations have been approached to investigate the idea, but only to affirm that it does not exist in "Hwang's nation." If you were an informant, it seemed impossible to live as a scientist or doctor in Korea. There were also concerns from my family. I was terrified thinking about my wife and child, who was only a few months old at the time. I was disappointed that the truth might be buried, blaming the incompetence of the Korean media and judicial institutions. On May 31, 2005, past 11 p.m., MBC's <PD Note> was conducting a special broadcast. <PD Note> is Korea's representative investigative news program and is famous for broadcasting news about corruption regardless of power, religion, and conglomerate for the past 15 years. I was watching the broadcast carefully and was moved by the words of PD Choi Seung-Ho, the anchor and executive producer.

> *There were times when we could not broadcast because we did not have the capability, but we never failed regardless of external pressure.*

After talking to my family that day, I added the following words along with the exact name, contact information, and information through the MBC PD Note Bulletin (Report No. 70433) during lunchtime on June 1, the following day:

> *It could be an international disgrace and I could suffer consequences, but I am writing this letter with the belief that the reputation gained by deception will be destroyed, and the truth will be revealed someday, so please do not forsake me and contact me.*

PD Note's producer, Choi did not miss my report, and PD Han Hak-Soo, who had related experience, was assigned as the person in charge. At that time, PD Han was in his late 30 s; he was a well-trained energetic journalist. My first meeting with him was like a spy movie, where I tested him with the following question: "What do you prioritize, the truth or the national profit?" PD Han answered without any hesitation that "the truth is the national interest." I was satisfied with this answer and reported the following facts:

1. There is a high possibility that the contents of the *Science* paper published in May 2005 are false. Verification is required. With most of the key personnel leaving the laboratory, the chances of success with two people are extremely low.
2. Even if only some stem cell lines may have been created, it is difficult to say that the teratoma formation experiment, which is an experiment to confirm self-replication and pluripotency in time, has been conducted. There must be a weak scientific process.
3. The eggs were illegally used in experiments for scientific papers in 2004 and 2005. The procurement of eggs was conducted at MizMedi Hospital and others. Moreover, Park and Kim, who have experience with fertilized egg embryonic stem cells, are dispatched from MizMedi Hospital every day.
4. In 2004, an experiment for the *Science* paper involved the use of eggs from two female graduate students, and a tragedy occurred in which one of them cloned her own egg.
5. In 2004, Hwang WS forced a consent form for egg donation from researchers, including female graduate students.
6. Hwang WS, who claims to be cloning a tiger from Mt. Baek-Du, uses pig eggs. The ultrasound picture of a pregnant tiger captured by ultrasound has not been genetically verified as a tiger. There is a high possibility that it was a pig.

7. There is no scientific evidence that the cloned cow Young-Long published in 1999 is a clone resulting from somatic cell cloning.
8. There is no scientific evidence that the cloned Korean cow Jin-Yi published in 2000 is also a cloned cow.
9. In 2003, the "mad cow" disease-resistant cattle were not genetically modified or cloned.
10. Actually, the real cloned cow was born in Hwang WS laboratory in 2003. Even though, there was no paper trail about this, it was confirmed through internal verification.
11. All of these scientific fabrications can be said to have been connived by insiders such as professor Lee, Kang, Lee C, Ahn, and Moon, as well as nearby scientists such as Yoon and MizMedi Hospital.

Earlier or around the same time, I reported this to the civic group the "People's Solidarity for Participatory Democracy, the Citizen Science Center," and Professor Sang-Ik Hwang of SNU College of Medicine, and we joined forces with MBC. Jae-Myung Lee, director of the Transparent Society Bureau of the People's Solidarity for Participatory Democracy, and Byung-Soo Kim of the Citizen Science Center protected the informants, and Professor Sang-Ik Hwang took on the role of academia, continuously helping the informants and the MBC PD Note.

2.3.3 Hwang WS's Resistance

PD Note investigated the truthfulness of my report for about 5 months and confirmed that there were numerous suspicions about Hwang's achievements. The authenticity of the stem cells, known as the core suspicion, had to be confirmed directly with Hwang WS, which was impossible to do without proof of Hwang's lies. On October 4, 2005, PD Han and PD Bo-Seul Kim directly visited SNU's College of Veterinary Medicine and requested verification. Professor Hwang WS, Lee BC, and Kang SG, along with the key participants, and 11 doctors and professors who were helping Hwang WS from the outskirts, were present [29]. Han PD asked about the verification details prepared thus far, but Hwang WS was unable to respond and showed signs of anxiety. However, he could not succumb to the intense pressure of Han PD and listen to the verification that was unfavorable to him. To reassure the external organizations seated next to him, Hwang WS pretended to address the verification and handed over what he called stem cells. However, in the end, the results of the genetic test were inconclusive, and it was strongly suspected that the method of adding a DNA-degrading enzyme was used.

Even with Hwang's conciliation, <PD Note> did not waver, and on November 22, when the first broadcast about the violation of human oocytes ethics was aired, Hwang WS declared that he would not comply with all the verification requests. From then on, Hwang WS mobilized the press, the Ministry of Science and ICT, the Blue House (the official residence of the president of Korea, which means the core power of the national administration), and the National Assembly to attack the informants, reporters, <PD

Note>, and MBC from all directions. The first was a media attack. From December 3, 2005, Hwang WS conspired with Professor Ahn, with the help of former journalist Yoon, the president of the news channel YTN, Mr. Pyo, Mr. Hong, head of the press, and Mr. Kim, a science reporter, to go to Pittsburgh, U.S. and begin attacking Park, saying that the coverage of PD Note was a violation of broadcasting ethics. Meanwhile, Yonhap News Agency's journalist Kim, Chosun Ilbo's Lee, Dong-A Ilbo's Kim, JoongAng Ilbo's Hong, Kyunghyang Shinmun's Lee, KBS' Hong, and science reporters who were close to Hwang WS (the so-called Hwang's scholarship) were mobilized to attack PD Note and search for the informant. Consequently, I was fired from the hospital that was affiliated with the Ministry of Science and ICT. I could no longer live at home because reporters and supporters of Hwang WS guarded the front of the house and even entered the house. At that time, the director of the Transparent Social Bureau of the People's Solidarity for Participatory Democracy civic group, Jae-Myung Lee, and Byung-Soo Kim, deputy director of the Citizens Science Center, protected and sheltered us. Ahn, a professor at Seoul National University, and Roh, chairman of MizMedi Hospital, accused me of being a whistleblower and a criminal who violated Medical Law by disclosing patients' information.

The indiscriminate bombing of power continued. At that time, Mr. Yoo SM, who was the Minister of Health and Welfare under the Roh Moo-Hyun government, said in a lecture at Chonnam National University on December 7, 2005, *"It is absurd to try to verify Professor Hwang's research."* Mr. Lee HC, the 36th Prime Minister, declared at a Cabinet meeting on December 6, 2005 that it was, *"A situation that damaged the credibility of our academia and adversely affected the morale of scientists by excessively covering and trying to dig up scientific results in the research stage."* In addition, Mr. Sohn HK, the governor of Gyeonggi Province at the time, urged the public to attack the informants and MBC at the groundbreaking ceremony of "the Hwang Woo-Suk Bio-Organ Research Center" on December 8, 2005, saying, *"People who oppress and harm Hwang Woo-Suk must be rejected and isolated."* The indiscriminate bombing of power continued. Mr. Sohn eventually committed a historical crime by providing Hwang with taxes in Gyeonggi Province for several years, laying the groundwork for a falsified researcher's return.

After wrapping up all the attacks and organizing them to some extent, Hwang WS was hospitalized at SNU Hospital and began asking for sympathy from the people who attacked companies that provided advertisements to MBC, which resulted in not having a single company advertisement. In the end, MBC President Mr. Choi MS surrendered the white flag by indefinitely suspending the broadcast of the PD Note and apologizing to the public.

2.3.4 The Revealed Truth and the Conclusion of SNU's Investigation Committee

On December 4, 2005, PD Choi and PD Han were subject to disciplinary measures and did not know when they would be fired. As the informant, I was kicked out from the hospital where he worked and wandered the streets to avoid Hwang's supporters and reporters. At that point, the previously unknown truth began to emerge. At dawn on December 5, a ray of truth finally shone brightly. A potato research scientist who used the pseudonym "Anonymous" on the bulletin board of BRIC, a scientific community run by POSTECH, posted a title quoting the group Queen's song *"The show must go on."* They noted that when you examine the 2005 paper, there were two fabrications in the picture. Shortly thereafter, a scientist using the pseudonym "A-Reung" announced that he had found a manipulation in a DNA gene fingerprint photo. These two facts were conclusive and undeniable evidence of the fabrication of papers. There were reports that manipulations had been found not only by young Korean scientists but also by American and Japanese scientists. This reversal could not have been announced by the Hwang WS-friendly media. However, a few journalists, including Pressian's journalist Yang-Gu Kang, Hankook Ilbo's Hee-Won Kim and Hee-Jung Lee, Yonhap News Agency's Han-Ki Seo, and The Hankyoreh's Yang-Joong Kim, started to disclose the truth, and cracks began to show in their solidarity with Hwang WS.

Professor Moon SY, co-corresponding author of 2004 *Science* paper, made a quick judgment in this unexpected situation. He suspected Hwang WS, who could not provide an accurate answer to the allegations. On the night of December 14, 2005, Roh SI, chairman of MizMedi Hospital, visited Hwang WS, who was hospitalized at SNU Hospital. At this meeting, Roh asked, *"Do you currently have stem cells?"* and the words that came out of Hwang's mouth were truly shocking: *"There are currently not a single stem cell line"*.

After hearing this, Moon decided to abandon the sinking ship with Dr. Roh, and the next day, called for a press conference on December 15, 2005. Dr. Roh announced, "Hwang Woo-Suk currently has no stem cell line." [12] This announcement reversed the one-sided situation at once. On the same night, anchors Mr. Um KY and Miss Kim JH of MBC's < 9 o'clock News Desk> said in a trembling voice, *"Oh, everyone! How can I deliver this news? Professor Hwang Woo-Suk said that there were no stem cells."* Leaving a news scene that was still under discussion, MBC aired the second episode of <PD Note> Hwang, which had been postponed indefinitely, "Why <PD Note> Demanded Re-Verification," hosted by Choi Jin-Yong, Director of Current Affairs and Culture.

Under these circumstances, SNU was unable to defend itself against the University's case of the *Science* paper falsification. Young professors affiliated with the College of Natural Sciences visited the president's office to protest. Before this incident, SNU President Chung convened a meeting of the deans. At the meeting, only Kyu-Chang Wang, Dean of the College of Medicine, and Jung-Hye Roh, Dean of Academic Affairs, insisted that the verification should be conducted; all the other deans opposed the verification. [13] There is also testimony from SNU President Chung that even the Ministry of Education blocked Hwang's investigation [30]. However, things changed dramatically after December 15, and the verification could no longer be postponed. President Chung set up an investigation committee at SNU

[12] Young-Joon Ryu heard this directly from Shin-Yong Moon.

[13] Young-Joon Ryu heard this directly from Gyu-Chang Wang and Jung-Hye Roh.

on December 16 and launched the investigation. It was an independent special organization composed of nine experts, headed by Professor Chung, a professor at SNU College of Medicine. They were professors Oh of the College of Pharmacy, Kim of the College of Dentistry, Park of the College of Law, Lee of the College of Agricultural Life Sciences, Ryu of the College of Veterinary Medicine, Director of Research Jeong, Lee of Hanyang University, and Chung of Yonsei University. Professor Ryu of the College of Veterinary Medicine resigned voluntarily for unknown but predictable reasons, leaving 8 active members.

On January 10, 2006, the Investigation Committee submitted a 49-page report titled "Professor Hwang Woo-Suk's Research Suspicion-related Investigative Committee Activities" which can be summarized as follows:

1. The 2005 *Science* paper was fabricated, and there is currently not a single stem cell.
2. The 2004 *Science* paper was also fabricated, and there is no evidence to corroborate somatic cell cloning. There is one stem cell line by parthenogenesis.

The scientific findings on the falsification of two *Science* papers were concluded with this report. To quell speculation among Hwang's supporters, the SNU Research Institute issued "Supplementary Data on the Results of the SNU Investigation Committee on Hwang Woo-Suk's Research" on May 1, 2006, explaining the origin of the 2004 paper's self-reproducing stem cells, or NT-1.

On January 16, 2006, Hwang WS held a press conference at the Scofield Auditorium of SNU College of Veterinary Medicine with the help of Professor Yang, Dean of the College of Veterinary Medicine. At the conference, answering the question, *"How about one stem cell or three stem cells?"* Hwang WS said, *"We do not have the source technology to make stem cells."* These remarks were unimaginable as a scientist. At that conference, he was greatly criticized for trying to increase the credibility of his remarks by lining up graduate students behind him. Hwang WS submitted his resignation to SNU on the same day, but the University dismissed Hwang WS along with Professor Kang SG based on the results of the investigation committee. Hwang WS, who said he would leave on his own, filed a lawsuit for dismissal and eventually lost the case in 2013, confirming his dismissal from the professorship. Professor Lee managed to escape, and the decision on his status was postponed until the outcome of the trial.

2.3.5 How Did a Scientific Study Fabrication Case End Up Standing Before the Court?

In the scientific field, research irregularities caused by researchers are beyond the academic community's judgment and rarely handled by judicial law. In fact, there was no basis for the rules within SNU to pass the case to a criminal investigation. As if Hwang WS had already known, he reported the case to the prosecution for his own immunity. In the instance of manipulation, which he did not corroborate, he directly pointed out the party and hoped he would evade punishment. After receiving the data from the SNU Investigation Committee, Deputy Prosecutor Mr. Lee of the

Seoul Central District Prosecutors' Office, who received the case, conducted a five-month investigation by mobilizing 9 prosecutors and over 50 investors, with Mr. Hong, the third special division chief, as the chief prosecutor. Seven raids were conducted, and 950 people were summoned to the Prosecutor's Office [31]. On May 12, 2006, the prosecution released a 144-page investigative report and indicted Hwang WS on charges of fraud, embezzlement, and violation of the Bioethics Act.

2.3.6 The Criminal Justice Process and Outcome

The first trial began on June 20, 2006 with the hearing of the 26th Division of Criminal Agreement (Chief Judge Mr. Hwang) in Court 417, West Building, Seoul Central District Court. Prosecutors demanded 4 years in prison for Hwang WS, 1 year and 6 months for both Professor Lee BC and Kang SG, 3 years for Kim Sun-Jong, 1 year for Dr. Yoon HS, and 10 months in prison and 2 years of probation for Dr. Jang, director of Hanna Obstetrics and Gynecology. After more than 43 trials over the course of 3 years and 4 months, Hwang WS was sentenced in the Supreme Court of Criminal No. 417 at 2 p.m. on October 26, 2009, in the 26th Criminal Division of the Seoul Central District Court (Chief Judge Mr. Bae). Hwang WS was sentenced to 2 years in prison and 3 years of probation. Kim Sun-Jong, who mixed and planted cells himself, was sentenced to 2 years in prison and 3 years of probation. Professor Lee BC was fined 30 million won, Professor Kang SG was fined ten million won, Dr. Yoon was fined seven million won, and Dr. Jang, who sold oocytes, was sentenced to 4 months in prison [32]. With the exception of Hwang WS, the other indicted accepted the results of the first trial and gave up their appeal, which confirmed the sentence [33].

The second trial was held on November 3, 2009, when the prosecution appealed without changing the indictment. On December 16, 2010, the Third Criminal Division of the Seoul High Court (Chief Judge Mr. Lee) acquitted Hwang on the charge of about 105 million won of research funds recognized as embezzlement and reduced his sentence to 1 year and 6 months in prison with 2 years of probation. The rest of the charges that he had been found guilty of remained unchanged. On February 27, 2014, the Supreme Court's trial confirmed the lower court's verdict, which convicted Hwang WS to the First Division of the Supreme Court (Chief Judge Mr. Ko) and sentenced him to 1 year and 6 months in prison and 2 years of probation. On the same day, Hwang WS filed a lawsuit against SNU to overturn the dismissal of his professorship, and the dismissal was declared a political party. This eliminates the possibility of Hwang WS returning to his post as a professor.

2.4 What Created the Scientific Research Manipulator

As seen earlier, Hwang WS, a graduate of the prestigious Daejeon High School and SNU, who later became a professor at the same university, became a national hero overnight by exploiting the gaps in the Korean media—he had learned about the gaps from an alumnus of the same high school. He was able to meet several political and government dignitaries and became a member of the Presidential Advisory Council on Science and Technology by deliberately misrepresenting facts about cloned cows, Young-Long and Jin-Yi. Continued demands for verifications from academics were deflected by provision of research grants; those demands by the media were forcefully blocked. Furthermore, he received continuous public attention and government support for the deception that he had successfully cloned a tiger from Mt. Baek-Du and a cow that was resistant to the mad cow disease. He even eyed the Nobel Prize by fabricating a thesis on human cloned embryonic stem cells.

In this case, as in the case of Dean Mr. Yang of the SNU College of Veterinary Medicine, there was an alliance of Korean *"Loyalty."* As graduate students at the time, many people directly or indirectly participated in his fraudulent acts—some even became professors, such as professor of the SNU College of Agriculture and Life Sciences, Professor of the Chungnam National University College of Veterinary Medicine, Professor of the Chungbuk National University College of Veterinary Medicine, and Professor of the Gachon University Department of Life Sciences. They attempted to save Hwang WS even after his thesis was found to be fabricated, just to repay the favor they had received from him. Meanwhile, as I saw at the SNU dean's meeting at the end of 2005, there were testimonies proving that most of the college deans opposed the verification within the SNU, which should have been the arena for such verification, and that even the Ministry of Education blocked the verification. Thus, it can be seen that the Korean academic community does not have the ability to self-correct.

What about the media? In 1999, the news of the birth of Young-Long was accompanied with unfiltered videos of Hwang WS shouting "success!" and receiving the new-born calf on a farm. The media, generally, did not demand DNA fingerprinting between somatic cells and cloned animals, even a few reporters asked, were lied—Hwang WS responded that it was lost, but no media house doubted or criticized his behavior. It was widely rumored that Hwang WS had given his credit cards to a few reporters (so called Hwang's scholarship) that he favored [34, 35]. The rumor that researchers at Hwang's lab delivered beef to their set every year during the holidays turned out to be true. It was highly unlikely that a journalist who had graduated from the SNU or had ties to the Daejeon area would write a critical article about Hwang. Moreover, Hwang's remark, *"I put with the Taegeukgi—the Korea national flag—in the heart of the United States,"* in a presentation at the American Association for the Advancement of Science (AAAS) in the United States in 2005 [36] was enough to stimulate the nationalism of reporters and color the media with totalitarianism.

Korean conglomerates scrambled to invest in Hwang WS with high anticipation. SK and Nonghyup invested more than two million dollars, and a few other companies provided him with more than three million dollars as private support. No sooner did he receive the money, then he embezzled it. However, at the trial, the investors submitted that this amount was a purposeless donation, saving Hwang WS from being charged with fraud. Why did they do this? Scholars analyzing the Hwang's scandal noted that he was an influential person, given his proximity to high-ranking government officials and politicians and his regular private drinking sessions with powerful individuals [37].

3 Democracy, Research Ethics, and Pandemic in Korea

As described above, Hwang WS scandal occurred only 10 years after the establishment of the democratic government in South Korea. It was shown that in an immature democratic system, scientific fraud can easily occur in a structure such as SNU where power is centralized. Conversely, the strong solidarity and cooperation of the advocates of democracy finally exposed the injustice. The Hwang WS case is directly and substantially related to the establishment and revision of the "Bioethics and Safety Act," in South Korea. We have also seen a confrontational relationship between the government seeking for authority of science and the citizens seeking for protection of rights of research participants and research integrity. Unfortunately, the Act is not a sufficient deterrent to recurring scientific misconduct. I will provide a brief overview of historical development of the Korean democratic system, establishment and revision of the Korean Bioethics and Safety Act, as well as the current situation of research ethics and integrity throughout the period of COVID-19 pandemic.

After the completion of the World War II and Japanese occupation, through the disruptive Korean War from 1950 to 1953 that divided the north and south of the fatherland peninsula, Korean people have clearly been continuing resistance against abnormal affairs of government and political power. In 1960, when there was fraud during the president election by the incumbent president, the people demonstrated by holding large national protests to create a new democratic government. However, soldier-president Park Chung-Hee broke down the people's desire for democracy, with aggressive promotion of economic growth under his dictatorship. Korean people protested again through "the Busan-Masan Uprising" (1979), which was suppressed by the military government. Soon after that, the solder-president was assassinated by the director of the Korean Central Intelligence Agency (KCIA), and the assassin was executed the following year. The subsequent "Gwangju Democratization Movement" (1980), where at least 54 citizens were killed and more than 500 wounded, was suppressed by the soldier-president Chun Doo-Hwan and opened the era of military dictatorship again, with his presidency from 1980 to 1987. Until 1992 when the time of the next (last) soldier-president Noh Tae-Woo ended, Korean people endured a long period with patience. When the first non-military civilian president Kim Young-Sam was elected, and his government began to eradicate the previous military dictatorship one by one. This included the persecution of former president Chun for the atrocities during his reign. Following the presidency of Kim Young-Sam, Kim Dae-Jung, a political leader of democratization, who were threatened with assassination and sentenced to death under the Park's presidency, was inaugurated as president, assuming office until 2003. Then Roh Mu-Hyeon, who supported the students' democratization movement as a human rights lawyer, was inaugurated as the next president. As such, Korean people have resisted and denied atrocity along with corruption of power and tried to correct it at all costs. This political experience was expressed in the spirit of resistance against corruption or injustice by the government-led research plan or huge vested interests like in such power structure as the SNU.

3.1 A Legal System for Maintaining Research Ethics

3.1.1 The Bioethics and Safety Act

Korea's Bioethics and Safety Act (hereinafter referred to as the Bioethics Act) is a law that was established in December 29, 2003, and came into effect on January 1, 2005, and during this period Hwan's scandal was revealed. This Act is the first and only Asian comprehensive legal framework for bioethics that contains prohibition of human cloning as well as permission and regulation of the creation of human embryos and cloning of human somatic cells for research purpose, and later revised to legal framework for research involving humans generally. However, it did not have enough power to prohibit scientific misconduct. This is a difficult challenge shared in the world, because of "academic freedom." However, in my opinion, while this freedom is one of the recognized human rights, it should be limited so as not to impede the public interest. Moreover, the practice of this right must in be pursuit of truth and must not be eroded by deception for individual profit.

After the announcement of the plan to enact the bill in January 2000, serious disagreements and controversies arose among scientific circles, industries, religious circles, and social groups. This reflects the robust and intense conflicts between Korean people's aspiration for democratization along with protection of individual human rights and the government-scientific-industrial complex that seeks for scientific and economic growth at the expense of these rights?

The Act prohibits all human reproductive cloning and some therapeutic cloning of somatic cells except for research into the treatment of rare and incurable diseases, and the permitted range is determined by the National Bioethics Committee. Additionally, the commercial distribution of sperm and eggs is prohibited, and the government manages the registration and management of embryo-generating institutions. Violation of this law and somatic cell cloning or heterogeneous implantation is punishable by up to 10 years in prison. Furthermore, it includes regulations on genetic testing, gene therapy, human resource bank, and Institutional Review Boards.

3.1.2 Hwang's Intervention in the Legislative Process

The government prepared and implemented the draft of this law from 2002 to 2005 when I was attending graduate school at the SNU. A working officer Kim of the Ministry of Health and Welfare, who drafted this law, met Hwang WS in person twice to discuss restrictions on who could study human embryo cloning. In essence, rules were designed to allow none but Hwang WS to conduct research. [14] In other words, there are interim measures for research on somatic cell cloned embryos in Paragraph 3 of the Addendum, but researchers who have been conducting research in the past could continue their research with the approval of the Minister of Health and Welfare only in the following cases.

1. Must have continued research on somatic cloned embryos for at least 3 years.
2. Must have a record of publishing research papers on somatic cloned embryos at least once in a related academic journal.

Dr. Park, who was then the director of the Maria Obstetrics and Gynecology Fertility Research Institute, was the first to establish human embryonic stem cell line from IVF embryo in Korea and was a tough competitor to Hwang WS. When Hwang WS learned that Dr. Park was applying for approval to perform the same research, he may have used his connections to introduce a condition that prevented Dr. Park from getting the approval.

3.1.3 Revision of the Bioethics Act After the "Hwang WS Scandal"

Although the gist of the Bioethics Act has not changed much, it has been reformed as a more just law after 20 revisions. The original, core framework of the Act was the regulations on embryonic research including human embryo cloning, in-vitro fertilization (IVF) embryo creation, disposal, human embryo research institute, as well as genetic cutting-edge technology including genetic testing, gene therapy, and gene banking. Since the 2012 revision, the scope was expanded to any research with interaction with human beings, as well as biobanks for human biological materials and data.Although it can be said that the Hwang WS scandal had a direct impact on the revision of this Act, people have sought in particular to prohibit coercive requests for egg donation. However, there is no specific stipulation in the Act that has direct impact on this. Nevertheless, addition of the technical limitation was with the objective of strengthening monitoring of human embryos and cloning to prevent the recurrence of similar types of unethical research. Despite these legislative efforts, the Act does not have the power to prohibit scientific misconduct.

3.2 Has Research Ethics and Integrity in Korea Improved?

3.2.1 Research Ethics and Integrity Have Degraded Into a Tool for Eliminating Public Office Candidates

According to National Research Foundation of Korea reports titled, "Investigation of Research Misconduct Detected between 2007 and 2012" and "Status of Domestic Research Ethics, 2013," 43% incidents of thesis misconduct in Korean universities

[14] Heard directly from Kim SS.

were confirmed as plagiarism, and research misconduct was most frequent in social sciences, engineering, humanities, arts, and physical education. The reports reveal that the Korean academic community is lenient toward research fraud and offenders commit fraud based on their personal relations with relevant officials. Such leniency and misuse of personal relations prevent expulsion from academia even when one violates "research truthfulness," leading to limits on punishment and serious damage to the academic community's ability to self-correct. This is accentuated by the research dishonesty witnessed at the Korea National Sport University (KNSU)—the only national sports university in Korea. Of the 251 articles published in various academic journals by 95 professors from KNSU since 1993, approximately half of them—that is, 120—republished their students' dissertations without including them as co-authors or citing them. Another professor conducted muscle biopsy on a student for a long period of time, which eventually led to nerve damage, forcing the student to abandon sports. A state audit by the Board of Audit and Inspection of the National Assembly was conducted, but it was buried into oblivion—probably under the weight of private connections.

Alternatively, research ethics are being used as a means of political struggles to disqualify candidates for public offices in the Korean context. Plagiarism of articles and violations of research ethics have been frequent topics in the National Assembly hearings that verify candidates for public offices nominated by the opposing party.

- In 2006, Mr. Kim, Deputy Prime Minister of Education, resigned voluntarily after 13 days in office owing to allegations of plagiarism.
- In 2014, Mr. Kim, Deputy Prime Minister of Social Affairs and Minister of Education, withdrew his nomination in 2014 owing to allegations of plagiarism.
- In 2015, Mr. Jeong, Minister of Health and Welfare, resigned after 7 days on suspicion of plagiarizing from his student.
- In 2016, Mr. Park, a researcher at the Ministry of Public Safety and Security, resigned after 7 days because of allegations of plagiarism.
- In 2017, Mr. Cho, Minister of Employment and Labor, resigned after 32 days on suspicion of plagiarism.

Additionally, allegations of thesis and research ethics violations were raised on Education Minister Ahn, Unification Minister Hyun, Education Minister Lee, Education Minister Seo, Security and Public Administration Minister Jeong, Labor Minister Lee, Culture Minister Kim, Police Chief Kang, Prime Minister Lee, Unification Minister Hong, Minister of Justice Kim, Deputy Prime Minister and Education Minister Lee, Education Minister Lee, Minister of Environment Minister Cho, Minister of Foreign Affairs Kang, Deputy Prime Minister of Social Affairs and Education Minister Kim, Minister of Health and Welfare Park, etc., but most of them were appointed [38].

Likewise, the fact that research ethics violations are a media topic only at political hearings for public officials, and little effort is made to deal with the problem in the academic context, is evidence that the level of research ethics compliance in Korea has not improved much after the Hwang's scandal.

3.2.2 Re-Entry of Hwang WS and the Korean Society's Reaction

On January 16, 2014, David Cyranoski, a reporter of the science magazine *Nature* published a special article on Hwang's Supreme Court ruling and his current status [39]. It tells the story of the fall of the Hwang's Empire in 2005 and the Supreme Court ruling in 2014, and also the news that Hwang WS is still struggling to make a comeback with patents for dog cloning and human embryo cloning, fully ignoring his failures and trying to fulfill his ambitions. The article did not praise Hwang WS at all. After the article was published, the Korean media immediately began to spread the news that the world-class scientific magazine *Nature* had paid attention to Hwang WS again and reported his comeback as a known fact without conveying the magazine's intention. This was an effort by reporters, who had fallen along with Hwang WS, to also make a comeback and it seemed they were using the article as an opportunity to make up for their mistakes. Recognizing this unusual reaction in Korea, *Nature* published an editorial to refute their claims on January 24, 2014, titled, *"Don't rush to rehabilitate Hwang"* [40]. In this editorial piece, the magazine claimed that its article of January 16 was misconstrued in South Korea as an endorsement of Hwang WS by the international scientific community. The magazine stated: *"Nature's profile of a former fraudster's attempts to regain respectability should not be taken as an endorsement of the researcher's claims."* However, the Korean media hardly introduced the magazine's refutation to the public, ensuring that the science con-artist continued living as a hero in Korea.

3.2.3 The Park Geun-Hye Administration and Hwang's Efforts to Recover

Park Geun-Hye (a daughter of a previous soldier-president Park Chung-Hee) was elected president of South Korea in 2013. The following is a summary of the facts revealed during my trial in 2018. Hwang WS got acquainted with Park, who was a student at the time, while lecturing in the AIP course at the SNU between 2002 and 2003. In 2004, he attended the wedding of Park's younger brother Park Ji-Man, and in December 2005, Park GH visited the SNU Hospital where Hwang WS was hospitalized. Around that time, Hwang WS even received a small holiday gift from Park through the late aide Mr. Lee [41]. On November 11, 2016, the *Kyunghyang Shinmun*[15] reported, "There are suspicions that Choi Soon-Sil, a non-executive power, has also interfered in state affairs in the health care sector. This is because the government has actively taken steps to ease restrictions on the use of unfrozen oocytes, the long-awaited project of CHA Hospital, the parent company of Cha-um Hospital, where Choi and President Park received treatment. In particular, in the process, the head of the Ministry of Health and Welfare, who opposed the deregula-

[15] The Kyunghyang Shinmun is a newspaper founded in 1946 by the Catholic Church. While the newspaper was forced to cease publication due to its criticisms against the Rhee administration, it was revived after the pro-democracy revolution of April 19, 1960.

tion, was replaced, raising suspicions of Mrs Choi's pressure" [42]. Additionally, Park instructed Mr. Ahn, the chief economic aide, to come up with countermeasures for the biotechnology sector, and on April 28, 2016, a policy discussion session was held under the theme," Problems of the Law and System Relating to Biotechnology Research" at the conference room of KTX Seoul Station with Mr. Jeong as the moderator, who was then secretary of the Blue House for Trade, Industry and Energy. Hwang WS was contacted by Mr. Jeong, the bio-nano manager of the Ministry of Trade, Industry and Energy, and attended the discussion; he suggested loosening the regulations [43]. On July 22, 2016, *Money Today* reported as follows: "CEO of H-Bion Hwang Woo-Suk, a speaker at the biotechnology breakfast lecture centered on cloning technology, held at the Bank Hall in Myeong-dong, Seoul, said, "I suggested to approve CHA Hospital's somatic cell cloned embryonic stem cell research plan during a conference held at the Blue House last month." (omitted) Hwang WS said, "I really respect and thank the government for reopening the doors for CHA Hospital." On December 7, 2016, *Rapportian* reported my comment that, "it has been a long time since the Blue House "Doorknob triumvirate" and former professor Hwang WS have been sighted together. I remember hearing about Choi Soon-Sil's existence for the first time in 2007 and then in 2009. The story was of her meeting with Hwang WS. At that time, I did not know who Choi was.'" On July 11, 2016, the Ministry of Health and Welfare conditionally approved the somatic cell cloning embryo research plan submitted by the CHA University.

Based on these facts, I commented opinion in an interview with CBS < *Kim Hyun-Jung's News Show* > on November 21, 2016, that: (1) Hwang WS is acquainted with President Park, (2) he attended a government meeting through the Blue House senior office, and (3) when he said that he suggested to grant permission for human somatic cell cloning research at CHA Hospital, he denoted that he was thinking of resuming his research through his personal connections of the past. Hwang WS rebutted by saying that (1) he never attended the meetings through the Blue House, (2) he never asked for grant of approval to CHA hospital's somatic cell cloned embryonic stem cell research, (3) he has no acquaintance with Choi and the "Doorknob triumvirate" who are highlighted as President Park's non-executive influential figures, and (4) he never had a private consultation with the president. Hwang WS pressed charges against me proceeded to sue. However, the court's final decision was not in line with Hwang's expectations. What I said can be interpreted as an explanation of what was already reported in the media, and Hwang WS and President Park were actual acquaintances and there was sufficient room to interpret it as a request for approval of CHA Hospital's somatic cell cloned embryonic stem cell research. Additionally, the court ruled that there was insufficient evidence to conclude that his claim of not being acquainted with Choi or the doorknob triumvirate was false and supported my claims. Despite Hwang's request and the prosecution's appeal, the court maintained the original judgment of April 18, 2019. The prosecution surrendered their appeal and the case ended with my victory. Thereafter, President Park was impeached and Hwang WS seemed to have lost the opportunity to make a comeback using his personal connections.

This Hwang WS accusations revealed that Hwang WS has not changed at all since 2005, and he tried to reclaim his position by repeating his old tricks. In fact, in an interview with *Nature* in 2014, he said that although he was currently cloning dogs and he wanted to research human cloning. It seems for him human embryo cloning is the only way out of the abyss of ignominy, so he can move his research site to a location where he could procure real oocytes. [16] According to a *Money Today* article published on February 28, 2011, Hwang WS said, "Libya has promoted this project to solve the incurable disease problem that frequently occurs among its people with stem cell technology and to make the stem cell field a national growth engine in addition to the oil industry. I also received 900 million won to start the work with." Later, the Libyan project failed because of Gaddafi being ousted from power.

3.2.4 Recurrence Possibility of Scientific Fraud

In Korea, there are still a significant number of people who support Hwang WS scientific fraud. It is also true that Hwang's lies and the media's delusion of a stem cell powerhouse still linger in the memories of middle-aged and elderly Koreans. However, this is a delusion that ignores not only the bioethical issue of the use of human oocytes, but also the scientific fact that the possibility of actual treatment is low. Such an environment where scientific research fraud occurs repeatedly, coupled with the absence of a punishment system, is sufficient to bring Hwang WS back onto the scene. Even if it is not for Hwang WS himself, there is a possibility that a new young Hwang WS will appear and reproduce the situation that is not different from that of 20 years ago. Lessons from the scandal are clear, but efforts to improve are lacking. Scientific research fraud cases are not limited to Korea; thesis manipulation is prevalent in the US, Japan, China, and in many European countries, and even institutional ethics review committees do not operate properly. There are too many unscrupulous researchers, who care little about individual conscience or the traditions of the research community.

4 Ethical Issues in Korea During the COVID-19 Pandemic

4.1 Ethical Issues Related with the COVID-19 Pandemic

Ethical issues that occurred in Korea during the COVID-19 pandemic such as mandatory collection of personal information, allocation of medical services, research ethics during a disaster situation and the ongoing dispute about the violation of

[16] In Southeast Asia, China, Libya, etc., there are no laws on human embryo cloning, or they are working with people who support it.

research ethics since 2005 will be discussed in this section. The transmission of COVID-19 reminded us with its impact being experienced globally. The virus that affected the whole world did not steer away from Korea, and it did not discriminate people based on one's economic status or social position. Up until the current date of August 12th, 2022, Korea was shown to have relatively effective preventive measures with a total of 17,795,357 confirmed cases and 23,745 deaths. However, ethical issues started to arise in various fields, with a clash between doctors' professional duties being one of them. A medical doctor has the obligation to cooperate with public health policies with his/her medical expertise and act as a main agent not only to implement preventive measures, but also respect the patient's autonomy and guarantee patient's rights. However, these obligations resulted in a conflict in principle theories during the COVID-19 pandemic.

According to Centers for Disease Control (CDC), public health surveillance is a continuous and systematic collection, analysis and interpretation of crucial health-related data for planning, implementing and interpreting public health practice [44]. Therefore, doctors had to follow the governmental executive law and to collect and analyze essential health-related data of the infected patients as main agents of the public health surveillance. [17] This created a point of conflict between a doctor's duty and respect and protection of patient's autonomy and rights. For instance, during the early stages of COVID-19 spread in Korea, there was a huge increase in confirmed cases in the city of Daegu due to large numbers I the population having visited Wuhan, China. The government then issued an administrative order for location tracking and collecting history of credit card usage to trace infected patients, which made it possible to retrieve information on the names of patients, detailed movement traces and even who they met. These measures showed great effects in stopping the source of infection but also led to discrimination and stigma onto the city as the origin of large-scale infection [45].

Although the individual resistance against collection of personal information was not small, the resistance was less compared to those in Europe or other Western countries. The authoritarian views of East Asian countries seem to have created a social consensus, in which the current situation of infection that has the possibility of causing direct harm to others can be an exception to tolerating reduction of individual rights. This was not only the case for COVID-19, but also during the SARS outbreak in Hong Kong. Nevertheless, these social consensuses are similar to the physical attacks and stigmatization of Asians residing in the United States and Europe when seen closely. Stigma and discrimination are often based on prejudices and straying from facts, and greatly influences the economically vulnerable groups such as migrant workers, the disabled and sexual minorities, and are even health-threatening.

Conversely, the issue of social influence on vaccination was even more critical. Social pressure to vaccinate individuals increased in order to achieve good community health. Due to limited government's medical resources such as critical care

[17] INFECTIOUS DISEASE CONTROL AND PREVENTION ACT.

beds, the importance of lowering the severity of the disease through higher rate of vaccination led to an explicit pressure from the government. These various sources of pressure on vaccination sparked disputes of violation of individual's right to "freedom of choice" and resistance against vaccination. Arguments against vaccination included short vaccine development period, mRNA method instead of the conventional production method, strong cytokine storm in young adults, sudden death from anaphylaxis and small governmental compensations for side effects from vaccines such as myelitis. Some university professors and researchers even came forward and argued the danger and ineffectiveness of vaccines, which then affected the public's opposition toward vaccines even more [46].

Priority of distribution in medical resources was actively discussed in the Korean Society of Clinical Ethics [47]. In the early stages of the pandemic when the supply of vaccines was not well distributed, the ethical principles for distribution of medical resources had not yet been established. Rather than providing equitable treatment opportunities, alleviating health inequality, and considering fairness and transparency, administrative decisions to handle urgent cases first caused confusion in the medical sites. Moreover, a specific and detailed plan to allow people to get vaccinated with sufficient awareness of vaccine options and side effects were lacking in the early stages. Although critical care beds and medical services were essential for treating critical patients, there was always conflict and concerns regarding whom and where the limited medical resources should be allocated on. Next, there was criticism on research ethics during disaster situations [48, 49]. In times of public health crisis, the development and research process of vaccines or cures often requires urgency. WHO recommended that the national research governance framework and international research communities to establish methods and frameworks that guarantees expedite ethical reviews during urgent situations [50]. In other words, rapid decisions, prioritized decisions, and convenient methods such as remote or e-mail were chosen for research planning. During this process, disaster areas and community members that had the potential to be a subject of study were exposed to higher risks.

Within this confusing situation in medical ethics, scientific misconduct has continued worldwide.

4.2 CV-Building and Scientific Misconduct for University Admission

There have been many scientific misconduct cases around the world during COVID-19 pandemic [51]. Scientific results continued to be published rapidly to overcome the unprecedented pandemic, but there was a situation where sloppy peer review and a performance-oriented system that took advantage of the crisis accelerated fraud. In South Korea, no notable fraudulent research related to COVID-19 was revealed, but in April 2022 serious fraud involving minors during research conducted during the pandemic emerged publically.

While these cases occurred during the COVID-19 pandemic, they were not directly related to pandemic situation, and similar situations recur continuously. On April 26, 2022, the Korean Ministry of Education discovered 96 co-authored theses among minors (high school students) and cancelled university admission of five convicted students. At first glance, the publication of a thesis in an international SCI-level journal (journal in the database "Science Citation Index," granted impact factor) by a minor was considered a significant scientific achievement. However, most of it was research dishonesty as students did not directly participate in the experiment or thesis writing. Their names were mentioned on the theses to receive additional points in university entrance exams, and a nexus between parents and private university prep institutions was unearthed.

These students are taking advantage of their parents, who allow their children or acquaintances to participate in research. Several private education providers concur that this phenomenon began during the Lee Myung-Bak administration (25 February 2008 to 24 February 2013). Let us consider an example [52]. At the G20 Seoul Summit, a group of 14 students called "WAVE" [53] created brochures about Korea for foreigners. They also published a monthly newspaper called the "WAVE Newspaper," which covered human rights issues in North Korea; the monthly publishing cost was five million won (about 4000 USD). These students advertised their social activities through newspapers and received several awards. An admissions expert said that if these students really published a newspaper independently, they can be admitted to university without even looking at their grades. Among them, four students were commissioned to serve as public relations ambassadors for the Gyeonggi Tourism Organization, and they presented their thesis as the first author, claiming that they had conducted an experiment at the SNU. Together with his aunt, sister, and cousin, one student represented Korea at the Special Olympics. These students also held a seminar sponsored by the *Chosun Ilbo* and the Saenuri Party Rep. Mr. Jeong and other members of the National Assembly delivered congratulatory speeches. Seven of them were admitted to prestigious universities: Yale (2 students), Harvard (1 student), Stanford (2 students), Seoul National (1 student), and U Penn (1 student). Thereafter, the publication of the WAVE [54] newspaper stopped. The experience they had accumulated was impossible to acquire independently, yet it played a decisive role in their university admission. According to the media coverage of the time, their parents were lawmakers, judges, lawyers, professors at medical schools, and owners of medium-sized enterprises. This raised the issue of violation of authorship ethics, emerging as a social concern highlighting the inequality of opportunity because of the so-called "*Parent chance*".

On May 19, 2019, KBS reported that Professor Lee BC at the SNU College of Veterinary Medicine had listed his son—a high school student—as a co-author of his thesis [55]. Professor Lee BC used this thesis [56] to secure admission for his son at the Kangwon National University College of Veterinary Medicine; he was accused of obstructing the execution of official duties and business by soliciting the evaluation committee through academic connections and regionalism. [18] Later, when his son graduated from the Kangwon National University, Lee admitted his son to his own laboratory to major in veterinary obstetrics at the SNU's graduate school. On April 26, 2022, the Ministry of Education cancelled admission of Lee BC's son to the Kangwon National University. In the process, Lee BC and his SNU college alumni, Professor L, Chairman of the College Transfer Committee of Kangwon National University, and Professor C of the Chungnam National University were implicated and indicted [57]—L and C were members of Hwang' research organization as mentioned earlier. This is clearly a reenactment of the Hwang's scandal. Although it has been some time since the Hwang saga surfaced, such crooked and network-oriented mindsets continue to harm research ethics.

4.2.1 Research Ethics Violations by Daughters of Two Former Minister of Justice

Mr. C, Professor of law at the SNU, served as the chief of civil affairs in the Office of President Moon from 2017 to 2019. He became the 66th Minister of Justice in 2019, but resigned after 67 days of appointment because of allegations leveled against his family [58]. The allegations included a research fraud charge against his daughter. She studied at the Hanyoung Foreign Language High School in Seoul and at the Department of Environmental and Ecological Engineering, College of Life Sciences & Biotechnology, Korea University. Thereafter, she graduated from the Pusan National University School of Medicine and became a doctor. Mrs. J, the wife of Minister C, was tried on charges of forging documents for her daughter's university admission and a total of seven documents used during the university admission process were confirmed to be false at the trial. These included documents about field work at the Dankook University Medical Science Research Institute and first author thesis, [59] field work at the Gongju University Biotechnology Research Center, internship at the SNU Public Interest Human Rights Law Center, internship at the Busan Aqua Palace Hotel, Dongyang University President's award certificate, third author of the Gongju University thesis, internship at the Korea Institute of Science and Technology (KIST), and assistant researcher and internship at the Dongyang University Language Education Center. Prosecutors and the court believed that former Dongyang University professor Mrs. Jeong was involved in forgery and on January 27, 2022, the Supreme Court sentenced her to 4 years in prison. Based on this ruling, the Pusan National University cancelled the admission of her because even though it was not the forged documents that got the daughter admitted to the university, the information provided for the admission was false [60].

A prosecutor led the investigation team in the case of former Minister of Justice and his wife. Ironically, the prosecutor became the incumbent Justice Minister, a thesis written by high school sophomore, his daughter also embroiled in

[18] Articles dated April 26, 2022, 96 cases of fraudulent co-authorship with minors found.

controversy over research fraud. She is currently attending the 11th grade at an international school in Incheon and engaged in a multitude of activities to build up her résumé to be favorable for entering a foreign university. She has published English e-books and has written several articles for online English magazines. What is unusual, however, is that eight papers were published in her name within a short period of 3 months. The *Hankyoreh* newspaper confirmed that one of the theses was written by a ghostwriter [61]. Additionally, it is suspected that the volunteer activities at the local children's center were also false. A civic group has filed a complaint against the current Minister, his wife, and the head of a local children's center to the Corruption Investigations Unit for forgery of private documents and obstructing the execution of official duties.

On June 14, 2022, the <*MBC PD Note* > broadcast the results of an investigation into the current Minister of Justice's family [62]. Not only the thesis written by the daughter, but that written by [63] the current Minister of Justice's wife's minor nephew is also suspected of research fraud. It was actually written by a Professor of the Yonsei University School of Medicine and published in 2019. This professor was the corresponding author, and the above nephew of the Minister of Justice's wife was the first author. She was a high school student in the United States at the time; she is currently attending the School of Dentistry at the University of Pennsylvania. The article's submission date was June 21, 2019, and the publication date was June 28. In this regard, the Korean–Americans living in the United States issued a statement that the essence of the incident was an organized crime carefully designed by the privileged class of Korea to enter a prestigious American university [64]. The Yonsei University Committee on Research Integrity initiated a preliminary investigation into this.

5 Concluding Remarks

In this chapter, the root cause of research misconduct and a system that aids and abets such malpractices were examined based on the case of thesis fraud by Hwang WS in 2005 and the thesis research fraud cases of underage persons in 2022. As the parties involved in the incident more than 20 years ago continue with research misconduct, and because similar transgressions emerge from others as well, it is evident that the absence of research integrity is pervasive and the situation remains unchanged [65].

The fundamental purpose of science is the search for TRUTH. Researchers publish articles to share their research results, which serve as scaffolding for others to build a solid academic structure. However, if these articles are written without scientific integrity and sincerity, subsequent research based on it will be shaky and the academic structure will collapse. Manuscripts do not always contain the truth, so verification by fellow researchers is required; this activity filters out the deception. However, a perusal of the research fraud cases worldwide reveals a nexus between unscrupulous scientists, who verify as genuine each other's false claims. Moreover, there are few social mechanisms to punish those who intentionally manipulate study results to achieve their ends. Hwang's scandal and cases of fraudulent theses to secure university admission have a time lag and different characteristics, but they are all fraud committed for personal gain. In many cases, the same person has been committing fraud for decades. This is done under the belief that it is more profitable to succeed even through a lie than to lose everything when the fraud is discovered. Moreover, the society tolerates success acquired through wrong means. Consequently, this greed for success finds expression worldwide.

Despite the law reform as a result of the 20 revisions of the Korean Bioethics and Safety Act catalyzed by the Hwang WS debacle, there remain loopholes allowing for certain types of scientific misconduct to continue unfettered. In some cases, such as the "Terranos" scam, conviction has resulted in justice being served, but they are rare, and others, such as Hwang WS, have found a fresh start elsewhere [66]. In the Hwang WS scandal, none of the key players involved in the scam went to jail, and they were able to move on and take responsibility for the damage they caused. In the end, most of these people ended up taking responsibility for their lives for the failure of other things due to their own wrong views, but this does not seem fair. At the very least, scientific fraud by scholars should be cracked down on in academia, and it would be the realization of justice if they were held accountable for causing social turmoil. Scientific fraud cases, which are caused by scientists lying, are causing people to make mistakes in judgment, leading to social division, extreme human losses, and high social costs. Nevertheless, there are clear rules to punish scientists who commit scientific fraud, and it encourages lying when they return to practice elsewhere. These are problems that we all need to address as a matter of urgency.

Acknowledgement As a whistleblower of the Hwang WS scandal, I thank everyone who has helped me through many trials over the past 20 years. In particular, I would like to thank producer Hak-Soo Han, the Seung-ho Choi of MBC PD Notebook. I thank Jae-Myung Lee and Byung-soo Kim of the People's Solidarity for Participatory Democracy. I thank Professor Sang-Ik Hwang of Seoul National University and Professor Han-Gyeom Kim of Korea University.

Conflict of Interest There is no conflict of interest related to this manuscript.

References

1. Cyranoski D. Whistle-blower breaks his silence. Nature. 2014;505:593–4.
2. Hwang WS, et al. Evidence of a pluripotent human embryonic stem cell line derived from a cloned blastocyst. Science. 2004; 303: 1669–1674. Retracted in: Kennedy D. Editorial retraction. Science. 2006;311(5759):335. Epub 2006 Jan 12.
3. Hwang WS, et al. Patient-specific embryonic stem cells derived from human SCNT blastocysts. Science. 2005; 308: 1777–1783. Retracted in: Kennedy D. Editorial retraction. Science. 2006;311(5759):335. Epub 2006 Jan 12.
4. Choe SH. South Korean scientist indicted in cloning scandal. International Herald Tribune. Sect. Health & Science. 2006. https://www.nytimes.com/2006/05/12/health/12iht-korea.1725751.html. Accessed Jul 2023.

5. Ahn MY, Normile D. Korean supreme court upholds disgraced cloner's criminal sentence. Science. Sect. News: Science Insider. 2014. https://www.science.org/content/article/korean-supreme-court-upholds-disgraced-cloners-criminal-sentence. Accessed Jul 2023.
6. Nature News. Woo Suk Hwang. Sect. Specials. 2006. https://www.nature.com/collections/szl-cbykgyl. Accessed Jul 2023.
7. Lee EY. Hwang Woo-Suk's full story of cloning the Baekdu mountain Tiger. Shin Donga, No. 557; 2006. p. 120–7. https://shindonga.donga.com/3/all/13/105185/1. Accessed Jul 2023.
8. Huh MS et al. Testimony of Byung-gul Paik, professor at Chonbuk National University College of veterinary medicine Shin donga, March 6, 2006. https://shindonga.donga.com/3/all/13/105235/1. Accessed Jul 2023.
9. Park JH. Dr. Hwang Woo-Suk's false testimony still influences the damage on livestock farms. Voice of Jeonbuk, Sect. SPECIAL PROJECT. 2020. http://www.jbsori.com/news/articleView.html?idxno=2413. Accessed Jul 2023.
10. Moon GY. Seoul National University undergraduate course must be eliminated, local universities survive ... The theory of the abolition of Seoul National University erupted inside Seoul National University. MK Article. Sect. Society. 2022. https://www.mk.co.kr/news/society/view/2022/01/28655/. Accessed Jul 2023.
11. Saegusa A. South Korean researchers under fire for claims of human cloning. Nature. Sect. News. 1998. https://www.nature.com/articles/25421. Accessed Jul 2023.
12. Han HS. How should I tell you this news?. Social Commentary. pp. 54.
13. Lim DW. "Pollution · blackout · loss of data..." professor Hwang said accident at every important moment. The Korea Economic Daily, December 19, 2005. https://n.news.naver.com/mnews/article/015/0000858666?sid=105. Accessed Jul 2023.
14. Seo HG. Where's young-long? Yonhap News Agency. December 22, 2005. https://n.news.naver.com/mnews/article/001/0001180671?sid=105. Accessed Jul 2023.
15. Kim Sung-Hoon's statement. MBC PD Note. 2005. https://www.youtube.com/watch?v=S4YJu2wm7LY. Accessed Jul 2023.
16. Kim JW. [people focus] Hwang woo-Suk (from Buyeo, professor of clinical veterinary medicine at Seoul National University). Chungcheong Today. 2004. https://www.cctoday.co.kr/news/articleView.html?idxno=53486. Accessed Jul 2023.
17. Son JK. Members of the ruling and opposition parties formed a support group for professor Hwang woo-Suk. ETNews, December 6, 2005. https://n.news.naver.com/mnews/article/030/0000126192?sid=105. Accessed Jul 2023.
18. Son BK. Hwang WS met DJ "I will fix you and your son". Oh My News. 2006. http://www.ohmynews.com/NWS_Web/View/at_pg.aspx?CNTN_CD=A0000304593. Accessed Jul 2023.
19. Even the Baekdusan tiger will be cloned. Chosun Ilbo, Social News. February 21, 1999, p. 23.
20. Cloned Baekdusan tiger to be born next year. Hankyoreh, Social News. December 29, 1999, p. 15.
21. A cloned Mt. Baekdu tiger, 'Oh heung' next year? SCIENG, http://www.scieng.net/sisatoron/10831. Published 2005. Accessed Jul 2023.
22. Korea's first cloned Baekdusan tiger to be born in July. Maeil Economy. April 17, 2000.
23. Campbell KH, McWhir J, Ritchie WA, Wilmut I.Sheep cloned by nuclear transfer from a cultured cell line.Nature1996; 380(6569):64–66.
24. Thomson JA, Itskovitz-Eldor J, Shapiro SS, Waknitz MA, Swiergiel JJ, Marshall VS, Jones JM. Embryonic stem cell lines derived from human blastocysts. Science. 1998;282(5391):1145–7. Erratum in: Science 1998 Dec 4;282(5395):1827. https://doi.org/10.1126/science.282.5391.1145.
25. Peace Broad Casting Open World Today, I'm Jang Sungmin [radio broadcast]; December 23, 2005.
26. Korea, Missed three times to get Nobel prizes. Sisa Journal. February 28, 2011. http://www.sisajournal.com/news/articleView.html?idxno=131502. Accessed Jul 2023.
27. Park KY. Evidence of a pluripotent human embryonic stem cell line derived from a cloned blastocyst. Science. 2004;303:1669–74.

28. Ky Young Park received 250 million won from Professor Hwang. SBS News. January 17, 2006.https://news.v.daum.net/v/20060117214509327?fbclid=IwAR1dg_VuJ9tk0l80AgC_O--NNBhst2AKB-AwruCQF1tP_c8E5cFsQ2ZVuXKE&fs=e&s=cl. Accessed Jul 2023.
29. Han HS. Everyone, How I should I deliver the News? Social Criticism. p. 289.
30. Kang YG. The Ministry of Education also blocked SNU's Hwang woo Suk investigation. Dent Press; 2006. https://www.pressian.com/pages/articles/32237. Accessed Jul 2023.
31. Lee NH. The story behind the investigation into the Hwang woo Suk case that ended after 5 months. Dong-A Ilbo. June 21, 2006. https://woman.donga.com/3/all/12/135418/1. Accessed Jul 2023.
32. JoongAng Ilbo. First trial ruling in 3 years and 5 months for 'manipulation of stem cell thesis'. October 27, 2009. https://www.joongang.co.kr/article/3844387. Accessed Jul 2023.
33. Seon JS. Stem cell thesis fabrication' researcher Seonjong Kim abandons appeal... Only professors Hwang woo-suk and Yun Hyun-Soo remain. Kookmin Ilbo, December 31, 2009. http://news.kmib.co.kr/article/view.asp?arcid=0002725162&code=11131100. Accessed Jul 2023.
34. Lee SJ. The country of Hwang woo-Suk. Seoul Korea, Bada publishers; 2006. Bada Publishers.
35. Shin HC. Hwang's credit card in veil. Sisa Journal April 3, 2006. http://www.sisajournal.com/news/articleView.html?idxno=104728. Accessed Jul 2023.
36. Human Embryonic Stem Cells May Promise Medical Advances. ScienceDaily article. February 12, 2004. https://www.sciencedaily.com/releases/2004/02/040212090522.htm. Accessed Jul 2023.
37. Kang YG, Kim BS, Han JG. Silence and enthusiasm: seven years of Hwang woo-Suk scandal. Humanitas. To protect the honor of individuals, this book does not directly refer to them by their full names except in special cases, but for those who wish to research them, the following videos are available. The man who faked human cloning. https://www.youtube.com/watch?v=ett_8wLJ87U. How to catch criminal cloner. https://www.youtube.com/watch?v=st6yupvNkVo Accessed Jul 2023.
38. Lee JE, Lee HE. Plagiarism of papers from high-ranking officials... Research ethics pushed aside by politics. University news Network. July 17, 2017. http://news.unn.net/news/articleView.html?idxno=176765. Accessed Jul 2023.
39. Cyranoski D. Cloning comeback. Nature. 2014;505:468–71. https://doi.org/10.1038/505468a.
40. Don't rush to rehabilitate Hwang. Nature. 2014;505:453. https://doi.org/10.1038/505453a.
41. Court statement of Hwang Woo-Suk in 2017. Seoul Eastern District Court 2017 Go-Dan 3879.
42. Jeong HW. Nam JW. Heo JM. After the president's comments, oocyte regulations were eased... Oppositions transferred to different posts. Kyunghyang Shinmun. November 11, 2016. https://www.khan.co.kr/politics/politics-general/article/201611110600085. Accessed Jul 2023.
43. Court statement of Hwang Woo-Suk in 2017. Refer to the written judgment of the Seoul Eastern District Court 2017 Go-Dan 3879.
44. Centers for Disease Control. Comprehensive plan for epidemiologic surveillance. Atlanta: US Department of Health and Human Services, Public Health Service; 1986.
45. Moon JY, et al. Social responsibilities of health care professionals during the COVID-19 pandemic: avoiding privacy infringements and the stigmatization of patients. Korean J Med Ethics. 2021;24(2):159–70.
46. You KH, et al. A review of the justifications for mandatory vaccination policies in response to the COVID-19 public health crisis. Korean J Med Ethics. 25(1):1–20.
47. Fang YH, et al. The ethics of COVID-19 vaccine allocation. Korean J Med Ethics. 2021;24(2):121–37.
48. Fang YH, et al. The pandemic of COVID-19 and the ethics of human challenge study. J Korean Bioethics Assoc. 2020;21(2):55–74.
49. Choi EK, et al. COVID-19 and research ethics. Korean J Med Ethics. 2021;24(1):1–16.
50. WHO Guidance on Research Methods for Health Emergency and Disaster Risk Management. 2021. https://apps.who.int/iris/handle/10665/345591. Accessed Jul 2023.

51. Piller C, Servick K. Two elite medical journals retract coronavirus papers overdata integrity questions. Science. 2020. https://www.sciencemag.org/news/2020/06/mysterious-company-s-coronavirus-papers-top-medical-journals-may-be-unraveling. Accessed Jul 2023.
52. Newstapa. Building ‘special’ qualifications of children of a privileged class. 2019. https://www.youtube.com/watch?v=EE8P1dTTN6Y. Accessed Jul 2023.
53. WAVE: World Association of Volunteering Elites. https://www.facebook.com/hafs.wave/. Accessed Jul 2022.
54. Jeong SY. Dr. Hwang Woo-Suk ‘Guilty’... 2 years’ imprisonment, 3 years’ probation. SBS News. October 26, 2009. https://news.sbs.co.kr/news/endPage.do?news_id=N1000662038. Accessed Jul 2023.
55. Choi YK. Professor Lee Byeong-Cheon, high school student’s son as the author of a thesis. KBS News. May 17, 2019. https://news.kbs.co.kr/news/view.do?ncd=4202836. Accessed Jul 2023.
56. Oh HJ, Lee TH, Lee JH, Lee BC. Trichostatin an improves preimplantation development of bovine cloned embryos and alters expression of epigenetics and pluripotency genes in cloned blastocysts. Theriogenology. 2012;74(11):1409–15.
57. Choi YK. Professor Lee Byung-Cheon in court... “It’s ironic to be prosecuted for reducing the pain of a test dog”. KBS News. October 9, 2020. https://news.kbs.co.kr/news/view.do?ncd=5021613. Accessed Jul 2023.
58. Kang BJ. What happened during the 67 days from the appointment of Cho Kuk as minister of justice to his resignation? HuffPost. October 14, 2019. https://www.huffingtonpost.kr/entry/story_kr_5da409fae4b06ddfc51c8a2e. Accessed Jul 2023.
59. Cho M, Hyun KS, Chung DC, Choi IY, Kim MJ, Chang YP. eNOS gene polymorphisms in perinatal hypoxic-ischemic encephalopathy [2]. Korean J Pathol. 2009;43(4):306–11.
60. Kim MI. Pusan National University cancels admission to Cho min medical school homeland “It’s harsh, application for suspension of execution”. Chosun Ilbo. April 4, 2022. https://www.chosun.com/national/national_general/2022/04/05/WTGKRZ3LRBBAZLXB5E7GXPQEDE/. Accessed Jul 2023.
61. Jeong HB, Kim GY, Bae JH. On H’s daughter’s thesis, Kenyan ghostwriter says, ‘I wrote it.’ Hankyoreh. May 8, 2022. https://www.hani.co.kr/arti/politics/politics_general/1041913.html. Accessed Jul 2023.
62. PD NOTE. Fairness and falsehood - ivy league and high school students. MBC Broadcast. June 14, 2022. https://www.youtube.com/watch?v=Yct32_lXw8k. Accessed Jul 2023.
63. Choi SA, Lee SM. Encapsulation of Streptococcus salivarius in double emulsion droplets as a method for increasing the efficacy of oral topical medications. Biomed J Sci Tech Res. 2019;19(2). https://biomedres.us/pdfs/BJSTR.MS.ID.003271.pdf.
64. Yoon GH. Korean American mothers “Han dong-hoon’s daughter’s negative specs, we are the victims”. Oh My News. May 17, 2022. http://www.ohmynews.com/NWS_Web/View/at_pg.aspx?CNTN_CD=A0002835955. Accessed Jul 2023.
65. Lee YK. Cloning whistleblower: little changed in S. Korea AP News. October 24, 2014. https://apnews.com/article/south-korea-hwang-woo-suk-archive-cloning-ae89cce79c5342d69b4bb2dd0667a123. Accessed Jul 2023.
66. King of Clones: The fall of Dr. Hwang. Netflix Documentary. (It started streaming on June 23, 2023).

Therapeutic Misconception as the Basis for Vaccine Nationalism of Japan: A Historical Reflection and Perspectives for Global Public Health

Chieko Kurihara and Takeo Saio

Abstract Japan's democratic Constitution was legislated under the guidance of the occupying authorities immediately after Japan's defeat in World War II. Among the authorities, the political influence of the United States (US) was particularly overwhelming. During the postwar reconstruction process, war criminals responsible for unethical human experiments were exonerated in exchange for providing their data to the US. Long after that, Japan still lacks public awareness on the essential nature of clinical trials as experimental activities. This was reflected in the responses of the Japanese government toward COVID-19 summarized as the "therapeutic misconception" (misinterpreting "research" as being a "treatment"). Furthermore, Japan purchased a huge amount of anti-SARS-CoV-2 vaccines, ensuring enough to vaccinate more than seven times the entire Japanese population, despite non-participation in large-scale placebo-controlled global clinical trials to prove the efficacy and safety of these vaccines. This "vaccine hoarding" is a typical example of "vaccine nationalism" that violates the principles of justice in bioethics.

This chapter describes the process of formation of Japanese research regulations and the characteristics of the Japanese COVID-19 response. The authors argue for the need to overcome health nationalism, and that international legal instruments with the purpose of global health governance along with a global norm of ethics in research must be established. To avoid exploitation and for the benefit of the most vulnerable, we should recognize that health is global public good as opposed to a commodity.

Keywords Democratic constitution · Health nationalism · Global public health
Global ethics in research

C. Kurihara (✉)
Kanagawa Dental University, Yokosuka, Kanagawa, Japan
e-mail: chieko.kurihara@nifty.ne.jp

T. Saio
Department of Internal Medicine and Psychiatry, Fuji Toranomon Orthopedic Hospital, Shizuoka, Japan

C. Kurihara et al. (eds.), *Ethical Innovation for Global Health*,
https://doi.org/10.1007/978-981-99-6163-4_6

1 Introduction

Japan's democratic Constitution was established in 1946 [1], immediately after its defeat in World War II in 1945. It has not been amended to date. The war crimes of human experiments by the Japanese Imperial Army have not yet been recognized by the Japanese government. Even without a historical reflection on these facts, Japan developed research regulations but failed to establish a comprehensive legal and ethical system to protect human rights as well as the dignity of research participants. In the absence of recognition of the intrinsic experimental nature of a clinical trial at the national level, the concept of "therapeutic misconception" [2] (misinterpreting "research" as being a "treatment") is dominant in the thinking of the Japanese government as well as in the Japanese research community. With such an erroneous understanding of the nature of clinical trials by the government and general public, we find the most notable characteristics of COVID-19 response in Japan as follows:

- The Government has promoted the off-label use of drugs developed and marketed by Japanese pharmaceutical companies which lacked an indication in COVID-19 and has recommended observational research to obtain information of the therapeutic outcomes.
- While Japan did not participate in large-scale global phase 3 placebo-controlled clinical trials of COVID-19 prevention vaccines, the Japanese government excessively ensured these vaccines to the point of vaccinating its entire population more than seven times over.

This chapter provides an overview of the Japanese legal and ethical environment for research and analyzes the characteristics of the Japanese government's response to COVID-19 in terms of drug and vaccine development, considering that it is contrary to the principles of bioethics, and in particular that of justice.

2 Democratic Constitution: Characteristics from the Perspective of Ethics in Research

The democratic Constitution of Japan was established, being led by the occupying administration of the General Headquarters, the Supreme Commander for the Allied Powers (GHQ/SCAP) [3]. It was not achieved by a civil revolution of Japanese people. Among the so-called Allied Powers, the influence of the United States (US) was particularly overwhelming [4]. War crimes committed by the Imperial Japanese Army were judged at the International Military Tribunal for Far East [5]. In this process, the numerous unethical medical and scientific human experiments were not pursued and the specific researchers not prosecuted. Besides, more than 40 perpetrators were found guilty at the "Khabarovsk Trial," a Soviet District Court (for use of biological weapons against China and Russia and for conducting unacceptable or unethical human experimentation) and in the "Yokohama Trials" in Japan, held before the Military Commission of the US Army (for medical experiments on US prisoners of war) [6]. However, these war criminals were later released as many of these crimes were exempted in exchange for providing the experimental results to the US, who ruled Japan after the war [7], and in preparation for the Cold War against the regimes of socialism and communism. This immunity from inhumane experiments was particularly applied to those medical science criminals, who held power during the war. And in the post-war democratic regime they were given high positions in the medical community and in the pharmaceutical industries. Thus, the Japanese Government has not yet officially recognized the war crimes of human experiments. Such historical facts have been described in the commentary [6] of the Geneva Convention [8], which forms the core of international humanitarian law to regulate the conduct of armed conflict and seeks to limit its effects, especially, but not limited to the protection of civilian persons in times of war. Japan is a signatory to the Convention and the Japanese Government needs to focus on this.

This contrasts with the Nuremberg Trials, in which Nazi Germany's doctors who conducted the human experiments were convicted. The Nuremburg judgment provided the basis for the enactment of the Nuremberg Code [9], which listed 10 fundamental principles that must be observed to satisfy moral, ethical, and legal concepts to be applied to human experiments. Conversely, the Japanese Constitution stipulates the role of the emperor in Chap. 1 as the symbol of the State and the unity of the people, in Chap. 2 eternal renunciation of war (this is why it is called "Pacifist Constitution"), and then in Chap. 3 it states the fundamental human rights [1]. The occupying forces decided to maintain the imperial regime since the beginning of Japan's national history, in order to peacefully control the democratic society. This story symbolizes the characteristics of the Japanese people who leave the past behind, obey authorities, and value peaceful harmony, rather than condemning violations of human dignity and human rights [10].

With such a national history and paternalistic culture of Japanese doctor–patient relationship [7], the legal framework for medical care and medical research has left much to "doctors' discretionary power" and "academic freedom," which tend to take precedence over guaranteeing the rights of patients and research participants. In the Japanese research environment "therapeutic misconception" is dominant. Thus, "clinical trials" and "research involving humans" have often been rephrased and misinterpreted as "treatment." In 1979 Japan ratified the International Covenants on Human Rights, adopted at the United Nation in 1966 [11]. However, the legal system to implement Article 7 of this covenant to prohibit scientific experimentation without consent is limited [12]. The Japanese system of protecting research participants was developed in response to external pressures and scandals of scientific malpractice and misconduct, without an official and systematic reflection on its own history. Against this background, the "therapeutic misconception" has been particularly evident in the government's response to the COVID-19 pandemic.

3 Development of Framework of Ethics in Research Involving Humans

3.1 Development History

In the late 1990s, legislation in Japan on clinical trials was initiated in response to the external pressure to harmonize regulations on drug development aimed at regulatory authorization, common in global markets. In 1990, the International Council for Harmonisation (ICH, formerly the International Conference on Harmonisation) began among trilateral regulators and industries of European Union (EU), US and Japan (member states have been recently expanded) [13]. Among its several guidelines, Good Clinical Practice (GCP, or ICH-GCP) designated as a general standard for clinical trials was agreed upon in 1996 (updated in 2016 [14]). The legislation in Japan of GCP in 1997 [15] as a domestic ordinance was essential to ensure that the regulators in US, EU, and other regions could approve clinical trial results generated in Japan. Given that there was a lack of reflection upon scientific experiments on humans as well as an absence of recognition of the experimental nature of clinical trials, it is not surprising that Japan has given the GCP-regulated clinical trial the name "Chi-ken," implying therapeutic nature. "Chi" means therapy and "ken" means experiment or trial. This word "Chi-ken" is used not only for therapeutic studies but also for healthy volunteer studies. This GCP Ordinance under the Pharmaceuticals and Medical Devices Act [16] (PMD Act) applies only to clinical trials aimed at product approval and to post-marketing studies. Other types of clinical trials, e.g., for academic purposes only, are outside the scope of the PMD Act.

Also in the late 1990s, in response to the Human Genome Project in the US [17] and the international agreement of banning human cloning [18, 19], the Japanese government developed regulations for some specific areas of genetic and embryonic studies. While human cloning was prohibited by law in 2000 [20], legal guidelines for human cloned embryo production [21] and non-legally binding ethical guidelines for research of human genome analysis [22], embryonic stem cell [23], gene therapy [24], epidemiology [25], and general clinical research [26] were issued from 2001 to 2003. The guidelines for epidemiological and clinical research were combined in 2014 and further combined in 2021 with the guidelines for genetic analysis research [27]. Some patients or citizen groups collaborating with members of the Diet have proposed legislation for Acts to protect patient's rights in medical practices [28] or research participants' rights [12, 29], but none of these has been legislated.

When some guidelines were merged, another set of regulations was created. In 2017, the Clinical Trials Act [30] was established in response to the scandal arising in 2012 of inappropriate relationship between a pharmaceutical company and academic researchers [31]. Several results from multicenter phase 4 clinical trials of certain hypertension drug initiated by corporate-funded academic researchers have been retracted from prestigious journals [32–34]. These trials were conducted under ethical guidelines, outside the GCP regulations. A pharmaceutical company statistician suspected of manipulating data to show debatable effects of a pharmaceutical company's drug was hounded and arrested but eventually acquitted in the highest court [35]. Responding to this scandal, the Clinical Trials Act was established to regulate clinical trials to evaluate safety or efficacy of medicinal products but not aimed at marketing authorization. This Act does not cover "observational studies" analyzing the outcomes of drug therapy using unapproved or off-label products.

When the Clinical Trials Act was passed, the Diet requested in a Supplementary Resolution to respect the provisions of the International Covenant on Human Rights [11] that prohibits scientific experiments on humans without consent and to establish measures to protect the dignity and rights of study participants, including those of surgical studies, which is outside the scope of this Act [36]. However, such legislation has not yet been implemented.

3.2 *Application of Research Ethics Norms in Reality*

Because of the set of research regulations described above, research on COVID-19 related products can be conducted within one of the following frameworks in Japan.

- GCP-level clinical trials under the PMD Act aimed at product approval.
- "Research" under the Clinical Trials Act to evaluate safety and efficacy but without intention of submission for marketing approval.
- To provide therapy using unauthorized or off-label drugs and then conduct observational research using the medical record of the therapy.

During the process of developing these regulations, there was no reflection on war crimes or systematic review of past unethical studies. ICH-GCP [14] and the World Medical Association (WMA)'s Declaration of Helsinki (DoH) [37] always have been considered the ethical reference standards in clinical trials and in other research involving humans. The Council for International Organizations of Medical Sciences (CIOMS) research ethics guidelines [38] have been cited on limited occasions for international collaborative research.

The Clinical Trials Act [30] and Ethical Guidelines [27] adopted "eight principles" with minor modifications from those proposed by Emanuel et al. [39, 40]. In this process there has been no reflection on the discussions in the US to clarify the boundary of "practice" and "research" neither on the three principles (respect for persons, beneficence, and justice) from the Belmont Report [41]. There was also no reflection on the fact that principles of Emanuel et al. were proposed in response to the controversies over exploitative research sponsored by industrialized countries carried out in developing countries. Their principles appear to be a procedural description of the three principles of the Belmont Report, adding high priority to "social value." [42] The Japanese research community interprets "social value" just in terms of generating meaningful contribution to health science, without recognizing its implication in fair allocation of risks and benefits of research and in avoiding exploitation of vulnerable people in the world.

4 Therapeutic Research During the COVID-19 Pandemic

4.1 Failure in the Development of a Japan-Originated Drug

After the reported outbreak of SARS-CoV 2 in December 2019 in China, it was only at the end of January 2020 that the World Health Organization (WHO) declared it a "Public Health Emergency of International Concern" (PHEIC) [43], and in March, a "pandemic" [44]. The epicenter of the infection shifted from China first to Europe and then to the US. Each country had declared a state of emergency and its public health authorities had restricted citizen's lifestyles, such as travel, physical distancing, washing of hands, and use of masks. The mass media reported day by day the collapse of overburdened health systems. Japan declared the first state of emergency in April 2020, with political considerations weighing heavily over the ongoing plan for Tokyo Summer Olympics [45], which eventually ended up being postponed to the summer of the following year, 2021. The noteworthy event for the Japanese during this period was the border control carried out on the luxury cruise ship Diamond Princess, which docked in the port of Yokohama on February 3, 2020. All 3711 passengers and crew, including 1341 Japanese, were trapped on board, and disembarking was only completed on March 1 of 2020. Of the 1011 passengers (Japanese and foreigners) disembarked between 19 and 23 February and for whom a 14-day health observation was carried out, it was confirmed that 712 were infected and 13 died [46]. This event raised substantial public debate but no official review was provided from an ethical as well as public health perspective.

During this period, then Prime Minister Shinzo Abe advocated carrying out of "observational studies" gathering data on the off-label therapeutic use (repositioning or repurposing) of "favipiravir" (RNA synthase inhibitor) and other drugs developed by Japanese companies [47, 48]. Observational studies to collect data from medical records regarding the outcome of providing "optimal care" to patients are out of the scope of Clinical Trials Act and GCP. It is covered by non-legally binding ethical guidelines. This means that government promoted people's misinterpreting "research" (experimental use of off-label drug) as being a "treatment," which means "therapeutic misconception." Favipiravir was first approved in 2014 for new or re-emerging influenza virus infections, and two million doses were stockpiled by the government but had not been used in practice because there was no outbreak of this infection in Japan [49]. The government revealed a plan to use 700,000 doses of this drug for COVID-19 patients.

In March 2020, observational studies collecting therapeutic outcomes under the ethical guidelines, followed by interventional studies of favipiravir under the Clinical Trials Act, were launched, both involving medical research institutions across Japan, with government support. Thereafter, at the end of March, the drug manufacturer initiated clinical trials under the GCP Ordinance. The Prime Minister pushed the Ministry of Health, Labour and Welfare (MHLW) to grant it approval for COVID-19 [50]. This was similar to the then US President Donald Trump pressurizing the US Food and Drug Administration (FDA) to approve chloroquine and hydroxychloroquine (antimalarial drugs), as a "game changer" [51]. In early May, a citizen's group raised concerns about the teratogenecity of favipiravir and sent its opinion to the MHLW [52]. The Japan Medical Association (JMA) also criticized government and stressed the need for randomized-controlled trials (RCTs) to prove safety and efficacy [53]. The Japanese Pharmaceutical Manufacturers Association (JPMA) was similarly critical as the JMA and provided recommendations to improve the Japanese drug approval system in case of an emergency [48]. Eventually, none of these studies could successfully prove the safety or efficacy of this drug [48, 54].

4.2 *"Special Approval" of a US-Originated Drug*

Japanese pharmaceutical companies were unable to contribute to the development of antiviral for COVID-19 during this initial phase of pandemic, while Japanese people benefited from an antiviral developed in the US. In late February 2020, a randomized placebo-controlled trial of remdesivir (a nucleic acid analog) for hospitalized patients began in multiple institutions across the US (Clinical trial identifier: NCT04280705, 1062 patients were randomized). European, Mexican, and Asian including Japanese institutions joined this global study. The trial was sponsored by the National Institute of Allergy and Infectious Diseases (NIAID), which is a part of the US National Institutes of Health (NIH). Interestingly the first study participant was a US citizen who was a passenger of and discharged after quarantine from the cruise ship Diamond Princess, which docked in Yokohama [55]. Remdesivir was developed by a US company for Ebola hemorrhagic fever but it was not as effective as its comparators. Favorable results were thereafter shown in animal experiments for severe acute respiratory syndrome coronavirus (SARS-CoV) and the Middle East respiratory syndrome coronavirus (MERS-CoV).

The interim analysis of 1059 COVID-19 patients showed faster recovery in the remdesivir group (active drug, 10 days; placebo, 15 days) [56]. US FDA granted an emergency use authorization (EUA) on May 1 of 2020 and new drug authorization (NDA) was granted on October 22, 2020 [57]. On May 7 Japan granted "special approval" [58] for the drug. This "special approval" status of Japanese PMD Act is an exceptional approval, in case of emergency, of a drug already approved in another country, which can be granted without sufficient information required in usual approval. Because the frontline hospitals for COVID-19 treatment are not always research hospitals, there were many difficulties in participating in the global trial [59].

Meanwhile, at the end of March 2020, the WHO launched a large-scale international collaborative study called Solidarity Therapeutics Trial. WHO provided a master protocol of RCT for hospitalized patients assigned to four potential antiviral repurposed drugs (remdesivir, hydroxychloroquine, lopinavir/ritonavir, and interferon regimens) and investigators at the forefront of treatment around the world could participate and share the results. The interim analysis of the outcome was released by WHO in October 2020 [60], and subsequently the results of this open-label RCT sponsored by a French institution (NCT04315948) were published concluding that these antivirals "*had little or no effect on overall mortality, initiation of ventilation, and duration of hospital stay*." [61] 11,330 patients in 30 countries were randomized; however, Japanese institutions did not participate in this study.

4.3 Emergency Approval of a Japan-Originated Drug with "Social Value"

With the progress of vaccination in the world since the beginning of 2021, as well as changes in the clinical characteristics of viral variants, e.g., the spread of Omicron strains with higher prevalence but lower rates of severity, the pandemic situation changed from mid-2021 to the beginning of 2022. At the end of July 2022, WHO reported that Japan had the world's highest number of new weekly COVID-19 cases [62, 63]. One of the explanations could be that Western countries ended their systems of mandatory reporting of cases of infection around the first half of 2022, while Japan continued with mandatory reporting [64].This highest number of weekly reported new cases continued for several weeks through to the end of 2022 [65].

In May 2022, the Japanese PMD Act was revised and a new system of "emergency approval" was established [66]. Unlike the "special approval" system, this new system allows that, in case of emergency, the world's first approval could be granted in Japan even if the product has not yet been approved abroad. This emergency approval requires "confirmation" of safety but requires only "inference (estimation)" of efficacy. After this approval, efficacy needs to be proven within 2 years through a conformity study. The following month, a Japanese company submitted an application of its drug to treat COVID-19, named "ensitrelvir" (protease inhibitor, oral antiviral), utilizing this new approval system, showing the results from the phase 2b study. However, it was not approved. One prominent flaw of this study was that it used "co-primary endpoints" [67] which are not recommend by the US FDA guidelines for COVID-19 prevention and treatment clinical trials [68]. The next study named "Part III" used the time to symptom resolution as an endpoint and was approved in November 2022. It achieved a reduction of the symptomatic period from 8 to 7 days. The government agreed to purchase two million doses of the drug [69].

During discussions with a group of patients and the public (series of web meetings learning about the Declaration of Helsinki, resulting in the manuscript of Chap. "Our "WMA Declaration of Helsinki": Opinions and Proposals from Patient and

Public for Research Ethics"), one member from the public said that it is questionable that a large amount of state money was being spent on relieving the symptoms of mildly ill patients a day earlier, while there was a social problem in Japan that large numbers of children could not sufficiently access to meals, because of expanding poverty and lack of state support [70].

5 Vaccine Development in COVID-19 Pandemic

5.1 *Panic Purchasing and Discard of Western-Origin Vaccines*

In December 2020, three anti-SARS-CoV-2 vaccines developed by European and US companies and research institutes achieved efficacy rates of as high as 95% (mRNA vaccine developed jointly by Pfizer and BioNTech) [71, 72], 94.1% (mRNA vaccine developed jointly by Moderna and the US National Institute of Allergy and Infectious Diseases) [73, 74], and 70% (adenovirus vector vaccine co-developed by Oxford University and AstraZeneca) [75] (Table 1). The surprising speed of this achievement in less than a year can be partially attributed to Operation Warp Speed [76] under the Trump administration of the US and especially to the cumulative knowledge acquired, for instance, to develop a vaccine for the 2014 Ebola outbreak. The US government invested USD 17 billion and removed all possible administrative and procedural barriers. In addition, there was strong collaboration among industry, government, academia, as well as general public. The then US President promoted this project referring to the Manhattan Project, an atomic bomb development program during World War II, in terms of exceptional collaboration among stakeholders, including the strong leadership of military [77]. Soon after the release of interim analysis, regulators of Europe, US, and other countries granted emergency use approval to these vaccines. This was followed by many countries launching their mass vaccination programs, as well as local vaccine development.

At this time of worldwide vaccine approvals, the death toll from COVID-19 was about 1000 per million in countries in North America, South America, and Europe, while it was about 30 in Japan and less than 100 in the Association of Southeast Asian Nations (ASEAN) [78, 79]. Due to this low prevalence and mortality, Japan did not participate in the global phase 3, large-scale, placebo-controlled clinical trials of these vaccines (Table 1), but signed purchase agreements for these three vaccines in quantities much greater than its entire population as of the first half of 2021 [79].

The panic purchasing of vaccines by the Japanese Government was, in fact, typical "vaccine nationalism," [80] which violates human rights in the context of global health [81]. Japan granted "special approval" to vaccines produced by of Pfizer/BioNTech in February 2021, and vaccines of Moderna and AstraZeneca both in May 2021. Japan's total population is approximately 120 million. By March 2022, the amount of vaccines contracted from Pfizer/BioNTech and Moderna was enough

Table 1 Phase 3 placebo-controlled trials of the vaccines which Japan panic purchased without participation

Study phase/publication	Endpoint	Tested intervention	Control	Efficacy	Trial identifier, country
BNT162b2 Generic name: Tozinameran Bland name : Comirnaty Developed by BioNTech + Pfizer					
Phase 1/2/3, global		**BNT162b2 (21,720)**	**Placebo (21,728)**	**Efficacy**	NCT04368728 **US, Argentina, Brazil, Germany, South Africa, Turkey**
NEJM DOI: 10.1056/NEJMoa2034577	**PCR Positive:**	8	162	**95%**	
		(7 days after the second dose)			
	Severe:	1	9		
		(any time after the first dose)			
Real World Data, Israel		**Vaccinated (596,618)**	**Unvaccinated (596,618)**		**Israel**
NEJM DOI: 10.1056/NEJMoa2101765	**Infection**	**4460**	**6100**	**94%**	
	Symptomatic	**2389**	**3607**	**94%**	
	Hospitalization	**110**	**259**	**87%**	
	Sever	**55**	**174**	**92%**	
	Death	**9**	**32**	**84%**	
Real World Data, US		**Vaccinated (2961)**	**Unvaccinated (989)**		**US**
MMWR DOI: 10.15585/mmwr.mm7013e3	**PCR-confirmed**	**8** (Partially)	**161**	**80%**	
	PCR-confirmed	**3** (Fully)	**161**	**90%**	
mRNA-1273 Bland name : Moderna COVID-19 vaccine Developed by Moderna + NAID					
Phase 3, US		**mRNA-1273 (14,134)**	**Placebo (14,073)**	**Efficacy**	NCT04470427 **US**
NEJM DOI: 10.1056/NEJMoa2035389	**Symptomatic**	**11**	**185**	**94.1%**	
	Severe:	**0**	**30**		
ChAdOx1 nCoV-19 Bland name : Vaxzevria Developed by University of Oxford + AstraZeneca					
Phase 2/3, global		**ChAdOx1 nCoV-19 (5807)**	**Control (5829)**	**Efficacy**	NCT04324606 **UK (phase 1/2)*3**
Lancet DOI: 10.1016/S0140-6736(20)32661-1	**Symptomatic**	**30**	**101**	**70·4%%**	NCT04400838
	(subgroup)	**LD and SD (1367)*1**	**Control (1374)**		**UK (phase 2/3)*2**
	Symptomatic	**3**	**30**	**90.0%**	NCT04444674 **SA (phase 1/2)*3**
	[Notes only for ChAdOx1 nCoV-19] *1 : Supecific subgrup which achieved 90% efficacy that received low dose (LD) and standard dose SD). Other groups received two SDs or LD and SD. *2 : Studies included in the above analysis. *3 : Studies included in the above publication in Lancet. *4 : Large scale placebo-controlled study not included in the above publication in Lancet.				NCT04536051 **Brazil (phase 3)*2** NCT04516746 **US (phase 3)*4**

This table is updated from authors' previous analysis in the publication of Ref. [79]

to provide up to the fourth dose to the entire population [82]. Nevertheless, some members of the Diet raised concerns that, at that time, the government had taken budgetary measures of 2.4 trillion yen (approx. 18.23 billion USD), more than 2% of the national budget, to purchase 882 million doses of vaccine [83]. This volume is more than enough for all citizens to be vaccinated with seven doses [84, 85]. Despite repeated questions from those Diet members about the unit price of vaccines, the government has not disclosed it due to confidentiality agreements with the companies. Subsequently, the Board of Audit of Japan indicated uncertainty regarding the basis for calculating the resources corresponding to this total amount of vaccines [86]. As of January 2023, 81.4% of the Japanese population had received at least one dose, 80.4% had completed two doses and 68.0% had completed three doses, which means a total of 378 million doses [87] (as of December 2021, 77.5% of the population had already completed two doses and 79% had received one dose [88]).

Vaccination in Japan proceeded preferentially with Pfizer/BioNTech's, followed by Moderna's. AstraZeneca/Oxford's vaccine stalled due in part to reports of rare side effects of blood clots [89]. Half of the contracted amount of AstraZeneca/Oxford's vaccine was cancelled; two-thirds of its supplied vaccines were distributed abroad and a quarter was discarded (Table 2) [90]. In addition, more than a third of the Moderna vaccine provided was discarded (Table 2) [91]. The scrapping of these two vaccines was announced to be due to expiration dates.

Meanwhile, Japanese contributions to global vaccination are mainly as follows (information up to January 2023, except where noted) [92, 93]:

1. Financial support of USD 1 billion to the COVAX Facility's Framework for Developing Countries through COVAX Advance Market Commitment (AMC).
2. A total of above-mentioned 44 million doses of AstraZeneca/Oxford vaccines were provided to 32 countries, either directly to seven countries or through COVAX to 25 countries.

Table 2 The amounts of the vaccines that the Japanese government contracted and of supply, donation, discard, and cancellation

	Pfizer	Moderna[a]	AstraZeneca[b]	Novavax	Total
The amount of contract[c]	399	213	120	150	882
Supplied	NA[d]	143	57.7	NA	
Domestically supplied	NA	96.9	0.2	NA	
Distributed to abroad	NA	0	**44**	NA	
Discarded	NA	**46.1**[e]	**13.5**	NA	
Cancelled	NA	**140**	**62.3**	**141.76**[e]	

(million doses)

[a]Announcement from the MHLW on 10 Feb 2023 [91]
[b]Announcement from the MHLW on 30 Sept 2022 [90]
[c]A report by Board of Audit of Japan in March 2023 [86]
[d]NA means not available
[e]According to a news report on 18 Mar 2023 [82], discarded doses of Moderna are 63.9 million. Cancelled doses of Novavax are according to this news report

3. As part of the "Last Mile Assistance" program, grant aid was provided to support the development of cold chain systems in developing countries to the last mile before they reach the local level. As of March 2022, a total of 77 countries and regions had already received approximately a total of 18 billion yen (134 million USD).

The case of Japan's panic purchasing of vaccines and supplying a specific type of vaccine to other countries as it was not being utilized in Japan due to rare side-effects needs to be further considered, not just as a matter of "vaccine nationalism," but also as a form of "vaccine diplomacy."

5.2 *Failure and Questions of Developing Japan-Originated Vaccines*

During the initial phase of pandemic, Japan-originated DNA vaccines (AG0301, AG0302) jointly developed by biotech and manufacturing companies and regional universities attracted public attention, gaining more than 7 billion yen government budget. Unfortunately, the development was discontinued in September 2022, because early phase studies could not show the expected levels of neutralizing activity and antibody titers. At the beginning of clinical development in the summer of 2020, prefectural and city governors announced that this experimental vaccine would be administered to medical staff at the university hospital. Such "forward-looking" statement before the approval of research review board was criticized by mass media [94]. This is another type of therapeutic misconception (misinterpreting "experimentation" as being a "prevention"). Also, this episode recalled notorious history of a Japanese company in early 1960s to abuse employees to involve in phase 1 study of experimental drug and caused several cases of serious side effects. For the COVID-19 vaccine, US FDA guidance [95] states that healthcare workers should be excluded from early phase trials. Meanwhile, phase 3 trials for repositioning purpose of existing vaccines were conducted in many parts of the world, involving healthcare workers as study participants (M-M-R II®: NCT04333732; BCG: NCT04327206).

During the late phase of pandemic, two of the Japanese large pharmaceutical companies, which started clinical development after the successful study results of Western-originated vaccines were shown, reached to marketing application at the time of late 2022 and the early 2023. These development programs include early phase trials in Japan as well as late phase trials in abroad, both including placebo-controls. It is difficult to define ethical, scientific, social values, and justifiability of these development programs, including possibility of exploitation of study participants, in this complicated situation of late phase of pandemic. However, it certainly needs exploration in terms of distributive justice.

6 Debates on Ethics of Resource Allocation

In addition to issues related to ethics in research and development, the situation of allocation of scarce medical resources was the focus of debate among Japanese bioethicists, as well as bioethicists in other countries. The socio-economic impacts of the COVID-19 pandemic, with, in particular, socially vulnerable populations being most disproportionately and disadvantageously harmed are well recognized. Among various problems, Japanese bioethicists specifically focused on the issues of allocation of artificial respirators in the context of its shortage. What was unique and often discussed about was that, although Japan having more doctors and hospital beds per population than other countries, the Government and the Japan Medical Association (JMA) failed to take optimal measures for resource allocation in emergencies leading to a shortage of medical resources and prolonged restrictions on citizens' behavior. The JMA argues that based on the data from the Organisation for Economic Co-operation and Development (OECD), a relatively larger number of beds were for psychiatric and convalescent patients, hence acute phase beds for COVID-19 patients were not that many when compared globally [96]. Japanese bioethicists, without entering into an in-depth discussion of this general issue of medical resources, focused on the dilemma of "triage" and the justifiability of withdrawing a respirator from a patient with lower life-saving potential in order to transfer it to another patient with higher life-saving potential.

A proposal of justification for withdrawing artificial respirators in such a way was issued by a voluntary group of Japanese bioethicists in March 2020 [97], at the early stages of the pandemic. This was shortly after the Italian Society for Anaesthesia, Analgesia and Intensive Care (SIAARTI) made international headlines with its guidelines allowing restrictions to the provision of treatment based on the patient age [98]. At that time, the number of cases and deaths in Japan was much lower than in other countries. On November 20 of the same year, the Japanese Society of Intensive Care Medicine (JSICM) issued a proposal for laxer criteria to justify stopping a patient's treatment to save another patient, to make it easier to do so for those who were incapable making decisions and where there were no relatives or proxies [99].

The authors of this chapter disagreed with these recommendation [100–102]. In addition, a member of the Diet who is a patient of amyotrophic lateral sclerosis (ALS) [103] and organizations of people with disabilities voiced objections. The Science Council of Japan (SCJ) Philosophy Committee took this issue seriously and convened a series of web-based symposiums and published proceedings, to which one of the authors contributed [102]. Many ethicists expressed concern about the content of the recommendations, sharing positions of related situations in other countries. Particular attention was paid to the recommendations of the German Ethics Council [104] which stated that triage was contrary to the German Constitution and principles of medical ethics. The debate has not been resolved in Japan and no specific official opinion has been put forward. We have not confirmed any case of discontinuing a patient's treatment to treat another patient based on the above two recommendations.

Emanuel et al. [105], one of the references to support the argument of Japanese bioethicists of removing the artificial respirator, states that "*many guidelines agree that the decision to withdraw a scarce resource to save others is not an act of killing and does not require the patient's consent,*" citing number of references [106–110]. However, these cited articles strictly consider the conditions of acceptability of such decision, and in some of them we did not find relevant description that supports the argument [109]. These utilitarian positioning have also been countered with the argument that using utilitarianism will result in the disadvantaged being further devaluated due to lack of access, due to social determinants of health [111, 112]. Emanuel, known for his contribution to the US Obama administration's healthcare reforms, published in 2014 his opinion entitled "Why I hope to die at 75" [113] arguing that living too long without contributing to society is a waste of resource, which is also revisited during the COVID-19 pandemic.

The chapter authors organized a series of webinars, as a part of activities of the COVID-19 Task Force of the Japanese Association of Bioethics in collaboration with the Brazilian Society of Bioethics (SBB). In the first one [114], held in October 2020, Dirceu Greco, then president of the SBB gave a lecture to introduce SBB's recommendation [115] on the issue of allocation of limited resources. The next webinar, held in June 2021, was to discuss ethics in research where many of the authors of this book participated [116]. In these two webinars, the following points were identified as lessons learnt for bioethicists in Japan and in the world regarding the issue of resource allocation.

1. In Brazil, where COVID-19 was much more serious than in Japan, the Brazilian Society of Bioethics took the lead in issuing a recommendation that respect for human rights be given the highest priority in the allocation of resources during a pandemic. It allowed triage as priority setting only after careful consideration according to a decision-tree. However, it did not include statements that justified the discontinuation of treatment of a patient with less potential for saving life to be exchanged for another patient with greater life-saving potential [114, 115].
2. Tammam Aloudat, a Syrian clinician and humanitarian activist who has been involved in medical care in conflict zones and other disaster settings argued that utilitarianism was an unacceptable framework [117]. This statement was as a reminder that the idea of allocating medical resources by withdrawing from those with lower life-saving potential in order to save others with greater life-saving potential could undermine their humanitarian mission of working for patients in disaster situations.

Thus, we can summarize that the issue of health resource allocation in Japan during the COVID-19 pandemic was a failure of optimal allocation of excessive resources, along with the debate on utilitarian arguments of withdrawing resource from the vulnerable and switching to the privileged. This is against the principle of justice.

7 Discussion

7.1 Global Health Governance

In a public health crisis, we often face conflicting situations hinging on the protection of individual human rights and public health measures to prevent infection. In a global crisis triggered by a pandemic, conflict exists not only between the individual and the public. Conflict also emerges between the missions of sovereign states to protect their own nations and other obligations of states, such as solidarity, to contribute to global health.

We have seen various types of activities such as the Solidarity trial and COVAX intended to protect people in resource-poor nations, including the world's most vulnerable. The United States Operation Warp Speed placed top priority on protecting its own nation, which secondarily contributed to global health in terms of successful vaccine development. However, this huge amount of government funds invested for a specific aspect of public health has had limited return in terms of global public health. The profit-oriented commercialism of corporations and vaccine nationalism of high-income countries resulted in bilateral negotiations between the corporations and some national governments. As a result, it was high-income countries that urgently purchased vaccines, which caused insufficient allocation of vaccines through the COVAX initiative.

Similarly, Japanese vaccine policies were based on the selfishness of a single country and failed to contribute to solving the world's common problem, despite its great economic power as well as its excessive resources in medical practice and for research. This suggests a need for international legal instruments with the aim of establishing adequate global health governance mechanisms [118, 119] to overcome health nationalism [120].

A prominent challenge was the proposal from South Africa and India [121] to the World Trade Organization for a waiver from certain provisions of the Agreement on Trade-Related Aspects of Intellectual Property Rights (TRIPS Agreement), in order to increase the possibility of production and supply of COVID-19-related products by non-patent holder manufacturers. The June 2022 TRIPS Ministerial Council decision allowed a waiver only for COVID-19 vaccines [122]. This was criticized to be too narrow [123], and discussion on the expansion of the range of waiver was extended in December 2022 [124]. This proposal by South Africa and India must be included as an element of the global health governance framework for preparedness for future possible pandemics.

7.2 Global Norm for Ethics in Research

Along with global health governance, a legally established global standard of ethics in research involving humans based on internationally agreed human rights norm [11] is needed. This should contribute to the protection of the most vulnerable people from exploitation. In addition, it must serve to ensure the "social value" of the research, that is, whether the research question is in line with the common global goals, as outlined in the United Nation-designated Sustainable Development Goals (SDGs) [125].

The CIOMS ethical guidelines [38] are the first international guideline for ethics in research involving humans to define "social value" as a criterion for evaluating research. Traditionally, however, these principles have operated to protect each research participant in a single project [42]. Thus we need a legally binding norm for ethics in research involving humans, integrating existing principles such as not only CIOMS [38], WMA's DoH [37], but also the Declaration on Bioethics and Human Rights by the United Nations Educational, Scientific and Cultural Organization (UNESCO) [126] and others which promotes the design, evaluation, and conduct of research involving humans, toward common global goals, including global public health [127].

7.3 Ethical Foundation of Health Governance and Ethics in Research

Changes of current theories of health governance and ethics in research are required, including the change of current beliefs in some societies where health systems are equated to commercial commodities. The COVID-19 pandemic has revealed the structural deficiency of democracy in conjunction with capitalism [128] associated with extreme liberalism and excessive selfishness in the field of health. In a democratic system, the right of individuals to the pursuit of happiness and freedom and the right of states to self-determination have been assured by the constitution and the legal system. Thus, to ensure global public health and the right to health for all, it is unacceptable for individuals and states to exploit the vulnerable in their pursuit of greed. This is what we find when reflecting on Japanese war crimes in the old regime and now in the government's COVID-19 response in a democratic society. Our common worldwide experience has taught us that we must achieve global public health with the recognition that health is not a commodity for sale in the marketplace, but a "global public good" [129].

The first step toward achieving this is for countries to establish adequate, well-financed public health systems, e.g., National Health Systems of the United Kingdom or Brazil, where everyone is equally guaranteed free access to adequate essential healthcare. Furthermore, a global philosophical foundation must be established for common national public health policies that incorporate equitable resource allocation mechanisms to achieve global health.

8 Conclusion

Prominent features of the Japanese response to the COVID-19 pandemic can be summarized as the "therapeutic misconception" with the futile promotion of observational study of off-label use of a drug. In addition, without participation in the large scale placebo-controlled clinical trials to achieve the best-proven anti-SARS-CoV-2 vaccines, Japan panic purchased these vaccines seven times in excess required to vaccinate the entire population. This reinforces the need for international legal instruments with the aim of establishing global governance in health, together with global norm of ethics in research, to ensure the social value of research in a global perspective.

Robust ethical foundations must be established to support such norms. Health nationalism must be overcome through alteration of current theories for management of global public health. This will require the recognition that health cannot be commodified. It is a global public good, for the benefit of the most vulnerable. Exploitation is a flagrant injustice that must be both avoided and eradicated.

Conflict of Interest There is no conflict of interest to be disclosed related to the content of this manuscript.

References

1. Japanese Law Translation. The Constitution of Japan. Constitution November 3, 1946. https://www.japaneselawtranslation.go.jp/ja/laws/view/174. Accessed 12 Mar 2023.
2. Appelbaum PS, Roth LH, Lidz CW, Benson P, Winslade W. False hopes and best data: consent to research and the therapeutic misconception. Hastings Cent Rep. 1987;17(2):20–4. PMID: 3294743.
3. Britannica. Supreme Commander for Allied Powers: military office. Accessed 12 Mar 2023.
4. Kimijima A. Peace in East Asia and the Japanese constitution: a reexamination 60 years after its making. Ritsumeikan Kokusaikenkyu. 2009;21(3):539–49.
5. International Military Tribunal for the Far East. Special proclamation by the Supreme Commander for the Allied Powers at Tokyo January 19, 1946; charter dated January 19, 1946; amended charter dated April 126, 1946. Tribunal established January 19, 1946. https://www.un.org/en/genocideprevention/documents/atrocity-crimes/Doc.3_1946%20Tokyo%20Charter.pdf.
6. International Committee of the Red Cross. Convention (III) relative to the Treatment of Prisoners of War. Geneva, 12 August 1949. Commentary of 2020 Article 13: Humane treatment of prisoners. 2020. C.3.b Medical and scientific experiments, section 1601, 2, 4. https://ihl-databases.icrc.org/applic/ihl/ihl.nsf/Comment.xsp?action=openDocument&documentId=3DEA78B5A19414AFC1258585004344BD#73. Accessed 12 Mar 2023.
7. Kimura R. Asian perspectives: Experimentation on human subjects in Japan: Bioethical perspectives in a cultural context. In: Bankowski Z, Levine RJ, editors. Ethics and research on human subjects: international guidelines. Proceedings of the XXVIth CIOMS Conference. Geneva, Switzerland 5–7 February 1992. Geneva: CIOMS; 1993. p. 181–7.
8. International Committee of the Red Cross. Convention (III) relative to the treatment of prisoners of war. Geneva. 12 Aug 1949. https://ihl-databases.icrc.org/applic/ihl/ihl.nsf/INTRO/375?OpenDocument.

9. International Military Tribunal. Trials of war criminals before the Nuremberg Military Tribunals under Control Council law no. 10 Nuernberg, October 1946–April 1949, vol. 2. p. 181–2. (The Nuremberg Code.)
10. Kurihara C, Cho K, Toohey RE. Core ethical values of radiological protection applied to Fukushima case: reflecting common morality and cultural diversities. J Radiol Prot. 2016;36:991–1003.
11. United Nations. International Covenant on Civil and Political Rights. Adopted by General Assembly resolution 2200A (XXI) of 16 December 1966, entry into force 23 March 1976. https://www.ohchr.org/en/instruments-mechanisms/instruments/international-covenant-civil-and-political-rights.
12. Mitsuishi T, Nudeshima J, Kurihara C. Proposal of a draft human research participants protection bill: as the most essential legal basis for bioethical issues. Clin Eval. 2003;30(2, 3):369–95. Japanese. http://cont.o.oo7.jp/30_23/p369-95.pdf.
13. International Council for Harmonisation. History. https://www.ich.org/page/history. Accessed 12 Mar 2023.
14. International Council for Harmonisation of Technical Requirements for Human Use (ICH) Harmonised Guideline Integrated Addendum to ICH E6(R1): Guideline for Good Clinical Practice E6(R2). 2016. https://database.ich.org/sites/default/files/E6_R2_Addendum.pdf. Accessed 9 Nov 2016.
15. Ministerial Ordinance on Good Clinical Practice for Drugs. Ordinance of the Ministry of Health and Welfare No. 28 of March 27, 1997. Japanese.
16. Act on Securing Quality, Efficacy and Safety of Products Including Pharmaceuticals and Medical Devices. Act No. 145 of August 10, 1960. Japanese.
17. US National Institutes of Health, National Human Genome Research Institute. The Human Genome Project. https://www.genome.gov/human-genome-project. Accessed 12 Mar 2023.
18. Denver summit of the eight. Communiqué Denver, June 22, 1997. http://www.g8.utoronto.ca/summit/1997denver/g8final.htm. Accessed 12 Mar 2023.
19. United Nations. International convention against the reproductive cloning of human beings. 2005. A/59/516/Add.1.
20. Act on Regulation of Human Cloning Techniques. Act No. 146 of 2000. https://www.cas.go.jp/jp/seisaku/hourei/data/htc.pdf.
21. Ministry of Education, Culture, Sports, Science and Technology. The Guidelines for Handling of a Specified Embryo. 2001. http://www.lifescience.mext.go.jp/files/pdf/30_82.pdf.
22. Ministry of Education, Culture, Sports, Science and Technology; Ministry of Health, Labour and Welfare; Ministry of Economy, Trade and Industry. Ethical Guidelines for Human Genome/Gene Analysis Research. 29 Mar 2001, Abandoned on 30 Jun 2021. Provisional Translation (as of July 2018). http://www.lifescience.mext.go.jp/files/pdf/n2181_02.pdf.
23. Ministry of Education, Culture, Sports, Science and Technology. Guidelines on the derivation and utilization of human embryonic stem cells. 25 Sept 2001. Japanese.
24. Ministry of Education, Culture, Sports, Science and Technology; Ministry of Health, Labour and Welfare; Guidelines for gene therapy. 27 Mar 2002. Japanese.
25. Ministry of Education, Culture, Sports, Science and technology; Ministry of Health, Labour and Welfare. Ethical guidelines for epidemiological Research 17 Jun 2002. Abandoned. English version (as of December 1, 2008). https://www.lifescience.mext.go.jp/files/pdf/n796_01.pdf.
26. Ministry of Health, Labour and Welfare. Ethical Guidelines for Clinical Research. 30 Jul 2003. Japanese. Abandoned.
27. Ministry of Education, Culture, Sports, Science and Technology; Ministry of Health, Labour and welfare; Ministry of Economy, Trade and Industry. Ethical Guidelines for Medical and Biological Research Involving Human Subjects. 23 Mar 2021. Official English title and only Japanese text available.
28. Association for the legislation of a Patients' Rights Act. Japanese. http://www.iryo-kihonho.net/. Accessed 12 Mar 2023.

29. Mitsuishi T, Nudeshima J, Kurihara C, Asano S, Fukushima M. Proposal of a draft human research participants protection bill 2007: the critical legal basis for bioethical issues: second report. Clin Eval. 2007;34(3):595–611. Japanese. http://cont.o.oo7.jp/fukushima/34_3p595-611.pdf.
30. Clinical Trials Act (Act No. 16 of April 14, 2017). https://www.mhlw.go.jp/file/06-Seisakujouhou-10800000-Iseikyoku/0000213334.pdf.
31. Kurihara C. Regulations and credibility of phase IV clinical trials: lessons from the Japanese valsartan scandal. Clin Res. 2014;54–60.
32. Mochizuki S, Dahlöf B, Shimizu M, et al. Valsartan in a Japanese population with hypertension and other cardiovascular disease (Jikei Heart Study): a randomized, open-label, blinded endpoint morbidity-mortality study. Lancet. 2007 Apr 28; 369(9571): 1431–9. Retraction in: Lancet editors. Lancet. 2013;382(9895):843.
33. Sawada T, Yamada H, Dahlöf B, et al. Effects of valsartan on morbidity and mortality in uncontrolled hypertensive patients with high cardiovascular risks: KYOTO HEART Study. Eur Heart J. 2009;30(2):2461–9. Retraction in: Eur Heart J. 2013 Apr; 34(14): 1023.
34. Kimura S, Sawada T, Shiraishi J, et al. Effects of valsartan on cardiovascular morbidity and mortality in high-risk hypertensive patients with new-onset diabetes mellitus. Circ J. Epub 2012 Sep12. Retraction in: Shimokawa H. Circ J. 2013; 77(2): 552a. Epub 2012 Dec 28.
35. PHARMA JAPAN. Japan top court rejects prosecutors' appeal in Diovan case; Novartis not guilty. 30 Jun 2021. https://pj.jiho.jp/article/244718. Accessed 12 Mar 2023.
36. House of Councillors, The National Diet of Japan. 193rd Ordinary Session of Parliament on Clinical Trials Act (draft). As of April 14, 2017. Japanese. https://www.sangiin.go.jp/japanese/joho1/kousei/gian/193/meisai/m193080190056.htm. Accessed 12 Mar 2023.
37. World Medical Association. WMA Declaration of Helsinki: Ethical principles for medical research involving human subjects. Adopted by the 18th WMA General Assembly, Helsinki, Finland, June 1964, and last amended by the 64th WMA General Assembly, Fortaleza, Brazil, October 2013. https://www.wma.net/policies-post/wma-declaration-of-helsinki-ethical-principles-for-medical-research-involving-human-subjects/.
38. Council for International Organizations of Medical Sciences. International ethical guidelines for health-related research involving humans. 2016. https://cioms.ch/publications/product/international-ethical-guidelines-for-health-related-research-involving-humans/.
39. Emanuel EJ, Wendler D, Grady C. What makes clinical research ethical? JAMA. 2000;283(20):2701–11.
40. Emanuel EJ, Wendler D, Killen J, Grady C. What makes clinical research in developing countries ethical? The benchmarks of ethical research. J Infect Dis. 2004;189(5):930–7.
41. The National Commission for the Protection of Human Subjects of Biomedical and Behavioral Research. The Belmont Report: Ethical principles and guidelines for the protection of human subjects of research. 18 Apr 1979. https://www.hhs.gov/ohrp/regulations-and-policy/belmont-report/read-the-belmont-report/index.html.
42. Kurihara C. Chapter 9. A new horizon in health research ethics: a view from the perspective of radiological protection. In: Zölzer F, Meskens G, editors. Research ethics for environmental health. 1st ed. London: Routledge; 2021. https://doi.org/10.4324/9780429318436.
43. World Health Organization. Statement on the second meeting of the International Health Regulations. Emergency committee regarding the outbreak of novel coronavirus (2019-nCoV). 30 Jan 2020. https://www.who.int/news/item/30-01-2020-statement-on-the-second-meeting-of-the-international-health-regulations-(2005)-emergency-committee-regarding-the-outbreak-of-novel-coronavirus-(2019-ncov). Accessed 12 Mar 2023.
44. World Health Organization. WHO Director-General's opening remarks at the media briefing on COVID-19–11 March 2020. 11 Mar 2020. https://www.who.int/director-general/speeches/detail/who-director-general-s-opening-remarks-at-the-media-briefing-on-covid-19%2D%2D-11-march-2020. Accessed 12 Mar 2023.

45. Reuters. Japan adds Tokyo, other areas to COVID-19 ‘quasi-emergency’ state. 9 Apr 2021. https://jp.reuters.com/article/us-health-coronavirus-japan/japan-adds-tokyo-other-areas-to-covid-19-quasi-emergency-state-idUSKBN2BW02Y. Accessed 12 Mar 2023.
46. Ministry of Health, Labour and Welfare. List of press release SARS-COV-2 (charter flight, cruise ship). Japanese. https://www.mhlw.go.jp/stf/seisakunitsuite/bunya/0000121431_00108.html. Accessed 12 Mar 2023.
47. Press conference by Prime Minister Abe. 28 Mar 2020. Japanese. https://www.kantei.go.jp/jp/98_abe/statement/2020/0327kaiken.html. Accessed 12 Mar 2023.
48. Ohshima H, Aoyagi M, Tajima M, Nomura T, Kato Y, Koike R, Suzuki Y, Yamashita S, Ikeda T, Aoki H, Clinical Evaluation Expert Committee, Drug Evaluation Committee in Japan Pharmaceutical Manufacturers Association. Considering the future of clinical research at the turning point of implementation of the new legislation in Japan: consideration of regulatory approval system in Japan and ‘Emergency Approval System’ enforced by the amendment of the Pharmaceutical and Medical Device Act in 2022 due to the COVID-19 pandemic. Clin Eval. 2022;50(2):259–73. Japanese. http://cont.o.oo7.jp/50_2/p259-73.pdf.
49. The Mainichi. Japan keeping antiviral drug despite unproven COVID efficacy after spending $136 mil. 25 Apr 2022. https://mainichi.jp/english/articles/20220425/p2a/00m/0na/033000c. Accessed 12 Mar 2023.
50. Arai J, Mitsutake R. Avigan set for approval this month in Japan as coronavirus drug. Nikkei Asia. 5 May 2020. https://asia.nikkei.com/Business/Pharmaceuticals/Avigan-set-for-approval-this-month-in-Japan-as-coronavirus-drug. Accessed 12 Mar 2023.
51. Olorunnipa T, McGinley L. Drug promoted by trump as coronavirus ‘game changer’ increasingly linked to deaths. The Washington Post. 15 May 2020. https://www.washingtonpost.com/politics/drug-promoted-by-trump-as-coronavirus-game-changer-increasingly-linked-to-deaths/2020/05/15/85d024fe-96bd-11ea-9f5e-56d8239bf9ad_story.html. Accessed 12 Mar 2023.
52. YAKUGAI Ombudsperson “Medwatcher Japan”. Opinion on Avigan (regarding COVID-19). 1 May 2020. English version. https://www.yakugai.gr.jp/en/topics/topic.php?id=988. Accessed 12 Mar 2023.
53. COVID-19 Expert Committee of the Japan Medical Association. Urgent recommendations for the development of therapeutic agents in the event of a novel coronavirus pandemic. 18 May 2020. Japanese. https://www.covid19-jma-medical-expert-meeting.jp/topic/4549. Accessed 12 Mar 2023.
54. Doi Y, Hibino M, Hase R, Yamamoto M, Kasamatsu Y, Hirose M, Mutoh Y, Homma Y, Terada M, Ogawa T, Kashizaki F, Yokoyama T, Koba H, Kasahara H, Yokota K, Kato H, Yoshida J, Kita T, Kato Y, Kamio T, Kodama N, Uchida Y, Ikeda N, Shinoda M, Nakagawa A, Nakatsumi H, Horiguchi T, Iwata M, Matsuyama A, Banno S, Koseki T, Teramachi M, Miyata M, Tajima S, Maeki T, Nakayama E, Taniguchi S, Lim CK, Saijo M, Imai T, Yoshida H, Kabata D, Shintani A, Yuzawa Y, Kondo M. A prospective, randomized, open-label trial of early versus late Favipiravir therapy in hospitalized patients with COVID-19. Antimicrob Agents Chemother. 2020;64(12):e01897–20. PMID: 32958718; PMCID: PMC7674035. https://doi.org/10.1128/AAC.01897-20.
55. National Institutes of Health, National Institute of Allergy and Infectious Diseases. NIH Clinical Trial of Remdesivir to Treat COVID-19 Begins. 25 Feb 2020. https://www.niaid.nih.gov/news-events/nih-clinical-trial-remdesivir-treat-covid-19-begins. Accessed 12 Mar 2023.
56. Beigel JH, Tomashek KM, Dodd LE, Mehta AK, Zingman BS, Kalil AC, Hohmann E, Chu HY, Luetkemeyer A, Kline S, Lopez de Castilla D, Finberg RW, Dierberg K, Tapson V, Hsieh L, Patterson TF, Paredes R, Sweeney DA, Short WR, Touloumi G, Lye DC, Ohmagari N, Oh MD, Ruiz-Palacios GM, Benfield T, Fätkenheuer G, Kortepeter MG, Atmar RL, Creech CB, Lundgren J, Babiker AG, Pett S, Neaton JD, Burgess TH, Bonnett T, Green M, Makowski M, Osinusi A, Nayak S, Lane HC. ACTT-1 study group members. Remdesivir for the treatment of Covid-19—final report. N Engl J Med. 2020;383(19):1813–26. Epub 2020 Oct 8. PMID: 32445440; PMCID: PMC7262788. https://doi.org/10.1056/NEJMoa2007764.

57. U.S. Food and Drug Administration. NDA 214787. https://www.accessdata.fda.gov/drugsatfda_docs/appletter/2020/214787Orig1s000ltr.pdf. Accessed 12 Mar 2023.
58. Ministry of Health, Labour and Welfare. Press Release: On the Special exceptional approval of Remdesivir under the Pharmaceuticals and Medical Devices Act. 7 May 2020. Japanese. https://www.mhlw.go.jp/content/000628076.pdf. Accessed 12 Mar 2023.
59. Ohmagari N. The 12th Annual Meeting of the Japanese Association of Pharmaceutical Medicine 2021—a new style of pharmaceutical medicine (President: Kotone Matsuyama) <Keynote lecture>Clinical Practice for COVID-19. Clin Eval. 2022;49(3):343–60. Japanese.
60. World Health Organization. Solidarity Therapeutics Trial produces conclusive evidence on the effectiveness of repurposed drugs for COVID-19 in record time. 15 Oct 2020. https://www.who.int/news/item/15-10-2020-solidarity-therapeutics-trial-produces-conclusive-evidence-on-the-effectiveness-of-repurposed-drugs-for-covid-19-in-record-time. Accessed 12 Mar 2023.
61. WHO Solidarity Trial Consortium; Pan H, Peto R, Henao-Restrepo AM, Preziosi MP, Sathiyamoorthy V, Abdool Karim Q, Alejandria MM, Hernández García C, Kieny MP, Malekzadeh R, Murthy S, Reddy KS, Roses Periago M, Abi Hanna P, Ader F, Al-Bader AM, Alhasawi A, Allum E, Alotaibi A, Alvarez-Moreno CA, Appadoo S, Asiri A, Aukrust P, Barratt-Due A, Bellani S, Branca M, Cappel-Porter HBC, Cerrato N, Chow TS, Como N, Eustace J, García PJ, Godbole S, Gotuzzo E, Griskevicius L, Hamra R, Hassan M, Hassany M, Hutton D, Irmansyah I, Jancoriene L, Kirwan J, Kumar S, Lennon P, Lopardo G, Lydon P, Magrini N, Maguire T, Manevska S, Manuel O, McGinty S, Medina MT, Mesa Rubio ML, Miranda-Montoya MC, Nel J, Nunes EP, Perola M, Portolés A, Rasmin MR, Raza A, Rees H, Reges PPS, Rogers CA, Salami K, Salvadori MI, Sinani N, Sterne JAC, Stevanovikj M, Tacconelli E, Tikkinen KAO, Trelle S, Zaid H, Røttingen JA, Swaminathan S. Repurposed Antiviral Drugs for Covid-19 - Interim WHO Solidarity Trial Results. N Engl J Med. 2021;384(6):497–511. Epub 2020 Dec 2. PMID: 33264556; PMCID: PMC7727327. https://doi.org/10.1056/NEJMoa2023184.
62. World Health Organization. Weekly epidemiological update on COVID-19. 27 July 2022. https://www.who.int/publications/m/item/weekly-epidemiological-update-on-covid-19%2D%2D-27-july-2022. Accessed 12 Mar 2023.
63. The Asahi Shimbun. WHO: Japan tops the world in new weekly cases of COVID-19. 28 Jul 2022. https://www.asahi.com/ajw/articles/14681705. Accessed 12 Mar 2023.
64. TOKYO web. Japan has the world's highest number of corona cases in the world, 970,000 per week, according to WHO announcement, expert: 'Surge during overseas convergence'. 28 Jul 2022. Japanese. https://www.tokyo-np.co.jp/article/192380. Accessed 12 Mar 2023.
65. The Sankei News. For the sixth week in a row, Japan has the world's highest number of corona infections and the second highest number of deaths. 15 Dec 2022. Japanese. https://www.sankei.com/article/20221215-6PP2SXBELNJTXFNH5FBVPZLCI4/. Accessed 12 Mar 2023.
66. The Japan Times. Japan's parliament OKs bill for emergency approval of drugs and vaccines. 13 May 2022. https://www.japantimes.co.jp/news/2022/05/13/national/vaccine-drug-emergency-approvals/. Accessed 12 Mar 2023.
67. Saio T. On the procrastination of approval determination of ensitrelvir: consideration on the composite endpoint. Clin Eval. 2022;50(2):275–80.
68. U.S. Department of Health and Human Services, Food and Drug Administration, Center for Drug Evaluation and Research (CDER), Center for Biologics Evaluation and Research (CBER). Assessing COVID-19-related symptoms in outpatient adult and adolescent subjects in clinical trials of drugs and biological products for COVID-19 prevention or treatment Guidance for industry; Sept 2020. https://www.fda.gov/media/142143/download.
69. The Japan Times. Shionogi says Japan government to buy 1 million more COVID pill doses. 13 Dec 2022. https://www.japantimes.co.jp/news/2022/12/13/national/science-health/shionogi-covid-pill-purchase/. Accessed 12 Mar 2023.

70. The Mainichi. Children's cafeteria in Japan urges gov't to address poverty and lack of support. 23 Oct 2021. https://mainichi.jp/english/articles/20211022/p2a/00m/0na/048000c. Accessed 12 Mar 2023.
71. Polack FP, Thomas SJ, Kitchin N, Absalon J, Gurtman A, Lockhart S, Perez JL, Pérez Marc G, Moreira ED, Zerbini C, Bailey R, Swanson KA, Roychoudhury S, Koury K, Li P, Kalina WV, Cooper D, Frenck RW Jr, Hammitt LL, Türeci Ö, Nell H, Schaefer A, Ünal S, Tresnan DB, Mather S, Dormitzer PR, Şahin U, Jansen KU, Gruber WC. C4591001 clinical trial group. Safety and efficacy of the BNT162b2 mRNA Covid-19 vaccine. N Engl J Med. 2020;383:2603. Epub ahead of print. https://doi.org/10.1056/NEJMoa2034577.
72. Food and Drug Administration. Letter of Authorization. 2021. https://www.fda.gov/media/144412/download.
73. Baden LR, El Sahly HM, Essink B, Kotloff K, Frey S, Novak R, Diemert D, Spector SA, Rouphael N, Creech CB, McGettigan J, Kehtan S, Segall N, Solis J, Brosz A, Fierro C, Schwartz H, Neuzil K, Corey L, Gilbert P, Janes H, Follmann D, Marovich M, Mascola J, Polakowski L, Ledgerwood J, Graham BS, Bennett H, Pajon R, Knightly C, Leav B, Deng W, Zhou H, Han S, Ivarsson M, Miller J, Zaks T, COVE Study Group. Efficacy and safety of the mRNA-1273 SARS-CoV-2 vaccine. N Engl J Med. 2020;384:403. Epub ahead of print. https://doi.org/10.1056/NEJMoa2035389.
74. Food and Drug Administration. Letter of Authorization. 2023. https://www.fda.gov/media/144636/download.
75. Voysey M, SAC C, Madhi SA, Weckx LY, Folegatti PM, Aley PK, Angus B, Baillie VL, Barnabas SL, Bhorat QE, Bibi S, Briner C, Cicconi P, Collins AM, Colin-Jones R, Cutland CL, Darton TC, Dheda K, Duncan CJA, Emary KRW, Ewer KJ, Fairlie L, Faust SN, Feng S, Ferreira DM, Finn A, Goodman AL, Green CM, Green CA, Heath PT, Hill C, Hill H, Hirsch I, SHC H, Izu A, Jackson S, Jenkin D, Joe CCD, Kerridge S, Koen A, Kwatra G, Lazarus R, Lawrie AM, Lelliott A, Libri V, Lillie PJ, Mallory R, Mendes AVA, Milan EP, Minassian AM, McGregor A, Morrison H, Mujadidi YF, Nana A, O'Reilly PJ, Padayachee SD, Pittella A, Plested E, Pollock KM, Ramasamy MN, Rhead S, Schwarzbold AV, Singh N, Smith A, Song R, Snape MD, Sprinz E, Sutherland RK, Tarrant R, Thomson EC, Török ME, Toshner M, Turner DPJ, Vekemans J, Villafana TL, Watson MEE, Williams CJ, Douglas AD, Hill AVS, Lambe T, Gilbert SC, Pollard AJ, Oxford COVID Vaccine Trial Group. Safety and efficacy of the ChAdOx1 nCoV-19 vaccine (AZD1222) against SARS-CoV-2: an interim analysis of four randomised controlled trials in Brazil, South Africa, and the UK. Lancet. 2020;397(10269):99–111. Epub ahead of print. PMID: 33306989; PMCID: PMC7723445. https://doi.org/10.1016/S0140-6736(20)32661-1.
76. U.S. government accountability office. Operation warp speed: accelerated COVID-19 vaccine development status and efforts to address manufacturing challenges. GAO-21-319. Published: Feb 11, 2021. Publicly Released: Feb 11, 2021. https://www.gao.gov/products/gao-21-319. Accessed 12 Mar 2023.
77. KFF: Filling the need for trusted information on national health issues. Trump Administration organizing Manhattan Project-style effort to speed coronavirus vaccine efforts; NIH announces initiative to Hasten Diagnostics Development. 30 Apr 2020. https://www.kff.org/news-summary/trump-administration-organizing-manhattan-project-style-effort-to-speed-coronavirus-vaccine-efforts-nih-announces-initiative-to-hasten-diagnostics-development/. Accessed 12 Mar 2023.
78. Kurihara C, Saio T. Japan Medical Association COVID-19 Experts Conference. Ethics in COVID-19 prevention vaccine development: a milestone toward post-corona era. 15 Jan 2021. Japanese. https://www.covid19-jma-medical-expert-meeting.jp/topic/4068. Accessed 12 Mar 2023.
79. Kurihara C, Saio T, Matsuyama K, Imamura K. Opening remarks and situation in Japan and the world: proposal on ethics of placebo-controlled trials and post-trial access in the declaration of Helsinki. Clin Eval. 2021;49(Sup38):16–23. http://cont.o.oo7.jp/49sup38/49sup38contents_e.html.

80. Duke Global health Innovation Center. New study shows rich country shopping spree for COVID-19 vaccines could mean fewer vaccinations for billions in low-income countries. 2 Nov 2020. https://dukeghic.org/wp-content/uploads/sites/20/2020/11/COVID19-Vax-Press-Release__28Oct2020-1.pdf. Accessed 12 Mar 2023.
81. Dhai A. Ethics in vaccine allocation in developing countries. Clin Eval. 2021;49(Sup38):63–71.
82. Japanese government will purchase additional vaccines of 145 million doses. Kagaku-Kogyo-Nippo. 18 Mar 2022. Jaepanese. https://www.chemicaldaily.co.jp/%E6%94%BF%E5%BA%9C%E3%80%81%EF%BC%91%E5%84%84%EF%BC%94%EF%BC%95%EF%BC%90%EF%BC%90%E4%B8%87%E5%9B%9E%E5%88%86%E3%81%AE%E3%83%AF%E3%82%AF%E3%83%81%E3%83%B3%E8%BF%BD%E5%8A%A0%E8%B3%BC%E5%85%A5/. Accessed 12 Mar 2023.
83. The Mainich. Over 77.8 million COVID vaccine doses dumped in Japan, worth an estimate $1.6 billion. 2023. 18 Mar 2023. https://mainichi.jp/english/articles/20230318/p2a/00m/0na/010000c. Accessed 30 Mar 2023.
84. House of Councillors, The National Diet of Japan. Kamiya S. Question on the information disclosure on the contract for the purchase of COVID-19 vaccine. Question No. 35, 210th National Assembly (Extraordinary Session). 9 Nov 2022. Japanese. https://www.sangiin.go.jp/japanese/joho1/kousei/syuisyo/210/meisai/m210035.htm. Accessed 12 Mar 2023.
85. House of Councillors, The National Diet of Japan. Temporary Acting Prime Minister, Minister of State, Matsuno H. Reply to the question submitted by Mr. Sohei Kamiya, Member of the House of Councillors, concerning the information disclosure on the contract for the purchase of the COVID-19 vaccine. 210th National Assembly (extraordinary session), 18 Nov 2022. Japanese. https://www.sangiin.go.jp/japanese/joho1/kousei/syuisyo/210/touh/t210035.htm. Accessed 12 Mar 2023.
86. Board of Audit of Japan. Report pursuant to Article 30bis of the Act of the Board of Audit of Japan. 29 Mar 2023. Japanese. https://www.jbaudit.go.jp/pr/kensa/result/5/r050329.html. Accessed 30 Mar 2023.
87. Prime Minister's Office of Japan. COVID-19 vaccines. as of January 26, 2023. https://japan.kantei.go.jp/ongoingtopics/vaccine.html. Accessed 26 Jun 2023, to be updated accordingly.
88. National Institute of Infectious Diseases. On the COVID-19 vaccine. 17 Dec 2021. Japanese. https://www.niid.go.jp/niid/ja/2019-ncov/2484-idsc/10840-covid19-64.html. Accessed 12 Mar 2023.
89. The Mainichi. Japan ends distribution of AstraZeneca COVID-19 vaccine, discards 13.5 mil doses. 1 Oct 2022. https://mainichi.jp/english/articles/20221001/p2a/00m/0na/004000c. Accessed 12 Mar 2023.
90. Ministry of Health, Labour and Welfare. Termination of the provision of the COVID-19 prevention vaccine of AstraZeneca and its utilization. 30 Sept 2022. Japanese. https://www.mhlw.go.jp/stf/newpage_28291.html. Accessed 12 Mar 2023.
91. Ministry of Health, Labour and Welfare. Termination of the provision of the COVID-19 prevention vaccine (conventional strain vaccine) of Moderna Japan K.K. and its utilisation. 10 Feb 2023. Japanese. https://www.mhlw.go.jp/stf/newpage_30879.html. Accessed 12 Mar 2023.
92. Ministry of Foreign Affairs of Japan. Vaccine-related supports by Japan. 2023. Japanese. https://www.mofa.go.jp/mofaj/files/100221711.pdf.
93. Ministry of Health, Labour and Welfare. International contribution with a new coronavirus vaccine. Japanese. https://www.mhlw.go.jp/content/10900000/000950146.pdf.
94. Mainichi Newspaper. Focus: Corona DNA vaccine 'politically driven' clinical trial: baffling Osaka governor's pre-approval announcement. 5 Jul 2020. https://mainichi.jp/articles/20200705/ddm/002/040/038000c. Accessed 12 Mar 2023.
95. US Dept of Health and Human Services, Food and Drug Administration, Center for Biologics Evaluation and Research. Guidance for Industry: development and licensure of vaccines to prevent COVID-19. 2020 Jun.

96. Japan Medical Association. International comparison of the number of hospital beds. 22 Feb 2021. Japanese. https://www.med.or.jp/dl-med/teireikaiken/20210120_1.pdf. Accessed 12 Mar 2023.
97. Voluntary members of Bioethics-Medical ethics study group. A recommendation on the process for determining the allocation of artificial respirators during the COVID-19 outbreak. 30 Mar 2020. Japanese. https://square.umin.ac.jp/biomedicalethics/activities/ventilator_allocation.html. Accessed 12 Mar 2023.
98. Gruppo di lavoro, Marco Vergano, Guido Bertolini, Alberto Giannini, Giuseppe Gristina, Sergio Livigni, Giovanni Mistraletti, Flavia Petrini. Società Italiana di Anestesia Analgesia Rianimazione e Terapia Intensiva (SIAARTI). Raccomandazioni di etica clinica per l'ammissione a trattamenti intensivi e per la loro sospensione, in condizioni eccezionali di squilibrio tra necessità e risorse disponibili - versione 01 Pubblicato il 06.03.2020. https://www.quotidianosanita.it/allegati/allegato2675063.pdf. Accessed 12 Mar 2023.
99. Sawamura T, Norisue Y, Mima H, Ueta I, Shigemitsu H, Ohno M, Maki J, Ito K, Uemura S, Uesawa H, Gando S, Fujino Y, Nishida O. Withholding and withdrawing of the treatment for major surge in coronavirus disease 2019 (COVID-19) pandemic, from the viewpoint of healthcare resource rationing. J Jpn Soc Intensive Care Med. 2020;27(6):509–10. Japanese. https://www.jsicm.org/pdf/covid-19_iryohaibun_27_27_509.pdf.
100. Saio T. A critical appraisal of the recommendation on the allocation of artificial respirators during the COVID-19 outbreak in Japan. Clin Eval. 2020;48(1):161–6. Japanese. http://cont.o.oo7.jp/48_1/p161-6.pdf.
101. Saio T, Kurihara C. Demurral to the recommendation of withholding and withdrawing of the treatment of coronavirus disease 2019 (COVID-19). J Jpn Soc Intensive Care Med. 2021;28:297–9. Japanese. https://www.jstage.jst.go.jp/article/jsicm/28/4/28_28_298/_pdf.
102. Saio T. Chapter 3: Human dignity at stake: from clinical practice. In: Kagawa C, Doi K, editors. "Human Dignity" After the Corona Crisis (Science Council Series). Japanese. p. 87–113.
103. Funago Yasuhiko Official Site: Funago Y. Statement on the "selection of lives" following the spread of COVID-19. April 2020. Japanese. https://yasuhiko-funago.jp/page-200413/. Accessed 12 Mar 2023.
104. Deutscher Ethikrat. Solidarität und Verantwortung in der Corona-Krise. 27 Mar 2020. https://www.ethikrat.org/mitteilungen/mitteilungen/2020/solidaritaet-und-verantwortung-in-der-corona-krise/?cookieLevel=not-set.
105. Emanuel EJ, Persad G, Upshur R, Thome B, Parker M, Glickman A, Zhang C, Boyle C, Smith M, Phillips JP. Fair allocation of scarce medical resources in the time of Covid-19. N Engl J Med. 2020;382(21):2049–55.
106. Biddison LD, Berkowitz KA, Courtney B, De Jong CM, Devereaux AV, Kissoon N, Roxland BE, Sprung CL, Dichter JR, Christian MD, Powell T. Task force for mass critical care; task force for mass critical care. Ethical considerations: care of the critically ill and injured during pandemics and disasters: CHEST consensus statement. Chest. 2014;146(4 Suppl):e145S–55S.
107. Rosenbaum SJ, Bayer R, Bernheim RG, et al. Ethical considerations for decision making regarding allocation of mechanical ventilators during a severe influenza pandemic or other public health emergency. Atlanta: Centers for Disease Control and Prevention; 1 Jul 2011. https://www.cdc.gov/os/integrity/phethics/docs/Vent_Document_Final_Version.pdf. Accessed 12 Mar 2023.
108. Zucker H, Adler K, Berens D, et al. Ventilator allocation guidelines. Albany: New York State Department of Health Task Force on Life and the Law; November 2015. https://www.health.ny.gov/regulations/task_force/reports_publications/docs/ventilator_guidelines. Accessed 12 Mar 2023.
109. Cabinet Office, Department of Health. Responding to pandemic influenza: the ethical framework for policy and planning. London: UK Department of Health; 2007. https://www.hsj.co.uk/download?ac=1205809. Accessed 12 Mar 2023.
110. Hick JL, Hanfling D, Wynia MK, Pavia AT. Duty to plan: health care, crisis standards of care, and novel coronavirus SARS-CoV-2. NAM Perspect. 5 Mar 2020. https://nam.edu/duty-

to-plan-health-care-crisis-standards-of-care-and-novel-coronavirus-sars-cov-2/. Accessed 12 Mar 2023.
111. Reid L. Triage of critical care resources in COVID-19: a stronger role for justice. J Med Ethics. 2020;46(8):526–30.
112. Fallucchi F, Faravelli M, Quercia S. Fair allocation of scarce medical resources in the time of COVID-19: what do people think? J Med Ethics. 2021;47(1):3–6.
113. Emanuel EJ. Why I hope to die at 75. The Atlantic. 15 Oct 2014. https://www.theatlantic.com/magazine/archive/2014/10/why-i-hope-to-die-at-75/379329/. Accessed 12 Mar 2023.
114. Greco D, Kurihara C, Saio T, Nakazawa E, Organizers & Discussants. Kimura R, Reid L, Criscuolo D, Kerpel-Fronius S, Imamura K, discussants. Co-organized by the COVID-19 task force, Japan Association for Bioethics & Brazilian Society of Bioethics. COVID-19 and bioethics: Part1 ethical challenges and COVID-19. The recommendation no. 01/2020 of the Brazilian Society of Bioethics (SBB). Clin Eval. 2020;48(3):661–84. http://cont.o.oo7.jp/48_3/p661-84.pdf.
115. Brazilian Society of Bioethics. Brazilian Society of Bioethics (SBB) Recommendation No. 01/2020.
116. Proceedings: COVID-19 and bioethics Part 3: Pandemic and research ethics-Democracy, placebo and post-trial access. Clin Eval. 2021; 49(sup 38). http://cont.o.oo7.jp/49sup38/49sup38contents_e.html.
117. Aloudat T. Médecins Sans Frontières (MSF) access campaign for equitable access in the world. Clin Eval. 2021;49(Sup 38). http://cont.o.oo7.jp/49sup38/p61-109.pdf.
118. Dodgson R, Lee K, Drager N. Global Health Governance: a conceptual review. Written as part of a project funded by the Department of Health and Development. World Health Organization. February 2002. https://apps.who.int/iris/bitstream/handle/10665/68934/a85727_eng.pdf;jsessionid=BFF40A31A144AF07845A443C5271B425?sequence=1.
119. Kheir-Mataria WAE, El-Fawal H, Bhuiyan S, Chun S. Global Health governance and health equity in the context of COVID-19: a scoping review. Healthcare (Basel). 2022;10(3):540. PMID: 35327017; PMCID: PMC8949542. https://doi.org/10.3390/healthcare10030540.
120. European Commission. Statement by President von der Leyen following the Global Health Summit. 21 May 2021. https://ec.europa.eu/commission/presscorner/detail/en/statement_21_2622. Accessed 12 Mar 2023.
121. Council for Trade-Related Aspects of Intellectual Property Rights. Waiver from certain provisions of the TRIPS Agreement for the prevention, containment and treatment of COVID-19: Communication from India and South Africa; IP/C/W/669. 2 Oct 2020. https://docs.wto.org/dol2fe/Pages/SS/directdoc.aspx?filename=q:/IP/C/W669.pdf&Open=True.
122. World Trade Organization, Ministerial Conference Twelfth Session Geneva, 12–15 June 2022. Ministerial Decision on the TRIPS Agreement. Adopted on 17 June 2022; WT/MIN(22)/30 WT/L/1141. 22 June 2022. https://docs.wto.org/dol2fe/Pages/SS/directdoc.aspx?filename=q:/WT/MIN22/30.pdf&Open=True.
123. World Trade Organization. Decision text on extension of the 17 June 2022 ministerial decision to COVID-19 therapeutics and diagnostics: communication from the delegations of Argentina, Bangladesh, the Plurinational state of Bolivia, Egypt, India, Indonesia, Pakistan, South Africa and the Bolivarian Republic of Venezuela on behalf of the co-sponsors of the IP/C/W/669/REV.1 proposal; IP/C/W/694. 6 Dec 2022. https://docs.wto.org/dol2fe/Pages/SS/directdoc.aspx?filename=q:/IP/C/W694.pdf&Open=True.
124. World Trade Organization, Council for Trade-Related Aspects of Intellectual Property Rights. Paragraph 8 of the Ministerial Decision on the TRIPS Agreement Adopted on 17 June 2022: Report to the General Council; IP/C/95. 16 Dec 2022. https://docs.wto.org/dol2fe/Pages/SS/directdoc.aspx?filename=q:/IP/C/95.pdf&Open=True.
125. United Nations. Resolution adopted by the General Assembly on 25 September 2015—Transforming our world: the 2030 Agenda for sustainable development. 21 Oct 2015. https://www.un.org/en/development/desa/population/migration/generalassembly/docs/globalcompact/A_RES_70_1_E.pdf.

126. United Nations Educational, Scientific and Cultural Organization. Universal Declaration on Bioethics and Human Rights. 19 Oct 2005. https://www.unesco.org/en/legal-affairs/universal-declaration-bioethics-and-human-rights?hub=66535.
127. Kyobutungi C, Robinson J, Pai M. PLOS global public health, charting a new path towards equity, diversity and inclusion in global health. PLoS Glob Public Health. 2021;1(10):e0000038. PMID: 36962122; PMCID: PMC10021277. https://doi.org/10.1371/journal.pgph.0000038.
128. Piketty T. Capital in the twenty-first century. Cambridge, MA: Harvard University Press; 2014.
129. World Health Organization. Global Public Good for Health. 2002. https://apps.who.int/iris/bitstream/handle/10665/42518/9241590106.pdf.

Part II
Historical and International Perspectives on the Development of Ethical Principles in Research Involving Humans

The Declaration of Helsinki as a Living Document: Revisiting Its Principles in a Global Pandemic

Ramin W. Parsa-Parsi and Otmar Kloiber

Abstract As one of the core ethical documents of the World Medical Association (WMA), the Declaration of Helsinki (DoH) forms the foundation of ethics for medical research with human subjects. While ethical codes and guidelines are meant to stand the test of time, they must also be regularly revisited with consideration for new developments in medicine. Two of the issues that were most controversially discussed during the 2013 revision of the DoH—post-trial access and the use of placebos—are arguably more relevant than ever in light of the COVID-19 pandemic and the clinical trials for vaccines that were performed in an expedited manner. These and other issues must be examined regularly to ensure the continued applicability of this core WMA document.

Keywords World Medical Association · Research ethics · Medical ethics · COVID-19 · Placebo

The World Medical Association (WMA) is the global organization of physicians. One of its core missions is to ensure the highest possible standard of medical ethics. To achieve this objective, the WMA has adopted policies focused on a broad spectrum of ethical issues relevant to the medical profession. The declarations, resolutions, and statements published by the WMA are reviewed and revised at regular intervals to keep them up to date, to further improve them and incorporate new developments [1].

Within the WMA's policy apparatus, there are three documents that are heavily interrelated and form the backbone of the WMA's work: the Declaration of Geneva: The Physician's Pledge (DoG) [2], the International Code of Medical Ethics (ICoME) [3], and the Declaration of Helsinki (DoH) [4].

R. W. Parsa-Parsi (✉)
Department for International Affairs, German Medical Association, Berlin, Germany
e-mail: rparsa@baek.de

O. Kloiber
World Medical Association, Ferney-Voltaire, France

C. Kurihara et al. (eds.), *Ethical Innovation for Global Health*,
https://doi.org/10.1007/978-981-99-6163-4_7

The DoG was the first of these three documents to be adopted by the WMA's second General Assembly in 1948, just after the establishment of the WMA in 1947. Developed in the aftermath of the Second World War, it outlines the basic ethical principles for physicians and is intended to serve as a pledge for physicians upon admission to the medical profession. While minor amendments have been added over the years, primarily for clarification, the basic ethical principles of the DoG have not changed since 1948. The most recent revision in 2017 was probably the most significant. During the revision process, the principle of patient autonomy was introduced—a principle that had already been included in many other WMA documents, including the DoH. Another matter that was introduced was the responsibility physicians have to attend to their own health, well-being, and abilities in order to provide care of the highest standard [5].

Following the adoption of the DoG in 1948, the WMA Council installed a workgroup to develop an international code of medical ethics. The ICoME was adopted just a year later in 1949. It is the foundation of ethical principles defining the professional duties of physicians and is therefore relevant to all physicians in their daily practice. It outlines the physician's responsibilities vis-à-vis patients, society, as well as other physicians and health professionals. After a 4-year revision process, a revised ICoME was adopted unanimously by the WMA General Assembly in October 2022. The WMA went to great lengths to ensure the revised ICoME is a truly global code that is applicable to different cultures and political systems [1]. For example, the revision process was reinforced by a series of regional and international conferences in Brazil, Kuwait, Thailand, Nigeria, and the USA, along with a dedicated conference on conscientious objection in Indonesia, hybrid sessions at high-profile international bioethics conferences, and an online public consultation. The concepts of patient autonomy, physician well-being, physician conscientious objection, and equity and justice in health care found their way into the code, along with further elaborations on patient confidentiality and informed consent [1]. In addition, the newly added paragraph 36 in the revised ICoME clarifies that the physician "must support sound medical scientific research in keeping with the WMA Declaration of Helsinki and the WMA Declaration of Taipei" [3].

The third of these documents—the DoH—forms the foundation of ethics for medical research with human subjects. It is widely known and used by researchers, including non-physicians, around the globe, though it is primarily directed at the medical profession. The DoH expressly encourages other health professionals to follow the same principles.

Although these three documents serve three distinct purposes and can each stand on their own, there is a certain deliberate synergy among them. During their respective revision processes, each document was also reviewed for compatibility with the others. Revisions are also timed to ensure that time, resources, and attention can be fully dedicated to one document at a time.

While each of these documents must be considered in light of new developments in medicine, it is perhaps the DoH that is most impacted by the changing times. Ethical codes and guidelines are meant to stand the test of time, and each of the respective workgroups called upon to revise the DoG, ICoME, and DoH over the years has approached this process with due respect for the integrity of core ethical principles. Workgroups are generally advised to amend only that which is essential in these three documents and to avoid any major structural changes unless there is a good argument for doing so.

At the same time, new developments in medical research that were unimaginable when the DoH was first adopted have given rise to new ethical questions. Globalization of medicine has made it even more essential to ensure that these crucial documents are comprehensive and comprehensible to medical professionals throughout the world.

The DoH has been amended nine times since it was adopted in Helsinki during the WMA's 18th General Assembly in 1964. During the most recent revision of the DoH, which was adopted in 2013, two of the issues that were most controversially discussed were post-trial access and the use of placebos. In the revised 2013 version, the statement on post-trial access clarifies that "sponsors, researchers and host country governments should make provisions for post-trial access for all participants who still need an intervention identified as beneficial in the trial" and should inform the participants of this as part of the process of informed consent. This amendment is meant to ensure that patients are aware of the options for access to treatment after the trial before deciding whether or not to participate.

The DoH states very clearly that the goal of generating new knowledge in medical research "can never take precedence over the rights and interests of individual research subjects." This is very much in keeping with the duties and principles outlined in both the DoG and the ICoME. But the tension between knowledge acquisition and the rights and interests of research subjects is especially palpable when it comes to the use of placebo controls. In light of the many debates surrounding the topic of placebos, the WMA installed a dedicated workgroup during the last DoH revision to deal exclusively with this issue. An international expert conference was organized to discuss different approaches to the topic. While it was certainly challenging to come to a consensus, the DoH was ultimately amended to further protect study subjects, even above and beyond the protections that were already in place in earlier versions of the document. The key message is that "patients who receive any intervention less effective than the best proven one, placebo, or no intervention will not be subject to additional risks of serious or irreversible harm as a result of not receiving the best proven intervention."

But this is not the end of the story for these two sensitive issues. In fact, they are arguably more relevant than ever in light of the COVID-19 pandemic and the clinical trials for vaccines that had to be performed in an expedited manner in order to save time and potentially lives. These trials were undertaken in different parts of the world, particularly in those countries where patients had the highest risk of infection. It became especially tricky once many countries had approved a given vaccine for emergency use, but trials continued using vaccines with a placebo comparator even though an effective vaccine was already on the market. This would appear to be ethically questionable on the basis of the current version of the DoH, which begs the question: Should the DoH include a specific paragraph accommodating for these challenging global emergency situations?

Such consideration may also include the question of human challenge trials, which seem to be of particular interest when it comes to vaccinations.

There are further issues that must be examined in order to assess whether or not the current version of the DoH provides sufficient guidance for those involved in clinical trials. There are questions relating to post-trial access to care, compassionate use after a trial has been concluded or prior to market authorization. There is a growing field of patient-driven research and "Open Science" approaches. The DoH has to be tested against questions of inclusiveness in research when it comes to disadvantaged or traditionally excluded groups. This extends to the question of whether the far-reaching exclusion of "vulnerable groups" from research that can be conducted in other populations is not contradictory, when the "vulnerable group" is also the "concerned group", e.g. in the case of HIV-prevention research. The question that must be answered is whether new concepts and methods like real-world data use for control groups, or virtual patient data, but also more recent research designs like branched or adaptive trial designs, are sufficiently covered by the principles of the Declaration.

Furthermore, it would be remiss to discuss the DoH without addressing the need to critically assess the interface between this document and the Declaration of Taipei on Ethical Considerations regarding Health Databases and Biobanks (2016) [6] (DoT), as the requirements of the DoH are not fully consistent with the requirements and procedures laid down in the newer DoT [7]. The connection between the DoH and the DoT also points to the relevance of informed consent, its potential evolution and/or replacement by other mechanisms protecting individual rights.

In April 2022, the WMA Council installed an international and representative workgroup to review and revise the DoH again. The abovementioned topics will most likely be explored and debated once again in light of recent developments.

Conflict of Interest Dr Ramin W. Parsa-Parsi was a member of the WMA Council, chair of the 2013 DoH revision workgroup, and is currently a member of the workgroup responsible for the ongoing revision of the DoH. In this capacity he has received compensation from the WMA in the form of travel reimbursement. Dr Otmar Kloiber is Secretary General of the WMA.

References

1. Parsa-Parsi RW. The International code of medical ethics of the World Medical Association. JAMA. 2022;328(20):2018–21. https://doi.org/10.1001/jama.2022.19697.
2. The World Medical Association. The Declaration of Geneva. 1948, last amended 2017. Available at: https://www.wma.net/policies-post/wma-declaration-of-geneva/.
3. The World Medical Association. The International code of medical ethics. 1949, last amended 2022. Available at: https://www.wma.net/policies-post/wma-international-code-of-medical-ethics/.
4. The World Medical Association. The Declaration of Helsinki. Ethical principles for medical research involving human subjects. First adopted in 1964, last amended 2013. Available at: https://www.wma.net/policies-post/wma-declaration-of-helsinki-ethical-principles-for-medical-research-involving-human-subjects/.

5. Parsa-Parsi RW. The revised declaration of Geneva: a modern-day physician's pledge. JAMA. 2017;318(20):1971–2. https://doi.org/10.1001/jama.2017.16230.
6. The World Medical Association. The declaration of Taipei on ethical considerations regarding health databases and biobanks. 2002. Last revised 2016. Available at https://www.wma.net/policies-post/wma-declaration-of-taipei-on-ethical-considerations-regarding-health-databases-and-biobanks/.
7. Kurihara C, Baroutsou V, Becker S, Brun J, Franke-Bray B, Carlesi R, Chan A, Collia LF, Kleist P, Laranjeira LF, Matsuyama K, Naseem S, Schenk J, Silva H, Kerpel-Fronius S. Linking the declarations of Helsinki and of Taipei: critical challenges of future-oriented research ethics. Front Pharmacol. 2020;11:579714. https://doi.org/10.3389/fphar.2020.579714.

Hidden Medical War Crimes and the Emergence of Bioethics in Japan

Rihito Kimura

Abstract This paper provides three aspects of bioethical instances relevant to human subject research during and after the World War II (WWII). First instance is about the "Japanese Military's Medical Crime" during the Sino-Japanese War (1937–1945). This crime was verified, however, the "Top Secret" documentary evidences have confirmed that the United States (U.S.) Occupation Forces have classified this "Japanese Military's Medical Crime" in order to ensure the National Security of the U.S. The second instance is about the Head-Quarter of the Japanese Military's Biological Weapon Corps in China which still exists in transforming itself into the Museum. This Museum and its exhibit itself became the visionary and actual evidences to remind us the cruel inhuman experimentations. Third instance is the human subject research performed on the Atomic Bomb Survivors at Hiroshima and Nagasaki after the end of WWII. The researchers "put priority on examining the A-bomb survivors rather than giving them treatment." It is important to recognize these human subject research such as Japanese Military's Medical Crime should be integrated into the emerging Bioethics Education, so that we will be able to learn from the past and proceed to the future with sincere hope for not repeating this kind of human subject research again.

Keywords Human subject research · Medical war crime · Unit 731 · Biological weapons · Atomic Bomb Casualty Commission (ABCC)/Radiation Effects Research Foundation (RERF) · Hiroshima and Nagasaki

1 Introduction

This chapter analyzes three examples of bioethical problems concerning human subject research during and after WWII. I use the narrative approach in writing this paper, based on my own investigation of relevant declassified "TOP SECRET"

R. Kimura (✉)
Bioethics and Law, Waseda University, School of Human Sciences, Tokyo, Japan

C. Kurihara et al. (eds.), *Ethical Innovation for Global Health*,
https://doi.org/10.1007/978-981-99-6163-4_8

documents and bioethical commitments including some site visits such as to the former headquarters of Unit 731 in China [1] and the Radiation Effects Research Foundation (RERF) in Hiroshima, Japan.

2 Bioethical Deficit: Documentary Evidences at Suitland, Maryland, U.S.

One day in October 1985, I visited the United States (U.S.) National Archives at Suitland, Maryland and I came across declassified documents relating to Japan's Manchuria-based Kwantung Army's Biological Weapon Corps "Unit 731" led by Colonel (General) Shiro Ishii. The official name of Unit 731 was "The Epidemic Prevention and Water Supply Department" [2, 3].

There was great value to seeing the original "TOP SECRET" documents from the 1940s which were declassified in October 1985 just before I had visited the National Archives. One of the key excerpts that I would like to share was written as follows [4]: (BW means biological weapon, in the following description)

> APPENDIX "A" CONCLUSIONS
> 4. It is concluded that:
> a. Information of Japanese BW experiments will be of great value to the U.S. BW research program.
> b. The data on hand as substantially outlined in paragraph 3 of appendix "A" does not appear sufficient at this time to constitute a basis for sustaining a war crimes charge against Ishii and/or his associates.
> c. The value to the U.S. of Japanese BW data is of such importance to national security as far outweigh the value accruing from "war crimes" prosecution.
> d. In the interests of national security, it would not be advisable to make this information available to other nations as would be the case in the event of a "war crimes" trial of Jap BW experts.
> e. The BW information obtained from Japanese sources should be retained in Intelligence channels and should not be employed as "war crimes" evidence, unless evidence developed at the International Military Trial presents overwhelming reasons which this procedure can no longer be followed.

By reviewing the relevant pages of Unit 731 related "TOP SECRET" documents in detail at Suitland, I was able to recognize there are repeated comments which are confirming (1) the "great value" of Japanese BW experiments, (2) there need not be "war crime charges" against Japanese researchers, (3) the importance of Japanese BW data for national security, (4) the information should not be available to other nations.

In the document dated 29 September 1947 with the name of J.B. Cresap, Comdr., USN, Secretary also added the following Memorandum [5]:

1. The alternate CAD member cannot agree with the statement by the State Member in the enclosure to SFE 188/3 that "on the bases of facts brought out in the subject paper

(SFE 188/2) that it is possible that the desired information can be obtained from Colonel (General) Ishii and his assistants without these assurances." Appendix "C" SFE 188/2 (cable C-52423 dtd 6 May 1947 from CINCFE which message originally presented the problem for a decision) says in part, "Ishii states that if guaranteed immunity from 'war crimes' in documentary form for himself, superiors and subordinates, he can describe the program in detail." Paragraph 3 B of that message further states: "Additional data, possibly including some statements from Ishii probably can be obtained by informing Japanese involved that information will be retained in intelligence channels and will not be employed as "war crimes evidence." The message concludes by recommending the adoption of the method outlined in 3B. It is apparent from a reading of the entire message that it is the wish of CINCFE to make the most expeditious arrangements possible with the Japanese BW group, headed by Ishii, for the desired information and that in his opinion this is the least possible offer than can be successfully made.

2. It is recognized that by informing Ishii and his associates that the information to be obtained re BW will be retained in intelligence channels and will not be employed as war crimes evidence, this government may at a later date be seriously embarrassed. However, the Army department and Air Force members strongly believe that this information, particularly that which will finally be obtained from the Japanese with respect to the effect of BW on humans, is of such importance to the security of this country that the risk of subsequent embarrassment should be taken.
3. Further, it is the considered opinion of responsible American officials, both military and civil, who have had close personal contact with General Ishii and other members of the BW group that all the desired information in a detailed form will give the greatest value cannot be obtained unless this Japanese group is informed that the information will not be employed as war crimes evidence.
4. Therefore it is believed that, in the final analysis, the security of the United States is of primary importance.

The "war crimes" in these documents referred to war crime cases to be indictable in the International Military Tribunal for the Far East in Tokyo, which was installed by Douglas MacArthur, the Supreme Commander of Allied Forces in 1946. This International Military Tribunal's Article 1 stated that "There shall be established an International Military Tribunal for the Far East for the trial of those persons charged individually, as members of organizations, or in both capacities, with offenses that include crimes against peace." This was neglected in the name of "national security" and the "great value for U.S. BW operations." In fact, the U.S. authorities have never confirmed the documented assurance of the immunity for Ishii and his associates and the reluctance to prosecute. However, it seems to me that there have been gradual changes on the side of Ishii for their voluntary positive presentation of the various experimentation data, etc. as expressed in the following documents [6]:

<u>FACTS BEARING ON THE PROBLEM</u>

1. Part 2 of the cable cited in paragraph 1 (Enclosure) states that general Ishii, the Japanese authority of BW, will describe the Japanese BW program in detail if guaranteed immunity from "War Crimes" in documentary form for himself, superiors, and subordinates. Ishii and associates have to date, voluntarily supplied and are continuing to supply such information without a documentary guarantee of immunity.

2. Nineteen Japanese BW experts have written a 60 page report concerning BW research using human subjects. A twenty page report covering 9 years of research on crop destruction has been prepared. A report by 10 Japanese scientists on research in the veterinary field is being written. A Japanese pathologist is engaged in recovering and making photomicrographs of the selected example of 8000 slides of tissues from autopsies of humans and animals subjected to BW experiments. General Ishii is writing a treatise embracing his 20 years of experience in all phases of BW.
3. f. Experiments on human beings similar to those conducted by the Ishii BW group have been condemned as war crimes by the International Military Tribunal for the trial of major Nazi war criminals in its decision handed down at Nuremberg on 30 September 1946. This government is at present prosecuting leading German scientists and medical doctors at Nuremberg for offenses which included experiments on human beings which resulted in the suffering and death of most of those experimented on.

The documentary evidence of the U.S. military concealing the case of Japanese war crimes in the name of "national security" and their exclusive interests in BW led me to write my first book on Bioethics in Japan in 1987 [7]. My point was how a *bioethical deficit* and *double standard* were present in dealing with the medical crimes perpetrated by Nazi Germany and Japan.

In these original declassified documents, we recognize clearly the *legal crimes* against the principle of justice. I have analyzed the difference in the cases of medical war crimes during WWII between Nazi Germany and Japan and pointed out the ethical double standard taken by the U.S. for the contradictory outcome of differences. One of the most important reasons for dealing with these issues was "the extremely valuable" outcome of human experiments. And, as I mentioned before, U.S. wanted to keep them without disclosing them to other nations, particularly to the Union of Soviet Socialist Republics (U.S.S.R.) according to the following note by J.B. Cresap dated on 1 August 1947 [8].

> c. For all practical purposes an agreement with Ishii and his associates that information given by them on the Japanese BW program will be retained in intelligence channels is equivalent to an agreement that this government will not prosecute any of those involved in BW activities in which war crimes were committed. Such an understanding would be a great value to the security of the American people because of the information which Ishii and his associates have already furnished and will continue to furnish. However, it should be kept in mind that there is a remote possibility that independent investigation conducted by the Soviets in the Mukden area may have disclosed evidence that American prisoners of war were used for experimental purposes of a BW nature and that they lost their lives as a result of these experiments, and further, that such evidence may be introduced by the Soviet prosecutors in the course of cross-examination of certain of the major Japanese war criminals now on trial at Tokyo, particularly during the cross examination of Umezu commander of the Kwantung Army from 1939 to 1944 of which army the Ishii BW group was a part. In addition, there is a strong possibility that the Soviet prosecutors will, in the course of cross examination of Umezu, introduce evidence of the experiments conducted on human being by the Ishii BW groups, which experiments do not differ greatly from those for which this Government is now prosecuting German scientists and medical doctors at Nuremberg.

The Unit 731 related cases were not introduced in Tokyo International Military Court following the Top Secret Policy of U.S. "With respect to the issue of immunity from war crimes prosecution, immunity was provided in March 1948, but this is not backed up by written documents" [9].

One year after the end of Tokyo International Military Court, from Dec. 2nd till 30th in1949, the U.S.S.R. held the Military Court in Khabarovsk prosecuting 12 persons who were active at Unit 731. The document of this court trial, published in 1950, has received wide attention due to the detailed witness and record on the issues of Japanese BW [10].

In my book, I have also included my experience of participating in the public hearing of Disabled Veterans Hearing at The U.S. Congress on 17th, September 1986, which had been witnessed by Mr. Frank James [11]. He became a U.S. prisoner of War while he battled at Corregidor, Philippines and after his survival of the Bataan Death March, he was transported to China where he was forced to work at one of the Japanese Prisoner of War Camps in Mukden which had close contact with General Ishii's BW connection.

We have to recognize the Imperial Japanese Army's *bioethical deficit* as a War Crime, as evidenced by the existing "TOP SECRET" documents. Biomedical human subject research against Chinese and other victims before and during WWII became a tool for U.S. national security interests which justified giving immunity to Japanese military medical criminals for their crimes. Contrary to the situation in Nazi's medical war criminal cases in Germany, these Japanese medical war criminals survived in safety, with some even becoming leading figures in postwar Japanese society in the medical academic community, publishing businesses, and blood-product businesses. Former army medical officer (Lieutenant Colonel) related to Unit 731, Dr. Ryouichi Naito, for example, was the founder of Midori-Jyuji which was involved in the HIV infected blood scandal [12].

3 Inhuman Experiments: Audio-Visionary Evidence at the Former Unit 731 Head-Quarter in Harbin, China

On the 4th of Sep. 2013, I visited the site of the former headquarters of the Imperial Japanese Army Unit 731 at Ping-fan which is located on the outskirt of Harbin in the North-Eastern area of China. This notorious headquarters has now become the Harbin Unit 731 Museum (The Crime evidence exhibition hall of the 731 Japanese armies invading China) [13].

When I talked with Director Jin Chengmin of this Museum, I mentioned my continuous concern since my boyhood in the 1940s as I have heard some rumors about Chinese Captives for cruel human experimentation by the Japanese Military Medical Corps. I then handed him some of my 731 related papers and explained the research project of the Asian Bioethics Program at the Kennedy Institute of Ethics at Georgetown University [14].

I was guided by him to the second floor where I saw in the showcase a large rectangular photo of the actual use of military poison-gas smoke with some Japanese soldiers wearing gas masks. There were also various kinds of equipment and apparatus to wage poison-gas military operations displayed in the showcases on the same second floor. Coming out of this display room, we were guided to the corridor which both sides were filled with many plates of recording the dates and places written for the incidents of the poison-gas military operation of 1800 cases as confirmed to date. I came across to check the "Secret Documents; Reports of Incidents of Use of Gas by Japanese" at the National Archive in Suitland in 1985 which was declassified on 12th January 1980 and has been confirmed only 65 times between 1937 and 1944. It took more than 70 years to accumulate these facts and findings from local Chinese experts. One of the serious problems in China is the widely spread remaining Japan's military weapon waste such as unused poison gas, stock of explosives, etc. which causes various accidents among the local people even now [15–17].

Director Jin Chengmin showed me the map of the area of Unit 731. He then guided me through the remains of the main building facilities as well as the freeze-experimentation laboratory. He also pointed out the apartment buildings of former Unit 731corp staff members and the 200 acre airport area. In addition, I was able to investigate the remaining facilities of mice-cages and Anthrax Experimental Laboratory.

In each corner of the exhibition room, there are some touch panels to which we can touch and listen about the case in process of cruel human experimentation such as vivisection, frost-bite research, etc. done to captured Chinese, and Russians called *maruta* (logs). These victims were transferred from Harbin Police Bureau in Harbin to Unit 731 headquarters to be used for materials for inhuman biomedical experimentations. Now, we have to recognize the epoch-making historical evidence of Justice was made to know that this former Harbin Police Bureau building became "The Memorial Hall of The Martyrs" in honor of the victims and the struggle of the North-Eastern peoples of China against Japanese hegemony [18].

There are some moves and appeals by the Chinese Government and local people as well as international people for the appeal registration of this Unit 731 headquarters building and its campus to UNESCO's "World Heritage" toward future in following both accepted cases and places such as Auschwitz and Hiroshima.

It is our bioethical responsibility particularly as Japanese to remind ourselves of these inhuman research tragedies that have been committed by the Japanese Military against local people in China including some Russians and possibly U.S. prisoners of war.

The Japanese Military Medical Corps has not only expanded operations to various local areas in China but also throughout the Southeast Asian region. However, the Japanese Government officially responded that it has no resources to confirm the details of these activities during the War [19].

In order to recognize these facts, it is necessary to highlight the importance of bioethical learning from experiential audio-visual evidence of cruel and unacceptable medical crimes through field site visits, which have taught Japanese and other visitors about the unforgettable aspects of human subject research. These kinds of site visits would be a very useful and positive opportunity for Bioethics Education, for people now and future, to learn the historical background of medical crimes in this contemporary world [20].

4 The Tragedy of Hibakusha: Genetic Evidence at Hiroshima and Nagasaki, Japan

In 2009 and 2014, I was appointed by the Department of Energy of U.S. Federal Government as one of the Ethics Review Committee members of Radiation Effects Research Foundation (RERF) in Hiroshima and Nagasaki.

In 1975, RERF was reorganized jointly with U.S. Federal Law and Japanese law as the successor to the Atomic Bomb Casualty Commission (ABCC) and continued the research initiated by ABCC in 1947 which has been the longest history of any ongoing international research program. There are now three cohort studies going on as follows:

1. The Life Span Study investigates diseases that lead to the death of the A-bomb survivors (Hibakusha).
2. The Adult Health Study provides periodic health exams, health counseling, and health care counseling. The tissue samples were collected from those who volunteer when they are tested through this study.
3. The Genetic Studies investigate diseases in the children of Hibakushas and includes research on mortality, birth defects, sex ratio, cancer incidence, chromosome aberrations, biochemical study, molecular genetics, and health examinations.

The introductory part of the Summary Report in 2014 states that "This review was conducted as a follow-up to reviews conducted in June 2009 to evaluate the compliance of Radiation Effects Research Foundation (RERF) Human Subjects Research Protection Program with the requirement of 10 Code of Federal Regulations (CFR) 745 (Department of Energy (DOE) codification of 45 CFR 46 subpart A), DOE Order 443.1B, and Japanese Guidelines (Ethics Guidelines for Human Genome/Gene Analysis Research and the Ethical Guidelines for Epidemiological Research)."

The review was conducted by subject matter experts in their respective fields: both U.S. and Japan. The reviewers were selected by DOE, and three of them also participated in the 2009 DOE review. All reviewers provided individual written feedback to DOE, based on their observations, interviews, and reviews of documents.

The following are some critical parts of my findings and observations presented to the DOE on September 1st, 2014.

1. My first comment was about the importance of the "Informed Consent Form"(ICF) for Adult Health Study (AHS) at RERF. In 2009, ICF's wording was too complicated and technical to read and not clear enough to be understood. After four years, now, a colorful "easy to read" cover page with clear instructions and understandable sentences in Japanese was made for the revised ICF in 2014.
2. My second comment was about the benefit to the participants of AHS. ICF for AHS had some notes for future research about the preservation and use of blood

and urine samples. Currently, the ICF states that the results of any future research will not be reported to individual participants. However, I have pointed out the necessity of research work and related outcomes needing to be made available as the participants have the right to know that their samples were used in scientific research and its outcome.
3. My third comment was about the roles of the Institutional Review Board (IRB) at RERF. The three main functions of IRBs are to review proposals, decide on the proposed research, and educate administrators and staff members on the bioethical principles (both scientific and non-scientific) are needed not only for IRB members but also for all RERF staff members.
4. My fourth comment was the institutional positive trends of the RERF since 2009. From a bioethical perspective, the most important change has been the effort to share research information with the participants and the general public. The annual Open House, lecture series, visitors and outreach lectures on demand, publications such as "RERF Update," and current websites are very impressive and show commitment by RERF especially as I have been very impressive and show commitment by RERF to be more transparent. There should be more efforts in both countries to highlight this very unique cooperative work between the U.S. and Japan at RERF especially as is supported by public funding and taxes.
5. Finally, with the approach of 70th anniversary of Hiroshima and Nagasaki (2015), I noted my hope that more attention would be brought to RERF and that the Health, Welfare and Labour's Annual Report (Kosei-Hakusho to be published in 2015) would include the work of RERF.

During this Ethics Review Comittee meeting for RERF in 2019 at RERF, Hiroshima, I had been talking with my colleagues about the RERF's researcher's behavior in the beginning stage have given rather negative reaction among the Japanese A-Bomb survivors. After three years of our Committee in Hiroshima, Mr. Gosuke Nagahisa, a staff writer for Hiroshima Peace Media Center of The Chugoku Shimbun (Chugoku-Newspaper), a local newspaper in the Hiroshima region, has reported about the RERF/ABCC's ceremony to mark its 70th anniversary on June 19th of 2017.

Mr. Nagahisa has written an article about this ceremony in the Chugoku Shimbun as follows [21]:

> Otsura Niwa, the Chairman of RERF, made a speech at the ceremony and mentioned the fact that the forerunner of RERF had put priority on examining the A-bomb survivors rather than giving them treatment at the time of its establishment. Mr. Niwa expressed words of remorse for his organization's behavior in the past, saying, "We take that fact to heart and feel deeply sorry." He also extended his gratitude to the A-bomb survivors, who have cooperated with the studies conducted by RERF.
>
> RERF was initially established as the Atomic Bomb Casualty Commission by the US government in March 1947. During its early days, the organization's work was primarily for military purposes and people criticize the organization as engaged in the investigation of the A-bomb's and fix effects without providing any treatment to the survivors.

As Mr. Nagahisa briefly mentioned in above articles, the case of radiation epidemiology studies prioritized knowledge production from genetic data, following radioactive effects of Atomic Bombing survivors in Hiroshima and Nagasaki, without providing care or treatment to these Hibakushas. These studies on innocent victims could be seen as a "quite a large scale human experimentation as a result of the military operation for the explosion of Atomic Bombs." Many ABCC research staff members at that time were so eager to collect genetic data from the Hibakushas [22].

5 Concluding Remarks

The establishment of Bioethics, as supra-interdisciplinary studies, can only be achieved by recognizing and reflecting on the inhumane acts in our history. To this end, I have written this chapter by using a narrative approach to analyze three negative case studies of human subject research [23, 24].

In any kind of human subject research, respect for human dignity and personhood should be the fundamental and primary concern. And the most important bioethical principle in performing human subject research should also be based on these principles [25, 26].

In Japan, we must continue to integrate the narratives of our dark past into our future biomedical research to protect human dignity. I also strongly believe that we should not use any data acquired through inhuman situations of exploiting human subjects in the past. Human subject biomedical and scientific research in dehumanized situations must never again be repeated in the name of "National Security" or "Truth-seeking in medical research for present and future generations".

Conflict of Interest There is no conflict of interest to be declared.

References

1. Kimura R. Verbrechen gegen die Menschlichkeit. Die vergessene Geschichte Japans. In: Ethik und Medizin 1947–1997.Was leistet die Kidifiziierung von Ethik?, Troehler U, Reiter-Theil S, editors. p. 161–70, Wallstein; 1997. [Troehler U, Reiter-Theil S, Herych E, editors. English Edition: Kimura R. Crimes against humanity: the forgotten history of Japan. In: Ethics codes in medicine: foundations and achievements of codification since 1947. Ashgate; 1998. p. 119–26.]
2. Tsuneishi K. Unit 731 and the Japanese Imperial Army's Biological Warfare Program. Asia-Pacific J Japan Focus [Internet]. 2005 [cited 2022 Sep 3];3(11). Available from: https://apjjf.org/-Tsuneishi-Keiichi/2194/article.html.
3. Tsuneishi K. 731 Butai zenshi: Ishii Kikan to gungakukansan kyōdōtai. [The Whole History of The Unit 731:The Ishii Organization and Military-Academic, Bureaucratic and Industrial Complex]. Kōbunken; 2022.
4. State-War-Navy Coordinating Subcommittee for the Far East, SFE 188/3, 8 Sept 1947, Appendix A, War Department General and Special Staffs, Folder: SWNCC 351, Record Group 165, NARA.

5. State-War-Navy coordinating subcommittee for the far east, SFE 188/4, 29 Sept 1947, Note. by the Secretary, J.B. CRESAP Comdr., USN, Secretary.
6. State-War-Navy coordination subcommittee for the far east: Appendix "A." 1 Aug 1947, J.B. Cresap.
7. Kimura R. Inochi o kangaeru: baioeshikkusu no susume. (Thinking of life—introduction to bioethics). Tokyo: Nihon Hyōronsha; 1987. p. 266–8.
8. Cresap JB. State-war-navy coordination subcommittee for the far east: J.B. CRESAP Comdr., USN, Secretary, Appendix "B." Discussion, 2. c. 1 Aug 1947.
9. Tsuneishi K. Reasons for the failure to prosecute unit 731 and its significance. In: Tanaka Y, McCormack T, Simpson G, editors. Beyond victor's justice? The Tokyo war crimes trial revisited. Martinus Nijhoff Publishers; 2011. p. 205.
10. Materials on the trial of former serviceman of the Japanese army charged with manufacturing and employing bacteriological weapons. Foreign Moscow: Language Publishing House; 1950.
11. Kimura R. Inochi o kangaeru: baioeshikkusu no susume. (Thinking of life—introduction to bioethics). Tokyo: Nihon Hyōronsha; 1987. p. 267.
12. Matsushita K. Midori Jyuji to 731 Butai (Green Cross and Unit 731). Sanichi Shobo; 1996. p. 9–23.
13. Chengmin J. Collection of pictures on Japanese military germ warfare. Inner Mongolia Culture Press; 2010.
14. Kimura R. Contemporary Japan. In: Reich, W, editor. Encyclopedia of bioethics, rev. ed. New York, London: Macmillan Pub. Co.: Simon & Schuster Macmillan; Prentice Hall International; 1995. p. 1496–505.
15. Reports of incidents of use of gas by Japanese. U.S. National Archives; 1944.
16. Isono O. Nippon-gun Iki Kagakuheiki Higai (The damages caused by the abandonment of chemical-weapons of Japanese Army). In: Yoshinaka T, editor. Unit 731 and the University. Kyoto University Academic Press; 2022. p. 419–47.
17. Nishiyama K. Dokugasumonndai to Gunjin Igaku Bunkakai (The problem of poison gas and military medicine sub-committee). In: The Japanese Association of Medical Sciences. 100 years anniversary issue. 2022; p. 195–9.
18. Penney M. Unit 731 and Preserving the History of Wartime Medical Atrocities. Asia Pacific J Japan Focus [Internet]. 2012;10(54, No. 18). Available from: https://apjjf.org/2012/53/18.html.
19. Nishiyama K, editor. Sensō to Igaku (War and medical science). Bunrikaku; 2014. p. 135–6.
20. Kimura R. 'Dark Tourism' and bioethics education. Japan Mission J. 2019;73(3):200–5. Oriens Institute for Religious Research.
21. Nagahisa G. 'RERF Chairman expresses words of remorse and appreciation to A-bomb survivors at 70th anniversary ceremony [Internet]'. Hiroshima Peace Media Center. 2017 [cited 2022 Sep 3]. Available from: https://www.hiroshimapeacemedia.jp/?p=77599.
22. Ham P. Hiroshima Nagasaki: the real story of the atomic bombings and their aftermath. New York: Thomas Dunne books St. Martin's Press; 2014. p. 435–7.
23. Kimura R. Bioethik als metainterdisziplinaere Disziplin In: Medizin Mensch Gesellschaft, Band 11 Heft 4, Dezember 1986 (IV), Stuttgart: Ferdinand Enke Verlag.
24. Kimura R. Asian perspectives: experimentation on human subjects in Japan-bioethical perspectives in a cultural context. In: Bankowski Z, Levine RJ, editors. Ethics and research on human subjects international guidelines: Proceedings of the XXVIth CIOMS conference. 1st edn. Geneva: Council for International Organizations of Medical Sciences (CIOMS); 1993. p. 181–7. and Kimura. R. 'Bioethics in Japan'. In: Bankowski Z, et al. Poverty, vulnerability, and the value of human life: a global agenda for bioethics, In: XXVIIIth CIOMS conference, Geneva: Council for International Organizations of Medical Sciences. p. 72, 222–5.
25. Kimura R. Japan, Bioethics In, p. 1760. In: Bruce J, editor, chief. Bioethics. Macmillan reference USA. 4th edn, vol. 4. Gale Cengage Learning; 2014, p. 1757–66.
26. Baker R. Transcultural medical ethics and human rights. In: Troehler U, Reiter-Theil S, Herych E, editors. Ethics Codes in Medicine. Foundations and achievements of codification since 1947. Ashgate; 1998. p. 312–31.

From Nuremberg to Helsinki: Historicizing the Codification of Post-War Research Ethics

Ulf Schmidt

Abstract In 1947, the Nuremberg Doctors' trial promulgated the ten-point set of principles for the conduct of human experiments known as the Nuremberg Code; in 1964, the World Medical Association (WMA) adopted the first version of its Declaration of Helsinki (DoH). The continued relevance of both documents offers ample reason to reflect on the development of post-war research ethics, particularly in light of recent COVID-19 pandemic experience and military conflict. Historicizing the codification and implementation of modern research ethics provides valuable context for a question of growing contemporary importance, namely how best to protect human participants and vulnerable communities in an increasingly complex global research environment. Efforts to safeguard human participants in clinical trials have intensified since the first version of the WMA's Declaration and are now codified in many national and international laws and regulations. However, most researchers lack a comprehensive understanding of how the DoH originated, changed, and functions in today's world. Over half a century, this "living document" has been criticized and revised many times, but its standing as of one the most universally accepted ethical codes remains largely undisputed. At the same time, it is far from certain whether our existing global framework provides sufficient guidance for tomorrow's research practices.

This paper is a revised and expanded version of a talk presented to the joint session of the Bioethics Societies of Japan and Brazil, June 4, 2021. Selected parts of the paper have previously been published in [1]. Research for this paper, which is part of the project "Taming the European Leviathan: The Legacy of Post-War Medicine and the Common Good," has been supported by the European Research Council under the European Union's Horizon 2020 research and innovation program (grant agreement no. 854503).

U. Schmidt (✉)
Faculty of Humanities, Department of History, University of Hamburg, Hamburg, Germany
e-mail: ulf.schmidt@uni-hamburg.de

C. Kurihara et al. (eds.), *Ethical Innovation for Global Health*,
https://doi.org/10.1007/978-981-99-6163-4_9

Keywords Research ethics · Nuremberg code · Declaration of Helsinki · Human experiments · World Medical Association · International Committee of the Red Cross · World Health Organization

1 Introduction

In June 1964, the World Medical Association (WMA) adopted the first version of its Declaration of Helsinki (DoH) [2], one of the most important landmarks in the history of biomedical research ethics; this had been preceded, in 1947, by the judgment handed down at the Nuremberg Doctors' trial, which promulgated the ten-point set of principles for the conduct of human experiments known as the Nuremberg Code [3–8]. The historical and contemporary relevance of both documents ensures that it is right to reflect on the development of post-war research ethics, particularly in light of recent pandemic experience and military conflict in Europe and elsewhere.

The announcement by the drug companies Pfizer and BioNTech of their development of the first fully tested COVID-19 vaccine in December, 2021, was hailed as major scientific milestone which, according to experts, would challenge "our whole paradigm of what is possible in vaccine development" [9]. The international scientific community received the news with great enthusiasm. Developing vaccines and novel treatments for diseases such as malaria and tuberculosis was apparently now only a matter of time, as long as the political will was there to act in a "true global emergency" and sufficient resources were made available to medical science [9].

Others judged the pandemic response in less optimistic terms. In May, 2021, a group of researchers writing in *Lancet Global Health* concluded that the COVID-19 pandemic had fundamentally changed clinical research in global health, but not for the better. Their findings suggested that the medical community's response to the COVID-19 outbreak had been "inefficient and wasteful" [10]. In an attempt to develop vaccines and therapies quickly, it had not only cut corners by expediting ethical review processes but had also conducted an "overwhelmingly large number" of human trials of "questionable methodological quality." Furthermore, most COVID-19 clinical trials had been too small-scale to provide conclusive data [10]. Indeed, researchers of the Pan American Health Organization (PAHO), while noting that multiple COVID-19 trials had been carried out in low- and middle-income countries, such as in Latin America and the Caribbean, discovered a trend toward "small repetitive non-rigorous studies that duplicate efforts and drain limited resources without meaningful conclusions on the safety and efficacy of the interventions being tested" [11]. It appears that scientists around the globe, including those unfamiliar with human trials, had rushed to work in vaccine research and development. "Many people who aren't used to doing clinical research started investigating treatments. And many of the trials were never finished," one the authors of the *Lancet* study noted in an interview with National Public Radio [12]. In one of the most severe public health emergencies the world had seen, scientists and drug companies had shown an inability, and perhaps unwillingness, to coordinate clinical research, and had instead controlled access to data and sought commercial rewards. It was a damning verdict that mirrored an earlier assessment delivered by UN Secretary-General António Guterres in September, 2020: the COVID-19 pandemic had been a "clear test of international cooperation," which the world had failed because of a "lack of global preparedness, cooperation, unity and solidarity" [13].

For scholars engaged with the history of research ethics, these findings are not particularly surprising. Viewed within a global context, research on humans has at times taken place against a backdrop of economic crisis, struggling health and welfare systems, and regulatory uncertainty, with impoverished and vulnerable populations on the one hand, and a thriving pharmaceutical industry with its increased demand for clinical trials to test new medicines on the other. Profit margins on the global pharmaceutical market are astronomical, with the revenue generated increasing from around US$390 billion in 2001 to over US$1.1 trillion in 2016, which was almost the nominal GDP of Russia in that year [14]. The top three international pharmaceutical companies are Pfizer (US), Novartis, and Roche (both Swiss). In 2017, Pfizer sold prescription medicines worth US$45.3 billion and invested US$7.6 in Research and Development (R&D). Novartis made sales of US$41.8 billion and invested US$7.82 billion in R&D, and Roche sold US$41.7 billion and invested US$9.1 billion in R&D [15].

Moreover, for companies such as Pfizer, which teamed up with BioNTech (a German company) for vaccine development, the COVID-19 pandemic boosted annual revenues from US$41.7 billion in 2020 to US$81.3 billion in 2021 [16]. This amounts to reported operational growth of 92%, thus almost doubling, in a single year, the company's revenues from the sales of its COVID-19 vaccine (Comirnaty) and oral anti-viral treatment (Paxlovid). Pfizer's projected revenues for 2022 have been reported to be in the region of between US$98 billion and US$102 billion [16]. Global health emergencies, in other words, not only exacerbate existing health inequalities in some societies, but can offer major drug companies a chance to tap into unprecedented government resources (US$10 billion as part of the US Operation Warp Speed vaccine program alone [9]), conduct large numbers of sometimes questionable clinical trials in low- and middle-income countries, and, above all, expand their global market share.

These are staggering figures. It might be assumed that unambiguous, uniform regulation that ensures the safety of participants and continued public confidence in human research would be in everyone's interest. However, little could be further from the truth. As Tony Judt has poignantly pointed out in *Ill Fares the Land* (2010), we are living in a society that is defined by relentless consumerism and the pursuit of material interests [17]. Ideas and ideologies no longer seem to matter, and justice, fairness, truth, and the common good have ceased to be important. Our relativist, post-modern, post-Cold War world has become "post-ethical."

Historicizing the codification and implementation of research ethics in the twentieth and twenty-first centuries addresses a subject of great contemporary relevance. In an increasingly complex global research environment driven by industry and government funding, and in which decades-old geopolitical certainties can no longer be relied upon, how can we best protect human participants and vulnerable communities? ("Vulnerable" refers to people whose backgrounds could make them more likely to be exploited: the disabled, children, people in poverty/dependent relationships, etc. However, the "non-vulnerable" are obviously also entitled to the same level of protection as "vulnerable" populations. There is no difference in ethics or in law. But ethicists and lawmakers should pay particular attention to the vulnerable because they have in fact historically been exploited more often than the non-vulnerable. For a discussion about the over- or under-utilization of the concept of "vulnerable populations" see [18–21].) Efforts to safeguard human subjects in clinical trials have intensified since the first version of the Declaration of Helsinki in 1964 and are now codified in many national and international laws and regulations. However, most researchers lack a comprehensive understanding of how the DoH originated, changed, and functions in today's world. Over the course of half a century, this "living document" has been criticized, and revised, many times, but its standing as of one the most globally accepted ethical codes remains largely undisputed. At the same time, it is far from certain whether our existing global framework provides sufficient guidance for tomorrow's research practices. Experts are just beginning to realize the enormous implications of—and demands upon—our current system of research governance. Today, an expanded recension of the Declaration, updated in 2013 [2], attempts to offer ethical leadership in an increasingly confusing research environment. At times, the process of change has been as disconcerting for those advocating freedom for science as it has been for those trying to protect vulnerable communities in both developed and developing nations.

Although until recently our understanding of the history that shaped the creation of the DoH had been limited to selected policy documents from the 1950s, newly discovered material that complements existing scholarship allows us to see more clearly the contours of certain policy debates that drove, held back, and at times accelerated the drafting and re-drafting process. (For some of the burgeoning literature on the DoH see [5, 6, 8, 22–36].) When, why, and how did the revision process of the WMA's 1949 International Code of Medical Ethics [37] begin? Which key players argued for revision and change? To what extent were specific groups and vested medical interests involved in the drafting process? What factors persuaded the WMA to produce a jointly formulated Code of Medical Ethics in Wartime with other stakeholders in the mid-1950s [38]? To what extent did this ethics code impact on the creation of the DoH? What role did international organizations play in shaping debates and guidelines on human research ethics, in collaboration with (or sometimes separately from) the WMA? While our knowledge of the workings of the WMA's medical ethics committee has previously been limited, we are now—using a near-complete run of minutes and reports in the WMA's archives at Ferney-Voltaire (France)—able to reconstruct the internal debates about research ethics and human experimentation that took place in the first twenty-five years of the Association's operation. These records, in conjunction with files relating to codes of ethics, research practices, and armed conflict, offer additional contextual detail which allows us to substantially revise our understanding of the genesis of the DoH.

An important factor behind the proclamation of the DoH in 1964 was a specific tension between international health and legal organizations such as the World

Health Organization (WHO) and the French National Academy of Medicine and the Medico-Juridical Commission of Monaco (MJC), on the one hand, and the WMA on the other. The former wished to draft and, ideally, implement international medical law in the post-war period, while the WMA was the body tasked with protecting the interests of the medical profession. These interests often went hand in hand with those of pharmaceutical and other companies sponsoring the Association in its first decade (including some tobacco companies), which is why, from the mid-1950s to the mid-1960s, the WMA saw itself as "the only protection for [the pharmaceutical] industry" [39, 40]. The process of defining the limits to Helsinki, the conceptual boundary between what was deemed to be ethically permissible and what was deemed legally permissible in human experiments, is thrown into stark relief by looking at the political controversies and negotiated compromises on the subject of medical ethics. This chapter highlights the determination with which the WMA fought for its independent status against international organizations, political interference, and even the rule of international medical law. Understanding some of the debates and negotiations before and after the creation of the Declaration allows us to see more clearly the long-term limits and historical paradoxes of this all-important medical ethics code, as well as its far-reaching success in protecting patients and trial participants in different communities around the world.

2 Controlling the Codification of Medical Ethics

Ever since the emergence of experimental medicine during the Renaissance, there have been attempts to define and enforce the boundaries of ethical science. In the mid-nineteenth century, the French physician Claude Bernard made it plain that medical morality dictated "never performing on man an experiment which might be harmful to him to any extent, even though the result might be highly advantageous to science, i.e., to the health of others" [41]. The issue of consent in experimental, non-therapeutic research played a considerable role in medical science during the long nineteenth century [42–45]. Not all experiments on humans, whether therapeutic or non-therapeutic, required the consent of the subject, but most scientists accepted the need for volunteers, particularly when there was a possibility of harm. From 1830, English law was understood to require that a physician had to obtain the informed consent of the research subject, even if the experiment was for therapeutic purposes. Doctors failing to do so risked litigation [46]. Earlier, in 1767, an English court had ruled in *Slater v. Baker and Stapleton* that the defendants would be held liable because they had operated without the "patient's consent" and without telling the patient "what is about to be done to him" [47]. (This case is also cited in [48–51].) In 1900, the Prussian authorities ruled that human research was not permitted "if the human subject was a minor or not competent for other reasons," or had not given unambiguous and informed consent [44]. In the 1930s, the UK Medical Research Council (MRC) advised scientists to make sure that any experiment had been performed with the "full consent of the patient, given after proper appreciation of the risks involved, and that it had been performed with all due care and skill" [52] ("TNA" refers to the UK National Archives in Richmond, Greater London). German regulations from 1931 likewise stated that "experimentation shall be prohibited in all cases where consent has not been given" [53]. (These regulations are also discussed in [54].) Scientists and the authorities generally accepted that human research had to be ethical in order to be permissible long before the Nuremberg Code. This is not surprising, given that one of the universal principles of medical ethics is that the physician–scientist do no harm to either a patient or a research participant. Those investigators who wanted to search for new knowledge which would not necessarily benefit the participant were required to inform the subject about the risks involved and obtain the subject's consent. (On the other hand, if the research is aimed at benefitting the patient, the degree to which consent would have to be obtained, including the risks to be explained, would not necessarily have to be same as when the research has no measurable benefit to the participant. Nevertheless, those who are healthy, and subject themselves to experimental science for the benefit of others, should enjoy the greatest degree of protection, both morally and in law.)

The post-war proclamation of human and civil rights through the UN's Universal Declaration of Human Rights (1948) [55] and other international legal instruments signaled the intention of legal experts and advocacy groups to construct a broad legal and moral framework that would provide individuals with greater protection and enforceable rights against state and non-state actors. In response to large-scale human rights violations during the Third Reich, Article 1 of the West German Basic Law of 1949 [56] stipulated that the "dignity of man shall be inviolable"—*Die Würde des Menschen ist unantastbar*—and that this has to be respected and protected by "all state authority." In addition, Article 2 confirmed that every person has the "right to life and physical integrity." The Basic Law acknowledged "inviolable and inalienable human rights"—so-called *Grundrechte*—as the foundation of every community, which also applied to the field of medicine and biomedical research. Whereas the Universal Declaration of Human Rights confirmed in Article 5 that "no one" shall be "subjected to torture or to cruel, inhuman or degrading treatment or punishment," the UN's International Covenant on Civil and Political Rights of 1966 stipulated, more explicitly, that "no one shall be subjected without his free consent to medical or scientific experimentation" [57]. Despite such attempts to enshrine the protection of patients and participants in international law and, in some cases, national constitutions, encouraging post-war medical professionals to reflect on the existing and emerging risks involved in medical science nonetheless required great effort on the part of experts and organizations alike [58].

At the Nuremberg Doctors' trial, which opened in December, 1946, 23 doctors and officials were charged with war crimes and crimes against humanity for their involvement in unethical and often lethal experiments. As part of their judgment, the Nuremberg judges issued a set of ten principles for permissible research on human participants that came to be known as the Nuremberg Code [3]. Although many Anglo-American scientists initially regarded the Code, in the words of medical ethicist Jay Katz, as a "good code for barbarians but an unnecessary code for ordinary physician–scientists" [59], and raised concerns about the Code's broader professional and legal standing, the principles nonetheless served as important guidelines for ethical conduct in medical science in times of both war and peace, as the debates during the drafting process of the DoH in the late 1950s and early 1960s demonstrate.

Following the post-war condemnation of Nazi medical war crimes, the WMA not only reaffirmed its broad support for Hippocratic medical ideals and values in the Declaration of Geneva in 1948 [60], but also issued an International Code of Medical Ethics a year later that repeated basic ethical principles relating to confidentiality, beneficence, and non-maleficence (to do no harm) [37, 61, 62]. In 1951, WMA delegates rejected a proposal to turn the International Code into a more authoritative document through the inclusion of principles dealing with human experimentation. They were adamant that the International Code needed to remain a "broad statement of ethical principles" that could be reinforced, if necessary, by national laws and regulations. Under no circumstances were the Code's "brevity and simplicity" to be altered [63, 64]. To be acceptable to doctors around the world, the argument—then as now—was apparently that the document be brief and simple, otherwise it would either not be read or it would be rejected.

However, medical war crimes trials held after 1945 in the Allied zones of occupation contributed to a climate in which public debate about the role of research ethics became increasingly inevitable. In the wake of the revelations at post-war trials, including those held at Dresden, Hamburg, and Nuremberg, of ever more atrocities, national medical organizations, funding bodies, and other non-state actors rushed to disassociate themselves from Nazi medicine. The harsh sentencing of German scientists during the Struthof Medical Trials in Metz and Lyon in the early 1950s, for instance, prompted the French National Academy of Medicine and the MJC to take a firm stand on medical misconduct in the field of clinical research [65, 66]. However, in trying to salvage what was left of the reputation and moral integrity of the medical profession through the creation of clearer rules and regulations, interest groups such as the MJC found themselves on a collision course with the WMA.

In 1952, the WMA responded to reports that the MJC was planning to promulgate an international code of medical ethics as a means to inform the future creation of international medical laws. The battlelines were clearly drawn between the WMA, which, as the body representing the interests of the medical community, was fiercely opposed to greater state and legal regulation, and the MJC, whose membership represented international medical jurisprudence. During its 6th General Assembly, held that year at Athens, the WMA adopted a resolution to the effect that if others, including lawyers from MJC, persisted in drafting such a document, "this code will not be accepted by the medical profession of the world" [67]. The resolution was transmitted to the MJC, the WHO, the International Committee of the Red Cross (ICRC), the International Committee on Military Medicine and Pharmacy (ICMMP), and the International Labour Organization. WMA concerns also related to the role played by the ICRC in the field of international medical law, which raised the specter of the ICRC codifying international medical ethics standards through its various conventions [68]. A couple of years later, in another rear-guard action to prevent medical lawyers, legislators, and physicians—who were attending the first International Congress on Medical Ethics, funded by the French doctors' professional body, the Ordre des Médecins—from issuing internationally binding medical ethics guidelines, the WMA not only refused to send a representative, but reiterated its position that it was the "function of the medical doctors of the world to formulate any code of international medical law" [69].

3 The Road to Helsinki

The strategy temporarily averted external encroachment on what the WMA regarded as its area of competence, but, by insisting that it was the only legitimate organization with the moral authority and expertise to draft such a document, it found itself under pressure to produce a more authoritative text. Moreover, in continental Europe, the painful memory of the Holocaust showed no signs of waning. One by one, national medical associations began to review their guidelines about permissible experiments on humans. In 1952 the French National Academy of Medicine stipulated that experiments to produce new knowledge (rather than therapies) had to be conducted on "informed volunteers free to accept or reject" the intervention [70]; a year later, the Royal Netherlands Medical Association called upon the WMA to engage with the subject of human experimentation [71]. WMA delegates such as Lambert A. Hulst, a former member of the Dutch resistance movement and subsequent president of the WMA, began to draw attention to the risks involved in clinical trials. As brother-in-law of Leo Alexander, one of the key medical experts in the Nuremberg Doctors' trial, Hulst had gained first-hand insight into the central issues concerning the court in relation to permissible human experimentation [72]. Speaking for a majority of WMA delegates in the early 1950s, he condemned as "criminal acts" the Nazi physicians' use of test subjects under compulsion. Experiments had to be voluntary, he noted, and were permissible only if the participant was informed about his or her right to "consent or refuse." Hulst called upon the WMA to define more clearly the boundaries within which research could legitimately be performed [73, 74]. The Société de Médecine de Paris likewise condemned the use of test subjects under compulsion [73]. The debate received additional impetus after criticism was made of editors of scientific journals for publishing the findings of human experiments prematurely and without due regard for ethical considerations [75].

In 1954, the WMA's medical ethics committee, headed by the French physician and vice-president of the WMA council Paul Cibrie, agreed to "establish a set of rules which would control experimentation on humans" [75, 76]. While refraining from condemning "scientific crimes" such as those conducted by Nazi Germany, the committee focused on "genuine human experimentation" within a four-pronged framework relating to (a) the scientific and ethical qualifications of the experimentation, (b) caution and discretion in the publication of early results, (c) a distinction between experiments applied to healthy and unhealthy subjects, and (d) the requirement that research participants "must understand fully the risks involved" [77, 78]. In September, the committee added a fifth principle by introducing a conceptual distinction between "healthy subjects" and "sick subjects" [79, 80].

After Cibrie's presentation of the committee's initial report to the WMA council, meeting prior to the 8th General Assembly in Rome in October, 1954, controversy ensued about the need to inform healthy participants in experiments on humans. Susan Lederer's account of the meeting gives the impression that American scientists—foremost among them Austin Smith, editor of the *Journal of the American Medical Association* and a colleague of Cibrie's on the committee—initiated the objections against the requirement to fully inform healthy human subjects, since this would "seriously undermine research in the United States" [80]. Smith's reservations were grounded in post-war changes in the conduct and methodology of clinical trials to test new drugs and therapies. As far as Anglo-American and Western research practices were concerned, the use of placebos and control groups had become the methodological "gold standard" for scientists evaluating the efficacy and side-effects of pharmaceutical agents. The above narrative lends weight to Lederer's argument that American scientists' objections to the development of more rigorous ethical standards in the 1950s make them, by implication, partly responsible for a delayed and weakened DoH [24].

However, while it is not the intention of this chapter to reduce perception of the "American influence" on the origins of the DoH without justification, a close reading of the minutes of the October, 1954, meeting [81] reveals that the initial impetus for the protest came from European rather than American delegates, in particular the Danish physician Otto Rasmussen. He argued, without explaining why, that explicit reference to a "control group" in the proposed principle relating to healthy participants "could not be accepted because it would make scientific experimentation impossible," perhaps because "fully informing" healthy participants in the placebo arm of a study might have been seen at the time to negate the purpose of a control group. Rasmussen proposed that the wording about control groups be deleted, since the point about informing subjects as to the nature and risk of experiments was covered elsewhere in the draft rules on human experimentation. British physician and editor of the *British Medical Journal* (*BMJ*) Hugh Clegg agreed, arguing that, if the WMA adopted the relevant section, it would "abolish all properly controlled human experimentation." Such a reference would, in his view, expose the WMA to international ridicule in the "light of scientific methods of procedure." Smith joined the discussion only after his European colleagues had voiced their objections. Before moving to have the wording about control groups deleted from the document, he made it plain that it would be "impossible to follow the procedure [in respect of control groups]" in the USA. Hulst attempted to defuse the controversy by characterizing it as a linguistic "misunderstanding" resulting from the intention to inform healthy individuals as to whether they had received an experimental vaccine or not. According to Hulst, it was not appropriate to leave people under the impression that they had been vaccinated when there was a 50% chance of their having been placed in the control group and therefore of having received the placebo. However, Clegg, who had discussed the subject with Edward Grzegorzewski from the WHO, argued that the specific issue of vaccination rarely applied to adults, and since children could not give legally valid consent, he proposed the deletion of any reference to vaccination as well. The debate was not only a setback for Hulst, as one of the architects of the document, but also reflected the extent to which research scientists protected their professional interests through the amendment or deletion of specific articles that could be construed as limiting their scope of action, thus foreshadowing future negotiations about the language used to codify human experiments.

Another contentious issue related to the problem of involving prisoners of war in human experiments. Claude Pilloud from the ICRC argued that scientists should not conduct experiments on this group "under any circumstances" [82] and reminded delegates that Article 13 of the ICRC's 1949 Geneva Convention prohibited human experiments on prisoners of war [83, 84]. Other delegates concurred on the grounds that, as a "captive" population, prisoners could be subject to coercion or blackmail and were thus unable to provide valid informed consent [82, 84, 85]. Given the contested nature of the field—as differences between deontological traditions and research cultures came to the fore—it was one of the few areas of temporary agreement. While the Belgian delegate, Dr. P. Lifrange, argued along the lines of the French National Academy of Medicine by stating that trial subjects would have to be "fully informed and free to accept or refuse" to participate, the Indian delegate Dr. S.C. Sen felt that any principles relating to human experiments "must be made elastic" since there were times when "the participant in the experiment either cannot or should not be told" [84]. An amendment to the draft rules on human experimentation, proposed by Dr. F.J.L. Blasingame from the USA, meanwhile, highlighted the need to inform participants of the nature of and reason for the proposed experiment, as well as of any attendant risks [84]. Perhaps surprisingly, given Western researchers' reservations about the Nuremberg Code, the wording of the Blasingame amendment reflected key elements of the Code's informed consent principle. It appears that post-war Anglo-American scientists, including those working within highly secretive military environments such as Porton Down in the United Kingdom, rejected or adopted elements of the Code's principles and ethical provisions according to their immediate needs [86].

Adopted by the WMA in October, 1954, the "Principles for Those in Research and Human Experimentation" stressed, among other "scientific and moral" dimensions of experimentation, the conditions under which healthy and unhealthy subjects could take part, and the necessity for full disclosure of the nature, purpose, and risks of the experiment [87–90]. However, rather than providing further clarification about human experimentation, the debate around the 1954 Principles not only highlighted significant scientific, cultural, and religious differences among WMA delegates, but also threw into relief the highly complex field of post-war research ethics. Indeed, the WMA's activities coincided with the detailed considerations of the conditions for permissible experiments on humans being undertaken by national medical organizations and funding bodies, including the MRC in the United Kingdom and the newly founded National Institutes of Health (NIH) in the USA, which resulted in a series of research guidelines; yet, while informing discussions on clinical trials, none of these documents managed to sufficiently reassure the international health community [91–93].

4 Medical Ethics in the Time of War

While protecting the key areas of self-proclaimed competence in the field of civilian research ethics, the WMA simultaneously engaged in international diplomacy to strengthen the rights of patients and human experimentation participants in times of armed conflict. By liaising with non-governmental and humanitarian organizations—including the ICRC, the ICMMP, and the WHO—which were seeking to develop universal standards for the role of medicine in wartime, the WMA attempted not only to raise its international profile, but also to extend its influence to physician–scientists and research facilities on the other side of the ideological divide, i.e., in Central and Eastern Europe.

Following the end of World War II, the WMA and other international organizations had made it one of their key objectives to define the role of doctors and medical science in time of war. From 1948 onward, the WMA participated in collective attempts by the ICRC and the WHO to revise the Geneva Convention of 1929 [94] in order to ensure that medical personnel would enjoy the same protection as prisoners of war. In July 1949, the WMA's Secretary-General, Louis H. Bauer, summarized the existing challenges of the subject in a confidential report on "Medicine in the Time of War" [95]. Bauer's report was based on one by Dr. Dag Knutson (a Swede) on a paper about the creation of a new code of international medical law ("Pour créer un Droit International Médical") by General Jules Voncken, general secretary of the ICMMP and former Commanding Medical Officer of the Belgian army [96]. Voncken considered the Conventions of Geneva and The Hague to be "inadequate" [95, 97, 98] and, in addition to calling upon the medical profession to develop shared medical ethics principles or "a common formula" to safeguard some of the "fundamental rights of man," Voncken addressed the complex relationship between medicine and international law, particularly in the field of biological warfare. As a military medical expert with an international agenda, who had studied the proceedings of the Nuremberg trials, he rejected the interference of state agencies in medical and humanitarian affairs, and suggested that the idea of a "self-contained, self-indulgent, supreme State must be done away with." He was part of a growing chorus of post-war medical professionals who, rather than reflecting on the role and risk of modern medical science, blamed Nazi medical atrocities on the authoritarian nature of the state. To prevent future medical ethics violations, he proposed a new code of medical ethics consisting of eight titles and 31 articles, which was to cover the standing of the doctor, the duties of the state in relation to the medical profession, and the procedures to be followed in case of "abuse and infractions of the code." A new international body made up of lawyers, military and civilian doctors, and military representatives was to be tasked with upholding medical ethics in time of war and peace [95, 96].

In reflecting on Voncken's paper, the WMA acknowledged that the subject of medicine in time of war raised issues of a "far-reaching nature." These issues required collaboration between medical and legal professionals and experts in international law, if not a "standing medico-legal committee" [95]. Bauer's report reflected the realization within the medical community that post-war medical science was likely to be shaped by greater adherence to laws and regulations. One way of limiting the expected interference of national governments and the judiciary in the field of medicine, WMA officials believed, was by expanding the WMA's cooperation with international organizations and develop an internationally recognized medical ethics code.

However, the WMA was concerned about the role played by the ICRC in the field of international medical law, which was reminiscent of the ICRC's codification of international medical ethics standards through its 1949 Geneva Convention [83], ratified by over twenty countries. This explains why, in 1953, the WMA felt that the time had come to "confront" the ICRC with a medical ethics code of its own [68]. At its General Assembly, the WMA also adopted a resolution requesting that the ICRC "extend the protection of the Red Cross emblem" to include civilian doctors attending to the sick and wounded in time of war [99]. Nevertheless, such inter-agency rivalries paled in comparison to events at the United Nations, which placed the WMA in a vulnerable position.

By the mid-1950s, disagreements with various UN agencies and state governments which the WMA considered to be "hostile" toward the medical professions, especially those in Central Eastern Europe, added to a sense of urgency to join forces with the ICRC and the ICMMP in order to ensure that an International Code of Medical Law be based upon the WMA's International Code of Medical Ethics. The situation had arisen after the Secretary-General of the UN instructed the executive committee of the WHO to discontinue a study about international medical law. The matter was placed before an administrative committee made up of the directors-general of UN governmental agencies, and the WHO was invited only in an advisory capacity. This, according to the WMA, produced a "serious problem as medical representation in [*sic*] UN is extremely limited." Moreover, while the WHO was considered to be "friendly" toward WMA objectives, it was unclear how it would develop once the "Eastern countries" were readmitted, which could be as early as 1956, and "in all event much too soon" [100].

WMA officials made it clear in their 1955 report that an "International Code of Medical Law must be based on the [WMA's 1949 International] Code of Medical Ethics." Countries advocating "socialized medicine" in Eastern and Western Europe, the WMA argued, did not necessarily adhere to the principle of the "defense of the individual." In some countries, doctors acted in ways which were irreconcilable with the expected conduct of the medical profession. Indeed, some doctors were considered "government collaborators" who did not "abide by the ideals" outlined in the WMA's medical ethics code. Worse still, in many countries doctors were apparently "willing to sacrifice the individual human." In summing up, the WMA acknowledged that it had a major fight on its hands if it was to protect the "right of the individual human" [100]. WMA officials, adopting Churchillian war time language, issued a call to arms in order to defend the principles underpinning its medical ethics code:

> There will be blood, sweat, and tears in the days ahead if the medical profession chooses to accept this challenge and the medical associations should be alerted and informed of the coming danger and strife. The resolution [of the WMA's International Liaison Committee on the Status of Civilian Doctors in War Time] adopted [by the WMA] in Rome cannot be presented to the United Nations [100].

In order to pre-empt the creation of binding international medical law that would impact on the permissible conduct of the medical profession, the four organizations—the WMA, the ICRC, the ICMMP, and the WHO—jointly agreed a new code of medical ethics in wartime [38]. Adopted by the WMA's 10th General Assembly in Havana in October, 1956, and revised a year later in Istanbul, the "Regulations in Times of Armed Conflict and Other Situations of Violence" defined medical ethics in time of war as "identical with medical ethics in time of peace" [101, 102]. The same applied to the rules governing human experimentation. Doctors were "strictly forbidden … under all circumstances" to conduct human trials on prisoners and citizens of occupied territories, regardless of instructions issued by the authorities [101]. The 1956 Code not only engaged with human experimentation in wartime, but also accelerated the formulation of the Declaration of Helsinki.

5 A Special Code of Ethics

Attempts by the WMA to separate internal policy debates from the broader Cold War context proved futile in the mid-1950s. The prospect of a renewed conflict in Europe added a sense of urgency to the need to offer guidance to physicians in time of war, and to use the WMA's political leverage to keep the channels of scientific knowledge exchange open across different ideological and economic systems. In November, 1956, responding to rising geopolitical tensions after the brutal suppression of the Hungarian uprising by the Soviet military, the WMA characterized any attempts to "restrict free scientific interchange," whether aimed at doctors, publications, conventions, or conferences, as "contrary to scientific development and progress" [103]. It was not the first time the WMA would intervene in politics. As early as 1950, the WMA council had criticized Nazi Germany, and more recently the Soviet Union, of undermining the freedom of science by using specialized fields such as genetics, anthropology, and physiology for "political ends," and called upon its 500,000 members to promote the "free pursuit of scientific truth" [104]. However, some national associations feared that such an outspoken approach could prove counter-productive for colleagues in Central Eastern Europe. Moreover, with its headquarters located in New York, the Association risked being seen as reflecting the interests of the US medical community. Research ethics in the 1950s had become intensely political.

At a more philosophical level, WMA ethicists soon tied themselves in conceptual knots. By the late 1950s, the WMA medical ethics committee attempted, but largely failed, to define what constituted a human experiment. Answering the question of whether human experiments could be considered a part of medical activity turned out to be similarly complicated. If an experiment could be considered a medical act, this placed the WMA and its national associations in a position of responsibility, and possible legal liability, in terms of providing physician–scientists with clear guidance. On the other hand, if an experiment was not considered to be a medical act, the question arose as to whether doctors carrying out human experiments continued to be governed by the rules of medical ethics. If they were not, this conflicted with the widely held belief in the medical community that the actions of doctors were to be governed by the "ethical code at all times" [105]. In or around October, 1959, by way of a compromise to resolve some of the apparent contradictions, and after liaising with national and international legal experts, the committee formulated a self-imposed pledge regarding human experiments:

> The objective of human experimentation is to obtain information for application to other patients and [it] is not undertaken for the welfare of the human submitting to the experiments. Therefore, human experimentation is not governed by medical ethics but by legal enactments. However, since human experimentation is fundamental in the advancement of medical science, principles must be established which will determine the conditions that must be observed in making the experiment acceptable. These principles would form a special code of ethics applicable to human experimentation [106].

If there was ever a moment when the WMA expressed a firm commitment to the creation of "a special code of ethics" regarding human experiments, the above statement can be seen as such a moment and as one of the first concrete steps en route to the DoH. Arguing that human experiments were not a medical act and therefore beyond the remit of the WMA, while at the same time claiming authority and superior competence in the drafting of medical ethics codes, placed the WMA in an untenable position. Having engaged with the subject for many years without tangible results, the only viable solution left was the decision, probably taken reluctantly, to draft a new code (or codes) of ethics to adequately address the subject, and thus allow the WMA to regain the initiative in the debate.

Preparations for the new ethics code involved an orchestrated publicity campaign about human experiments. In March, 1960, the *World Medical Journal* published reports by experts from across the world, including France, India, Japan, Thailand, and Uruguay, on their "codes of ethics governing human experimentation." These were flanked by a series of carefully curated opinion pieces on the subject of human experimentation, which quoted the Nuremberg Code in full. Significantly, while the two introductory articles were written by Hugh Clegg and Henry K. Beecher from a distinctly Anglo-American and Western perspective, three of the articles were authored by representatives of different religious institutions and Judeo-Christian faiths in order to garner broad international support for the WMA's ethics project [107–111]. Recognizing the difficulties in formulating a "universal code of ethics" applicable to different cultures, Rev. Bosio M. Giuseppe from Rome, Pastor Jacques de Senarclens, director of the Protestant Center of Studies of Geneva, and Johannes Juda Groen, from the Hebrew University Hadassah Medical School and Hospital in Jerusalem, all stressed the importance of respecting and protecting the integrity of the human body to various degrees, especially through the consent principle and the minimization of risk. Groen also proposed that different codes of ethics might govern human experiments in different countries and cultures, and questioned the effectiveness of ethics codes more fundamentally, arguing that human experimentation would remain within ethical boundaries only "if the leaders of our profession set the standard by their actual behaviour" [109–112].

In April 1960, the WMA council held a lively discussion about the work of the ethics committee. Somewhat surprisingly, as noted by Jean Maystre, none of the published articles engaged with, or reflected, ethical governance systems in Muslim and Buddhist countries. In order to integrate a more global perspective into the drafting process, and thus make the code more acceptable, the medical ethics committee was asked to collate information about existing ethical practices in different ethnic and religious contexts, especially in Muslim and Buddhist societies, and to consider the ethical implications of cognitive enhancement and behavioral experiments. Exploring the relevant aspects of human space exploration was likewise part of their brief. They also wanted to study, and better understand, the deontological landscape in Europe and elsewhere, including the "philosophy or code governing communist investigators" [85, 113].

International pressure by senior WMA delegates to broaden the perspective on research ethics explains why the medical ethics committee, while considering it a blueprint, ultimately rejected the Nuremberg Code as an insufficient guide for its own code of ethics. According to committee chair Hugh Clegg, the aim was to draft a "code which could serve at least as a guide to doctors working in different conditions and in different countries" [114, 115]. However, individual principles of the Code nonetheless functioned as a reference against which the authors of the DoH drafted their alternative document. The requirement, stipulated in Principle 3 of the Nuremberg Code, for instance, that an experiment should be "designed and based on the results of animal experimentation," was deemed to be "too restrictive"; while prior animal experiments were desirable, the new code needed to cater for situations in which it would "prove impossible to try the experiment on animals" [85]. Furthermore, balancing general principles of ethics with specific public health practices soon proved difficult. Most WMA delegates agreed that the "ethical interest of the individual must never be sacrificed in the interest of society" [85], and yet such a statement conflicted with public health programs advocated by post-war governments across the ideological divide, which mandated compulsory vaccination for communicable diseases. In light of the recent COVID-19 pandemic experience, the debate reflected, and in many ways foreshadowed, the tension between respect for individual autonomy and social responsibility, and the need to develop a public health ethic in which these seemingly diametrically opposed paradigms could coexist in equilibrium [116].

In September 1960, prior to and during the meeting of the WMA General Assembly in West Berlin, the drafting of the DoH received a major boost. After years of deliberations, the WMA medical ethics committee produced two reports on the "implementation of a project to draft a code of ethics governing human experimentation" [114]. In addition to defining more clearly the distinction between therapeutic and non-therapeutic experiments, WMA delegates were told that the code would cover pharmaceutical, psychological, neurosurgical, and behavioral experiments which interfered with human personalities. A supplementary report offered further details for a "draft code" about the conditions which would need to be fulfilled before researchers could conduct a human experiment [117].

6 The Declaration of Helsinki 1964

A year later, in 1961, Clegg's medical ethics committee finally produced a "Draft Code of Ethics on Human Experimentation," which was published by the *BMJ* in October, 1962 [118, 119]. Born in 1900 at St Ives, Huntingdonshire, Clegg had joined the *BMJ* in 1931, and as editor of the journal from 1947 and through his position in the WMA he played a prominent role in shaping the development of medicine in Britain and the wider world. Sent to Germany to report on the Nuremberg war crimes trials after the Second World War, he not only knew about the importance of the Nuremberg Code, but he had, according to a colleague, "been obsessed with the need for informed consent in trials," and later collaborated with Dr. Tapani Kosonen, a Finn, to write parts of the DoH [120]. However, the suggestion that Clegg be seen as the "principal architect" of the DoH exaggerates his role in the drafting of the document and needs to be seen within the context of obituaries of leading medical practitioners [121–123]. The other two members of the committee who worked on different versions of the Draft Code of 1962 were A.P. Mittra from India and the Italian Antonino Spinelli, president of the WMA from 1954 to 1955.

While there is some discussion about whether the Nuremberg Code left a mark on the DoH, we can be more certain about its impact on the Draft Code. Some of its general principles are almost identical, if not in wording, then in meaning, to those set out in the Nuremberg Code, for example "that during the course of the experiment the subject of it should be free to withdraw from it at any time" [118], which reflects Principle 9 of the Nuremberg Code. Important provisions of the Draft Code, however, such as the prohibition on the use of prisoners of war, or of persons confined to prisons and mental institutions [118], were omitted from the 1964 Declaration. Scientists in the USA, in particular, having made extensive use of prison inmates in clinical trials during and after the Second World War, were concerned that additional safeguards for these populations could hamper US-led drug research conducted in federal penitentiaries, as well as in Guatemala and elsewhere [124]. In the USA, the scientific community signaled a greater willingness to reform only after revelations about the medical ethics violations perpetrated by the US Public Health Service in the Tuskegee syphilis study led to public and political criticism, ushering in a period of major regulatory change which saw the introduction of Institutional Review Boards in the USA, known as Research Ethics Committees in other countries [125–128].

In 1963, a year after the publication of the WMA's Draft Code, the *BMJ*'s editorial writer "Pertinax" discussed the issue of trial participants in dependent relationships following the dismissal by Harvard University of two scientists for performing an experiment on a student with psilocybin, which causes hallucinations. Scholars curious as to Pertinax's identity know now that he was none other than Hugh Clegg, who wrote a weekly editorial in the *BMJ* under the heading "Without Prejudice" [121]. While the experiment was part of the growing interest in psychedelic drugs in both military and civilian facilities, and a reflection of the emerging hippy culture taking root in university campuses across the USA, Pertinax argued that it highlighted the dangers of experimenting on people in a dependent relationship. He was much in favor of making the WMA draft ethics code as detailed as possible to ensure that the "enthusiastic research worker" would not be able to "persuade all and sundry with subtle casuistry that his actions are correct and essential. The great ploy is to say, 'It would have been unethical *not* to have done this experiment which is unethical.'" Still, Pertinax had been surprised by American attempts to influence the wording of the draft code so that experiments on prisoners would not be prohibited: "I am disturbed to learn that the World Medical Association is now hedging on its clause about using—or *not* using—criminals as experimental material." This was all the more astonishing, he noted, since the USA had taken the lead in developing medical ethics standards in response to the "ghastly lessons" of the Nuremberg trials. Moreover, American researchers seemed to have forgotten that Nazi physicians had conducted experiments on camp inmates who had been classed as "criminals," and while admittedly there were significant differences between American and Nazi experiments, especially in terms of obtaining consent, the Nuremberg Code specifically warned against the use of "captive populations" as trial participants. His final comment seemed to suggest that the lessons of Nuremberg, and the deeper meaning of the Nuremberg Code, had either long been forgotten by the American research community in the early 1960s, or no longer had much impact on them: "One of the nicest of the American medical scientists I know was heard to say: 'Criminals in our penitentiaries are fine experimental material—and much cheaper than chimpanzees.' I hope the chimpanzees don't come to hear of this" [129].

In June, 1964, after prolonged debate, the WMA adopted the Declaration during its General Assembly in Helsinki, Finland. The extent to which the issues of vulnerable trial participants had divided opinion among member associations is highlighted by the fact that consensus had still not been reached before the opening of the 18th General Assembly on June 13, which was to approve the WMA's new ethics code. On Sunday, June 14, shortly before the start of the plenary sessions, the WMA council, under the chairmanship of Filip Worré from Luxembourg, agreed to change the name of the WMA ethics code from "Ethical Principles Guiding Research Workers in Clinical Medicine" to "Recommendations Guiding Doctors in Clinical Research." This minor last-minute linguistic amendment had far-reaching implications, and reflected a shift in emphasis from ethical standards informing the work of research scientists in modern medicine to those guiding the work of doctors who were also conducting research [130, 131]. It allowed scientists who were not doctors but who were involved in human experimentation, for instance, those in top-secret military facilities, to claim that the terms of the Declaration did not in fact apply to them [35]. At the stroke of a pen, the progress made in several years of discussion about how best to protect the human rights of patients and experimental subjects had been delivered a major setback: the group to whom the principles actually applied was limited to those with a medical degree. It was an effective way of eliminating the possibility that the WMA ethics code might be used as a point of reference in arguing against research scientists in a court of law, in the USA or elsewhere.

On the same day, the WMA council took another key decision, namely to delete "clause (III 3c)" from the latest version of the Draft Code, which stated that "No clinical research should be undertaken when the subject is in a dependent relationship to the investigator," and to replace it with an addition (italicized) to "clause 4a": "The investigator must respect the right of each individual to safeguard his personal integrity, *especially if the subject is in a dependent relationship to the investigator* (French: *surtout si l'individu est en état de dépendance vis-à-vis de lui*)" [130, 131]. Within less than 2 h, the character of the WMA's ethics code had fundamentally changed. Experimental research on institutionalized children, asylum inmates, the psychologically and/or physically handicapped, or the elderly, while requiring due care and special attention, was no longer ruled out. The final version, dated June 18, 1964 [2], was also conspicuously silent on the subject of prisoners and other vulnerable populations. For the time being, research scientists could carry on as if nothing had changed.

So it was that in June, 1964, after years of debate, the WMA finally adopted parts of the Draft Code during its 18th General Assembly in the Finnish capital. The document became known as the Declaration of Helsinki [5, 6, 31, 33, 34]. As noted above, important provisions of the Draft Code, such as the prohibition on conducting human experiments on prisoners of war, whether military or civilian, or on persons confined to prisons and mental institutions, had been deleted from the Declaration [132]. The *BMJ* felt that "American influence" might have weakened the Declaration, given the United States' recent history of conducting clinical trials on prisoners [129, 133]. In the context of the WMA's "advertised financial crisis"—overcome only after the American Medical Association (AMA) and the WMA's US Committee pledged to fund the organization with a grant of US $500,000 over 5 years—it is difficult to avoid the impression that financial considerations may have played a part, albeit indirectly, in allowing the representatives of the AMA to succeed in watering down the original draft code of 1962 [134]. In the summary of its activities for the year 1964, the WMA devoted fewer than three lines to a document that it now considers to be one of its most successful declarations [135]. It did not anticipate the impact that the document would have on research ethics over the next 50 years [31].

6.1 Conclusion

In promulgating the Declaration, the medical community had succeeded in supplanting the Nuremberg Code with research guidelines that strengthened the position of physician–scientists. Endorsed by the AMA, the American Society for Clinical Investigation, and the American Federation for Clinical Research, the Declaration promoted an important shift in the quality of international ethical codes, from the rights of patients and the protection of human subjects in clinical trials, as expressed in the Nuremberg Code, to the protection of patient welfare through physicians' responsibility. In comparison to the way in which the Code had been perceived, from its early conception, the Declaration was more aligned to the research culture of its time, and yet it also shifted attention away from the central importance of informed consent. The authors of the Declaration effectively moved away from a language of rights and legal liability in clinical trials to a protective system emphasizing patient health and welfare. The duty and responsibility of the scientist retained primacy over more abstract concepts such as informed consent. Succumbing to pressure from US scientists, who wanted to protect large-scale trials for the testing and approval of new drugs, the WMA's negotiated compromise meant that the Declaration did not prohibit the use as experimental subjects of prisoners of war, penitentiary inmates, or those confined to mental institutions. The Declaration, for all its shortcomings, nonetheless marked a major sea-change in research ethics: it constituted a powerful expression of intent by the international medical community to protect the health and wellbeing of research participants through a reaffirmation of Hippocratic medical ideals, in particular the undertaking to do no harm.

Moreover, the importance of post-war research ethics in a vastly expanded medical industry had propelled the WMA onto the international diplomatic stage, where it served not only as a skilled broker protecting the interests of the medical profession, but also as a perhaps unintentional advocate of liberal political ideals. The Second International Congress on Neutrality of Medicine, held at the invitation of the French government in Paris in November, 1964, a few months after the WMA's adoption of the Declaration of Helsinki, was attended by representatives of thirty-eight governments and international organizations [136]. In addition to the WMA, the ICRC, and the MJC, socialist states such as Bulgaria and Yugoslavia attended in an official capacity; others, including Poland, Czechoslovakia, and the Soviet Union, were present as "observers." Chaired by the WMA's liaison officer Jean Maystre, one of the commissions set up by the Congress called for a resolution stipulating that "the basic mission of the medical profession is to insure the preservation of health and the saving of life" [136]. Rather than following the orders of the military or state agencies in wartime, doctors were to rely on their "conscience as [their] supreme guide." The WMA's uncompromising position reflected its desire to use the proclaimed principle of the freedom of the medical profession to limit the rise of state interventionist measures in Eastern and Western Europe. Given the geopolitical tensions in the wake of the Cuban missile crisis, the Congress highlighted the extent to which principles of medical ethics and freedom of science had assumed a distinctly political dimension, especially when discussed in relation to the no less political concept of "neutrality of medicine" in time of conflict.

In the decades after the promulgation of the Nuremberg Code (1947), the ethics of Western research culture had undergone a process of profound transformation. It was a period in which exposure of human and civil rights violations went hand in glove with a realization by some self-appointed leaders in the field that further resistance to public and political demands for change could only lead to incalculable damage to the medical profession [137]. Enormous investment in medicine, science, and technology by public agencies in North America and Western Europe had created a situation in which the available resources were greater "than the supply of responsible investigators" [126, 138]. By the beginning of the 1960s, in the light of ever more frequent revelations about unethical research on vulnerable populations, and after calls for greater regulation prompted by the widely publicized Thalidomide tragedy, it was increasingly difficult to oppose the reform of existing research practices: the political, legal, and financial stakes had become too high for the medical community [139]. At the same time, medical experts and international organizations such as the WMA and the WHO began to recognize that deontological debates had the potential not only to advance reform in the field of research governance, and thus protect medical scientists from public scrutiny and legal liability, but also to serve as a tool for international diplomacy which could reach beyond, and occasionally overcome, national regulations and systemic divisions in Europe and around the globe.

Acknowledgment Research for this paper, which is part of the project "Taming the European Leviathan: The Legacy of Post-War Medicine and the Common Good," has been supported by the European Research Council under the European Union's Horizon 2020 research and innovation program (grant agreement no. 854503).

Conflict of Interest There is no conflict of interest to be disclosed related to the content of this manuscript.

References

1. Schmidt U, Sprumont D, Frewer A. Introduction: the limits of altruism. In: Schmidt U, Sprumont D, Frewer A, editors. Ethical research: the Declaration of Helsinki, and the past, present, and future of human experimentation. Oxford: Oxford University Press; 2020. p. 1–43.
2. WMA. Declaration of Helsinki—ethical principles for medical research involving human subjects. Adopted by the 18th WMA General Assembly. Helsinki, Finland. June 1964. https://www.wma.net/policies-post/wma-declaration-of--helsinki-ethical-principles-for-medical-research-involving-human-subjects/.
3. US Government Printing Office. Nuremberg code. In: United States v. Karl Brandt et al., Trials of war criminals before the Nuernberg Military Tribunals under Control Council

Law no. 10. Vol. 2: The medical case. Washington: US Government Printing Office; 1949. p. 181–2.
4. Schmidt U. Justice at Nuremberg: Leo Alexander and the Nazi doctors' trial. Basingstoke: Palgrave Macmillan; 2004.
5. Schmidt U, Frewer A, editors. History and theory of human experimentation: the Declaration of Helsinki and modern medical ethics. Stuttgart: Franz Steiner Verlag; 2007.
6. Frewer A, Schmidt U, editors. Forschung als Herausforderungen für Ethik und Menschenrechte. 50 Jahre Deklaration von Helsinki, 1964–2014. Cologne: Deutscher Ärzteverlag; 2014.
7. Moreno JD, Schmidt U, Joffe S. The Nuremberg code 70 years later. JAMA. 2017;318(9):795–6. https://doi.org/10.1001/jama.2017.10265.
8. Schmidt U, Sprumont D, Frewer A, editors. Ethical research: the Declaration of Helsinki, and the past, present, and future of human experimentation. Oxford: Oxford University Press; 2020.
9. Ball P. The lightning-fast quest for COVID vaccines—and what it means for other diseases. Nature. 2020;589(7840):16–8. https://www.nature.com/articles/d41586-020-03626-1.
10. Park JJH, et al. How COVID-19 has fundamentally changed clinical research in global health. Lancet Glob Health. 2021;9:e711–20.
11. Carracedo S, et al. The landscape of COVID-19 clinical trials in Latin America and the Caribbean: assessment and challenges. Rev Panam Salud Publica. 2020;44:e177. https://doi.org/10.26633/RPSP.2020.177.
12. Silberner J. Scientists say the rush to do COVID research led to a whole lot of waste. WFDD. April 23, 2021. https://www.wfdd.org/story/scientists-say-rush-do-covid-research-led-whole-lot-waste.
13. Secretary-General highlights "essential" failure of international cooperation, in address to Security Council Meeting on post-coronavirus global governance. Press release SC/14312. https://www.un.org/press/en/2020/sc14312.doc.htm.
14. Schmidt U, Sprumont D, Frewer A. Introduction: the limits of altruism. In: Schmidt U, Sprumont D, Frewer A, editors. Ethical research: the Declaration of Helsinki, and the past, present, and future of human experimentation. Oxford: Oxford University Press; 2020. p. 1–43, Annex 1.1; data from https://www.statista.com/.
15. Schmidt U, Sprumont D, Frewer A. Introduction: the limits of altruism. In: Schmidt U, Sprumont D, Frewer A, editors. Ethical research: the Declaration of Helsinki, and the past, present, and future of human experimentation. Oxford: Oxford University Press; 2020. p. 1–43, Annex 1.3.
16. Pfizer reports 92% operational growth in full-year 2021 revenues. The company provided full year 2022 guidance for revenues in the range of $98bn to $102bn. Pharm. Technol. Feb. 9, 2022. https://www.pharmaceutical-technology.com/news/pfizer-full-year-2021-revenues/.
17. Judt T. Ill fares the land: a treatise on our present discontents. London: Allen Lane; 2010.
18. Schonfeld T. The perils of protection: vulnerability and women in clinical research. Theor Med Bioeth. 2013;34:189–206.
19. Napier S. Challenging research on human subjects: justice and uncompensated harms. Theor Med Bioeth. 2013;34:29–51.
20. Levine C. Research involving economically disadvantaged participants. In: Emanuel EJ, et al., editors. The Oxford textbook of clinical research ethics. Oxford: Oxford University Press; 2008. p. 431–6.
21. Bergemann L, Frewer A, editors. Vulnerabilität und Autonomie in der Medizin. Menschenrechte—Ethik—Empowerment. Human rights in healthcare, vol. 6. Bielefeld: Transcript Verlag; 2019.
22. Fluss S. How the Declaration of Helsinki developed. Good Clin Pract J. 1999;6:18–21.
23. de Roy PG. Helsinki and the Declaration of Helsinki. World Med J. 2004;50(1):9–11.

24. Lederer SE. Research without borders: the origins of the Declaration of Helsinki. In: Roelcke V, Maio G, editors. Twentieth century ethics of human subjects research: historical perspectives on values, practices, and regulations. Stuttgart: Franz Steiner Verlag; 2004. p. 199–217.
25. Carlson R, Boyd K, Webb D. The interpretation of codes of medical ethics: some lessons from the fifth revision of the Declaration of Helsinki. In: Schmidt U, Frewer A, editors. History and theory of human experimentation: the Declaration of Helsinki and modern medical ethics. Stuttgart: Franz Steiner Verlag; 2007. p. 187–202.
26. Sprumont D, Girardin S, Lemmens T. The Helsinki Declaration and the law: an international and comparative analysis. In: Schmidt U, Frewer A, editors. History and theory of human experimentation: the Declaration of Helsinki and modern medical ethics. Stuttgart: Franz Steiner Verlag; 2007. p. 223–52.
27. Frewer A, Schmidt U, editors. Standards der Forschung: Historische Entwicklung und ethische Grundlagen klinischer Studien. Frankfurt am Main: Peter Lang; 2007.
28. Goodyear MDE, Eckenwiler LA, Ells C. Fresh thinking about the Declaration of Helsinki. Br Med J. 2008;337:a2128.
29. Goodyear MDE, Lemmens T, Sprumont DS, Tangwa G. Does the FDA have the authority to trump the Declaration of Helsinki? Br Med J. 2009;338:b1559.
30. Hedgecoe A. "A form of practical machinery": the origin of research ethics committees in the UK, 1967–1972. Med Hist. 2009;53(3):331–50.
31. Schmidt U. "Holding one's breath": reflections on the origins of the Declaration of Helsinki. In: Ehni H-J, Wiesing U, editors. Die Deklaration von Helsinki: Revisionen und Kontroversen. Cologne: Deutscher Ärzte-Verlag; 2012. p. 1–17.
32. Wahl M, Schmidt U. Ärzte und Forschung hinter dem Eisernen Vorhang. Medizinethische Diskurse und die Deklaration von Helsinki in der DDR (1961–1989). In: Frewer A, Schmidt U, editors. Forschung als Herausforderungen für Ethik und Menschenrechte. 50 Jahre Deklaration von Helsinki, 1964–2014. Cologne: Deutscher Ärzteverlag; 2014. p. 71–86.
33. Schmidt U, Frewer A. Geschichte und Ethik der Humanforschung. 50 Jahre Deklaration von Helsinki. Zur Einführung. In: Frewer A, Schmidt U, editors. Forschung als Herausforderungen für Ethik und Menschenrechte. 50 Jahre Deklaration von Helsinki, 1964–2014. Cologne: Deutscher Ärzteverlag; 2014. p. 9–13.
34. Schmidt U, Frewer A. The Declaration of Helsinki as a landmark for research ethics—protecting human participants in modern medicine. In: Wiesing U, Parsa-Parsi RW, Kloiber O, editors. The World Medical Association Declaration of Helsinki, 1964–2014. 50 years of evolution of medical research ethics. Cologne: Deutscher Ärzteverlag; 2014. p. 56–7.
35. Schmidt U. Secret science. A century of poison warfare and human experiments. Oxford: Oxford University Press; 2015.
36. Schmidt U, Wahl M. Doctors and research behind the iron curtain: medical ethics debates and the Declaration of Helsinki in East Germany (1961–1989). In: Schmidt U, Sprumont D, Frewer A, editors. Ethical research: the Declaration of Helsinki, and the past, present, and future of human experimentation. Oxford: Oxford University Press; 2020. p. 167–89.
37. WMA. International code of medical ethics. Adopted by the 3rd General Assembly of the World Medical Association. London, England. Oct. 1949. https://www.wma.net/policies-post/wma-international-code-of-medical-ethics/international-code-of-medical-ethics-1949/.
38. WMA. Regulations in times of armed conflict and other situations of violence. Adopted by the 10th World Medical Assembly, Havana, Cuba, Oct. 1956. https://www.wma.net/policies-post/wma-regulations-in-times-of-armed-conflict-and--other-situations-of-violence/.
39. WMA Archives. Minutes of 9th annual meeting. April 16, 1956.
40. WMA Archives. Summary of activities: 1967. p. 59.
41. Bernard C. An introduction to the study of experimental medicine, transl. by Greene HC. New York: Schuman; 1949. [originally published as Introduction à l'étude de la médecine expérimentale. Paris: Baillière et Fils; 1865]. p. 101.

42. Sass H-M. Reichsrundschreiben 1931: pre-Nuremberg German regulations concerning new therapy and human experimentation. J Med Philos. 1983;8:99–111.
43. Grodin M. Historical origins of the Nuremberg code. In: Annas GJ, Grodin MA, editors. The Nazi doctors and the Nuremberg Code: human rights in human experimentation. Oxford: Oxford University Press; 1992. p. 121–44.
44. Vollmann J, Winau R. Informed consent in human experimentation before the Nuremberg Code. Br Med J. 1996;313:1445–7.
45. Hazelgrove J. The old faith and the new science: the Nuremberg Code and human experimentation ethics in Britain, 1946–73. Soc Hist Med. 2002;15:109–35.
46. Willcock JW. The laws relating to the medical profession: with an account of the rise and progress of its various order. London: J. and W.T. Clarke; 1830. p. 109–10.
47. Faden RR, Beauchamp TL, editors. History and theory of informed consent. Oxford: Oxford University Press; 1986. p. 117.
48. Katz J. Experimentation with human beings: the authority of the investigator, subject, profession, and state in the human experimentation process. New York: Russell Sage Foundation; 1972. p. 526–7.
49. Kennedy I, Grubb A. Medical law: text with materials. London: Butterworths; 1994. p. 526.
50. Berg JW, et al. Informed consent: legal theory and clinical practice. Oxford: Oxford University Press; 2001. p. 42.
51. Annas GJ. Doctors, patients, and lawyers—two centuries of health law. N Engl J Med. 2012;367(5):445–50, here p. 446.
52. TNA. TS27/398. [Maurice] Gwyer [Treasury Solicitor] to MRC. June 21, 1933.
53. Grodin M. Historical origins of the Nuremberg Code. In: Annas GJ, Grodin MA, editors. The Nazi doctors and the Nuremberg Code: human rights in human experimentation. Oxford: Oxford University Press; 1992. p. 121–44, here p. 131.
54. Bonah C, et al., editors. La médecine expérimentale au tribunal: implications éthiques de quelques procès médicaux du XXe siècle européen. Paris: Archives Contemporaines; 2003.
55. UN General Assembly resolution 217A. Dec. 10, 1948. Universal Declaration of Human Rights. https://www.un.org/en/about-us/universal-declaration-of-human-rights.
56. Basic Law of the Federal Republic of Germany, May 23, 1949. https://www.cvce.eu/de/obj/the_basic_law_of_the_frg_23_may_1949-en-7fa618bb-604e-4980-b667-76bf0cd0dd9b.html.
57. Office of the UN High Commissioner for Human Rights. International Covenant on Civil and Political Rights. Adopted and opened for signature, ratification and accession by General Assembly Resolution 2200A (XXI) of Dec. 16, 1966. Entry into force March 23, 1976, in accordance with Article 49. https://www.ohchr.org/en/professionalinterest/pages/ccpr.aspx, Article 7.
58. Schmidt U. The Holocaust and the medical professions. In: Jockusch L, Pendas D, editors. Cambridge history of the Holocaust. Vol. 4: Outcomes, aftermath, repercussions. Cambridge: Cambridge University Press; 2023.
59. Katz J. The Nuremberg code and the Nuremberg trial. A reappraisal. JAMA. 1996;276(20):1662–6.
60. WMA. Declaration of Geneva. Adopted by the 2nd General Assembly of the World Medical Association, Geneva, Switzerland. Sept. 1948. https://www.wma.net/policies-post/wma-declaration-of-geneva/.
61. WMA Archives. First decade report. Chapter 4. p. 2–3.
62. Frewer A. Human rights from the Nuremberg Doctors trial to the Geneva Declaration. Persons and institutions in medical ethics and history. Med Health Care Philos. 2010;13:259–68.
63. WMA Archives. First decade report. Chapter 4. p. 7.
64. Wiesing U. The future of the Declaration of Helsinki. Introduction—remarks about the next revision. Presentation. Rotterdam. June 26, 2012.

65. Bonah C, Schmaltz F. From Nuremberg to Helsinki: the preparation of the Declaration of Helsinki in the light of the prosecution of medical war crimes at the Struthof medical trials, France, 1952–4. In: Schmidt U, Sprumont D, Frewer A, editors. Ethical research: the Declaration of Helsinki, and the past, present, and future of human experimentation. Oxford: Oxford University Press; 2020. p. 69–100.
66. Maio G. Medical ethics and human experimentation in France after 1945. In: Roelcke V, Maio G, editors. Twentieth century ethics of human subjects research: historical perspectives on values, practices, and regulations. Stuttgart: Franz Steiner Verlag; 2004. p. 235–52.
67. WMA Archives. First decade report. Chapter 4. p. 4.
68. WMA Archives. Armed conflict. Report to the 8th General Assembly by Jean Maystre, Liaison Officer, International Red Cross. (Confidential). July, 1953.
69. WMA Archives. First decade report. Chapter 4. p. 9.
70. Frenkel D. Human experimentation: codes of ethics. In: Carmi A, editor. Medical experimentation: its legal and ethical aspects. Ramat-Gan: Turtledove; 1977. p. 131–2.
71. Lederer SE. Research without borders: the origins of the Declaration of Helsinki. In: Roelcke V, Maio G, editors. Twentieth century ethics of human subjects research: historical perspectives on values, practices, and regulations. Stuttgart: Franz Steiner Verlag; 2004. p. 199–217, here p. 202.
72. Schmidt U. Justice at Nuremberg: Leo Alexander and the Nazi doctors' trial. Basingstoke: Palgrave Macmillan; 2004. p. 220.
73. WMA Archives. Research. Dr. LA Hulst. Experiments on human beings. 38.3/53.
74. WMA Archives. Pol. DoH, Adopted versions, 1953 onwards.
75. WMA Archives. Research. Medical Ethics Committee. Minutes. 4.3/54.
76. Bonah C, Schmaltz F. From Nuremberg to Helsinki: the preparation of the Declaration of Helsinki in the light of the prosecution of medical war crimes at the Struthof medical trials, France, 1952–4. In: Schmidt U, Sprumont D, Frewer A, editors. Ethical research: the Declaration of Helsinki, and the past, present, and future of human experimentation. Oxford: Oxford University Press; 2020. p. 83.
77. WMA Archives. Research. Medical Ethics Committee. Minutes. 17.5/54.
78. Lederer SE. Research without borders: the origins of the Declaration of Helsinki. In: Roelcke V, Maio G, editors. Twentieth century ethics of human subjects research: historical perspectives on values, practices, and regulations. Stuttgart: Franz Steiner Verlag; 2004. p. 199–217, here p. 202ff.
79. WMA Archives. Research. Medical Ethics Committee. Minutes. Sept., 1954.
80. Lederer SE. Research without borders: the origins of the Declaration of Helsinki. In: Roelcke V, Maio G, editors. Twentieth century ethics of human subjects research: historical perspectives on values, practices, and regulations. Stuttgart: Franz Steiner Verlag; 2004. p. 199–217, here p. 203.
81. WMA Archives. Research. Medical Ethics Committee. Minutes. 4.1/55.
82. WMA Archives. Research. Medical Ethics Committee. Minutes. 4.11/54.
83. ICRC. Convention (III) relative to the treatment of prisoners of war. Geneva, Aug. 12, 1949. Article 13. Humane treatment of prisoners. https://ihl-databases.icrc.org/applic/ihl/ihl.nsf/ART/375-590017/OpenDocument.
84. WMA Archives. Research. Medical Ethics Committee. Minutes. 4.10/55.
85. WMA Archives. Research. Medical Ethics Committee. Minutes. 4.3/60.
86. Schmidt U, Secret science. A century of poison warfare and human experiments. Oxford: Oxford University Press; 2015. p. 352f.
87. WMA. Principles for those in research and experimentation. [Adopted by the 8th WMA General Assembly, Rome, Italy, Sept. 1954.]. World Med J. 1955;2:14–5.
88. WMA Archives. First decade report. Chapter 4. p. 5f.
89. WMA Archives. First decade report. Chapter 10.

90. Schmidt U, Sprumont D, Frewer A, editors. Ethical research: the Declaration of Helsinki, and the past, present, and future of human experimentation. Oxford: Oxford University Press; 2020; Appendices 1a and 1b.
91. Schmidt U. In the absence of alternatives: the origins and success of the Helsinki Declaration, 1947–1982. In: Schmidt U, Sprumont D, Frewer A, editors. Ethical research: the Declaration of Helsinki, and the past, present, and future of human experimentation. Oxford: Oxford University Press; 2020. p. 101–130, here p. 104f.
92. NIH. Nov. 17, 1953 ("Group consideration of clinical research procedures deviating from accepted medical practice or involving unusual hazard"). In: United States Advisory Committee on human radiation experiments. Final report, 1. New York: Oxford University Press; 1996. https://bioethicsarchive.georgetown.edu/achre/final/chap1_4.html.
93. TNA. FD9/855. Draft—[MRC] council memorandum "Experiments on man: conditions for conduct," including statement on the "condition on which experiments can be conducted on man." Jan. 2, 1956.
94. ICRC. Convention relative to the treatment of prisoners of war. Geneva. July 27, 1929. https://ihl-databases.icrc.org/ihl/INTRO/305.
95. WMA Archives. Armed conflict. Medicine in the time of war. (Confidential). July 8, 1949.
96. Voncken J. Pour créer un Droit International Médical [To create an International Medical Code of Law]. Bull Int Serv Santé Armées Terre Mer l'Air. 1949;3(4):57–74. (The author would like to thank Chief Warrant Officer Catherine Van Den Berghe of the International Committee of Military Medicine (ICMM) for her research on this paper and for supplying a copy thereof).
97. ICRC. Convention (X) for the adaptation to maritime warfare of the principles of the Geneva Convention [1864]. The Hague. Oct. 18, 1907.
98. Baer H-U, Sanabria J. History of the International Committee of Military Medicine. Mil Med. 2002;167:2–3.
99. WMA Archives. Research. The Red Cross Emblem. 15.5/54.
100. WMA Archives. Armed conflict. Report of the International Liaison Committee: The status of civilian doctors in war time. March, 1955.
101. WMA Archives. Armed conflict. Code of medical ethics in wartime. May 15, 1956.
102. WMA Archives. Research. International Liaison and Medical Ethics Committee: Draft regulations in time of armed conflict. May, 1957.
103. WMA Archives. Research. Resolution on free scientific exchange. Havana, Cuba. Nov., 1956.
104. WMA Archives. Research. Council meeting of the WMA at Copenhagen, Denmark. April 24–28, 1950.
105. WMA Archives. Research. Medical Ethics Committee. Minutes. 4.3/58.
106. WMA Archives. Research. Medical Ethics Committee. Minutes. 4.10/59.
107. Clegg H. Human experimentation. World Med J. 1960;7(2):77–9.
108. Beecher HK. Human experimentation—a world problem from the standpoint of a medical investigator. World Med J. 1960;7(2):79–80.
109. Giuseppe BM. Human experimentation—a world problem from the standpoint of spiritual leaders. World Med J. 1960;7(2):80.
110. de Senarclens J. Human experimentation—a world problem from the standpoint of spiritual leaders. World Med J. 1960;7(2):81–2.
111. Groen JJ. Human experimentation—a world problem from the standpoint of spiritual leaders. World Med J. 1960;7(2):82–3, 96.
112. Benbassat J. Johannes Juda Groen (1903–1990): a forgotten visionary in the history of medical education. Rambam Maimonides Med J. 2020;11(4):e0038.
113. Lederer SE. Research without borders: the origins of the Declaration of Helsinki. In: Roelcke V, Maio G, editors. Twentieth century ethics of human subjects research: historical perspectives on values, practices, and regulations. Stuttgart: Franz Steiner Verlag; 2004. p. 199–217, here p. 205f.

114. WMA Archives. Research. Medical Ethics Committee. Minutes. 17.2/60.
115. Lederer SE. Research without borders: the origins of the Declaration of Helsinki. In: Roelcke V, Maio G, editors. Twentieth century ethics of human subjects research: historical perspectives on values, practices, and regulations. Stuttgart: Franz Steiner Verlag; 2004. p. 199–217, here p. 206.
116. Moreno JD, Sándor J, Schmidt U. Autonomy and social responsibility: the post-pandemic challenge. Perspect Biol Med. 2022;65(3):426–41.
117. WMA Archives. Research. Medical Ethics Committee. Minutes. 17.2A/60.
118. WMA. Draft code of ethics on human experimentation. Br Med J. 1962;3212:1119.
119. Schmidt U, Sprumont D, Frewer A, editors. Ethical research: the Declaration of Helsinki, and the past, present, and future of human experimentation. Oxford: Oxford University Press; 2020, Appendices 2a–2c.
120. Reynolds LA, Tansey EM, editors. Medical ethics education in Britain, 1963–93. London: The Wellcome Trust; 2007. p. 6.
121. Clegg HA. Obituary. Br Med J. 1983;287:220–1.
122. Lock SP. Hugh Anthony Clegg. Munk's Roll. 1984;7:103–6.
123. Lock SP. Clegg, Hugh Anthony (1900–1983). In: Oxford dictionary of national biography online, vol. 2004. Oxford: Oxford University Press.
124. Presidential Commission for the Study of Bioethical Issues. Ethically impossible. STD research in Guatemala from 1946 to 1948. Washington: Presidential Commission for the Study of Bioethical Issues; 2011. https://bioethicsarchive.georgetown.edu/pcsbi/node/654.html.
125. Jones JH. Bad blood: the Tuskegee syphilis experiments. London: Collier Macmillan; 1981.
126. Schmidt U. Justice at Nuremberg: Leo Alexander and the Nazi doctors' trial. Basingstoke: Palgrave Macmillan; 2004. p. 284f.
127. Reverby SM. "Normal exposure" and inoculation syphilis: a PHS "Tuskegee" doctor in Guatemala, 1946–1948. J Policy Hist. 2011;23(1):6–28.
128. Reverby SM. Ethical failures and history lessons: the U.S. Public Health Service research studies in Tuskegee and Guatemala. Public Health Rev. 2012;34(13):1–18.
129. Pertinax. Without prejudice (editorial column). Br Med J. 1963;1(5345):1603; original emphases.
130. WMA Archives. Minutes of 51st council session. Helsinki, Finland. June 13–14, 1964.
131. Schmidt U, Sprumont D, Frewer A, editors. Ethical research: the Declaration of Helsinki, and the past, present, and future of human experimentation. Oxford: Oxford University Press; 2020, Appendix 2e.
132. Katz J. The consent principle of the Nuremberg Code: its significance then and now. In: Annas GJ, Grodin MA, editors. The Nazi doctors and the Nuremberg Code: human rights in human experimentation. Oxford: Oxford University Press; 1992. p. 227–39, here p. 233.
133. Lederer SE. Research without borders: the origins of the Declaration of Helsinki. In: Roelcke V, Maio G, editors. Twentieth century ethics of human subjects research: historical perspectives on values, practices, and regulations. Stuttgart: Franz Steiner Verlag; 2004. p. 199–217, here p. 210.
134. WMA Archives. Summary of activities: 1964. p. 37.
135. WMA Archives. Summary of activities: 1964. p. 36.
136. WMA Archives. Armed conflict. Report of the Committee on International Liaison. Jan. 29, 1965.
137. Schmidt U. Secret science. A century of poison warfare and human experiments. Oxford: Oxford University Press; 2015. p. 372f.
138. Beecher HK. Ethics and clinical research. N Engl J Med. 1966;274:1354–60, here p. 1354.
139. Schmidt U. Justice at Nuremberg: Leo Alexander and the Nazi doctors' trial. Basingstoke: Palgrave Macmillan; 2004. p. 264–97.

CIOMS Research Guidelines: Considering the Needs of Developing Countries

Lembit Rägo and Monika Zweygarth

Abstract This chapter gives an overview of the history and activities of the Council for International Organizations of Medical Sciences (CIOMS) in three interrelated areas: research ethics, medicines development, and medicines safety (pharmacovigilance). Three specific CIOMS guidelines are then briefly discussed.

The 2016 CIOMS *International ethical guidelines on health-related research involving humans* were prepared in collaboration with the World Health Organization (WHO). They state the core principles and provide detailed commentaries to facilitate their implementation, with a focus on resource-limited settings. Section 2 highlights selected key aspects that are relevant to recent developments in the global health environment.

Building on the 2016 CIOMS ethical guidelines, the 2021 CIOMS Working Group report on *Clinical research in resource-limited settings* describes the challenges and opportunities to address unmet health needs in these settings. Section 3 gives an overview of the report and of its recommendations to governments, researchers, and funders.

Patients, including those in developing countries, must be empowered to be active partners in advancing safe and effective treatments. Section 4 highlights some key aspects from the report of the CIOMS Working Group XI on *Patient involvement in the development, regulation, and safe use of medicines*.

Keywords CIOMS · Research ethics · Product development · Pharmacovigilance · Patients · Developing countries

L. Rägo (✉) · M. Zweygarth
Council for International Organizations of Medical Sciences (CIOMS), Geneva, Switzerland
e-mail: ragol@cioms.ch; Zweygarthm@cioms.ch

C. Kurihara et al. (eds.), *Ethical Innovation for Global Health*,
https://doi.org/10.1007/978-981-99-6163-4_10

1 Background

1.1 What Is CIOMS?

The Council for International Organizations of Medical Sciences (CIOMS) is an international non-governmental, non-profit organisation [1] in official relations with the World Health Organization (WHO) and is an associate partner of the United Nations Educational, Scientific and Cultural Organization (UNESCO). Founded in 1949, it currently includes 41 international, national, and associate member organisations which, through their own members—for example, the World Medical Association's national associations of physicians—represent a substantial proportion of the biomedical scientific community.

The mission of CIOMS is to advance public health through guidance on health research and policy including ethics, medical product development, and safety [2]. A major strength of CIOMS lies in its Working Groups, which bring together some of the world's foremost experts in their fields from across different sectors: regulatory authorities, health professional and scientific organisations, patient organisations, and the pharmaceutical industry. CIOMS provides a neutral 'think tank' forum in which stakeholders can discover common ground and develop documents that express strategic goals, specific process ideas, and consensus aspirations. The CIOMS Working Group reports and recommendations usually take 2–4 years to be finalised; the minutes of Working Group meetings are published on the CIOMS website. All CIOMS reports are freely available online; printed copies can be ordered via the CIOMS website at postal cost.

1.2 CIOMS Guidance on Research Ethics

The work of CIOMS in the area of ethics started with the first CIOMS Round Table Conference on *Biomedical science and the dilemma of human experimentation*, held in Paris in 1967 [3]. Since then, biomedical sciences, biotechnology and their applications in medical practice have progressed, confronting our societies with new ethical dilemmas. In response, international guidelines such as the Declaration of Helsinki [4] have been revised periodically, and so too have the CIOMS guidelines.

In the late 1970s, CIOMS began working with the World Health Organization (WHO) on ethics in relation to research. The initial objective was to prepare guidelines on how the ethical principles that should govern the conduct of biomedical research involving human subjects, as set forth in the Declaration of Helsinki, could be effectively applied, particularly in developing countries, given their socioeconomic circumstances, laws and regulations, and executive and administrative arrangements. This resulted in the publication of the 1982 *Proposed International Ethical Guidelines for Biomedical Research Involving Human Subjects* [5].

The period that followed saw rapid advances in medicine and biotechnology, the growth of multinational clinical trials, a move towards more research involving children and other special populations, a shift in attitudes towards regarding research in humans as beneficial rather than threatening, and the outbreak of the HIV/AIDS pandemic. During this period the Declaration of Helsinki was revised in 1983 and again in 1989. In response to these events CIOMS, in cooperation with WHO and its Global Programme on AIDS, issued its *International Ethical Guidelines for Biomedical Research Involving Human Subjects* in 1993 [6]. Recognising that ethical guidance was also needed for public health research, CIOMS published its *International Guidelines for Ethical Review of Epidemiological Studies* in 1991 [7].

This was followed by the 2002 CIOMS *International Ethical Guidelines for Biomedical Research Involving Human Subjects* [8], and in 2003 a revision of the 1991 guidelines on epidemiological studies was initiated, resulting in the publication of the revised CIOMS *International Ethical Guidelines for Epidemiological Studies* in 2009 [9]. The following year, the CIOMS Executive Committee decided to revise both the 2002 and 2009 guidelines, taking into account new developments in the research landscape and new revisions of the Declaration of Helsinki. The CIOMS *International Ethical Guidelines for Health-related Research Involving Humans* were prepared in collaboration with WHO and published in 2016 [10]. They have been translated into all six UN languages as well as Japanese, Korean, Portuguese, and Ukrainian; a Polish version was in preparation at the time of writing this chapter. The unique features of this guideline and selected aspects related to recent global developments are explored in Sect. 2.

Historically CIOMS has been dealing with a wide range of ethical issues and has issued recommendations that were pioneering and innovative at the time, e.g., on protection of prisoners against torture [11], medical genetics [12], and research involving animals [13, 14]. The aim of CIOMS ethics guidelines for research has always been to provide internationally vetted ethical principles together with detailed commentary on how these principles should be applied, with particular attention to low-resource settings. This pragmatic focus on feasibility in difficult circumstances has been appreciated, and the guidelines have been widely used around the world, including in low- and middle-income countries (LMIC).

1.3 CIOMS Activities in Medicines Development

The general topic of 'medicines development' or 'product development' was taken up by CIOMS in the early 2000s, when the need to prevent and treat HIV/AIDS, malaria, tuberculosis, and other diseases affecting developing countries required new and innovative interventions to be developed and tested. It was obvious that therapeutic effectiveness and risks of adverse reactions needed to be studied in relevant local populations, which often differ from the Western populations that have traditionally been participants in clinical trials. A joint CIOMS/WHO Working Group was formed in 2003 to advise on models and solutions to carry out clinical trials in resource-limited settings, with operational safety vigilance systems for medicines before and after their authorisation. The group did not complete its work but made public its draft report in 2005 [15]. In November 2017, CIOMS launched a new Working Group on this topic, leading to the publication of the consensus report on *Clinical research in resource-limited settings* in 2021 (see Sect. 3) [16].

A principle whose value has long been underestimated is that of involving patients in medicines development. Patients can be involved from the very outset, when the need for a new product is recognised, throughout the life cycle of a product until it is retired from the market. Previous CIOMS guidelines have recognised the role of patients as key stakeholders in the safe and effective use of medicines, and input from patients themselves was obtained for the first time by the CIOMS Working Group IX when formulating its 2014 guidance on *Practical Approaches to Risk Minimisation for Medicinal Products* [17]. The CIOMS Working Group XI on Patient Involvement in the Development, Regulation and Safe Use of Medicines was launched in April 2018. Its report was released for public consultation in March 2022, and was at the final stages of publication at the time of drafting this chapter (July 2022). Some points from this very comprehensive report are presented in Sect. 4. The group has also issued a statement on COVID-19, affirming that the patient voice must be integral to the scientific march in defeating the virus [18].

1.4 CIOMS Activities in Medicines Safety

Medicines developers and regulators have ethical and moral obligations to do their best to ensure the safety of medicines both during research and when authorised medicines are used in clinical practice. The vigilance towards a variety of medicines safety issues in a comprehensive manner is known as pharmacovigilance. WHO has defined the term pharmacovigilance as 'the science and activities relating to the detection, assessment, understanding and prevention of adverse effects or any other possible drug-related problems'. [19]

The first CIOMS Working Group (WG) on pharmacovigilance was set up in 1986 with the aim to explore means of coordinating and standardising international adverse drug reaction (ADR) reporting by pharmaceutical manufacturers to regulatory authorities, more precisely the post-marketing reporting of ADRs having occurred in one country to regulatory authorities of other countries where the drug was also to be marketed. This resulted in the CIOMS WG I reporting form, which is now the basis for reporting ADRs globally. Subsequent CIOMS WGs targeted additional emerging aspects of pharmacovigilance such as periodic drug safety update summaries, core clinical safety information on drugs, benefit–risk balance for marketed drugs and, more recently, signal detection, risk minimisation, and meta-analysis for drug safety. To date CIOMS has published more than 20 reports with recommendations on pharmacovigilance topics [20].

While CIOMS recommendations are not in themselves legally binding, they have had a profound impact on practice. In the 1990s, several CIOMS pharmacovigilance reports were taken up by the International Council for Harmonisation of Technical Requirements for Pharmaceuticals for Human Use (ICH) and became an integral part of daily practice of regulatory authorities and the pharmaceutical industry. In developing countries, WHO has made substantial efforts over the last two decades to strengthen national pharmacovigilance systems, taking up many principles of CIOMS pharmacovigilance guidance. These are now helping to ensure the safety of medicines and vaccines in LMICs, while also preparing the ground for the implementation of harmonised ICH guidelines as the ICH membership is expanding towards a global reach [21].

In addition, CIOMS has issued guidance on specific medicines safety issues. An example is drug-induced liver injury (DILI) [22, 23]. A risk of DILI is associated with a wide range of medicines, including, for example, the commonly used painkiller paracetamol, certain anti-tuberculosis medicines, and various herbal medicines [22], making DILI an issue in all settings, including in developing countries. Another organ-specific adverse reaction, namely severe cutaneous adverse reactions to drugs (SCARs) is currently being addressed by a CIOMS Working Group.

During the COVID-19 pandemic a CIOMS topic that was of immediate interest in all parts of the world was vaccines safety. The CIOMS guidelines on active vaccine safety surveillance and vaccine safety communication [24, 25] were taken up in guidance such as the WHO Global Vaccine Safety Blueprint [26] and the WHO safety surveillance manual for COVID-19 vaccines [27].

Other topics of ongoing CIOMS Working Groups are also very relevant to pandemic situations. For example, while evidence is still being gathered from clinical trials, how can real-world data (RWD) and real-world evidence (RWE) be used to guide effective, coordinated, ethical regulatory decision-making? The CIOMS Working Group XIII on the use of RWD and RWE in regulatory decision-making is discussing this topic. And how should the benefits and risks of medical products be evaluated in the face of pressing need and high uncertainty? In its statement on COVID-19 the CIOMS Working Group XII on Benefit–Risk Balance for Medicinal Products has emphasised the critical importance of good science, best practices, and proper communication during public health emergencies [28]. At the time of writing these two CIOMS Working Groups were about to release their draft reports for public consultation, with a view to finalise them for publication by the end of 2023.

2 CIOMS International Ethical Guidelines for Health-Related Research Involving Humans

2.1 Background

The 2016 revision of the CIOMS guidelines, the *International Ethical Guidelines for Health-related Research Involving Humans* [9], combines the topics of the 2002 guidelines and the 2009 CIOMS guidelines [7, 8], covering both biomedical research and epidemiological studies including biobanking and research with health-related data. The 2016 guidelines were developed in collaboration between CIOMS and WHO, and in close cooperation with the WMA. They are based on other authoritative ethical guidance documents, including the WMA's Declaration of Helsinki [4] and UNESCO's Universal Declaration on Bioethics and Human Rights [29], and also consider other documents from UNESCO, WHO, the Council of Europe, as well as various regional and international initiatives that had emerged or changed at the time when the guidelines were drafted.

The Declaration of Helsinki is a succinct, concise set of general and specific principles of approximately 2000 words. In contrast, the CIOMS 2016 guidelines count approximately 50,000 words and consist of 25 numbered guidelines on specific topics. Each of these starts with the core principles, followed by extensive, carefully worded commentaries with general considerations, justifications, and conditions of their application. They complement the Declaration of Helsinki, facilitating its implementation. The annexes to the CIOMS guidelines have practical tools, i.e., a list of items to be included in research protocols, and essential information to be provided to prospective participants for informed consent.

2.2 Selected Key Aspects

Recent global developments in the research landscape are mirrored in the changes from previous guidelines that have been adopted in the 2016 CIOMS *International ethical guidelines on health-related research involving humans*. These changes relate to five main areas: (1) an emphasis on the social and scientific value of research, (2) a non-geographic view of low-resource settings, which might also exist in middle- and high-income countries and may change over time, (3) a new guideline on community engagement, (4) a context-dependent definition of vulnerability in research, and (5) the concept of broad informed consent to the future use of data and samples collected from an individual [30]. Health emergencies are discussed in Guideline 20 on Research in disasters and disease outbreaks.

COVID-19 has continued to affect the global community on an unprecedented scale, showing that 'no one is safe until everyone is safe'. The principles of the CIOMS guidelines still hold, and adherence to them will indeed reinforce an effective response to counter this and future threats. Some key aspects are outlined below.

2.2.1 Responsiveness to Communities' Health Needs

Two conditions have emerged in international debates to avoid exploitative research: Firstly, the research must be responsive to health needs or priorities of the communities in which the research is conducted, and secondly, the new interventions developed during the research must be become reasonably available to them after the trial [31].

The CIOMS Guideline 2 supports these two conditions. It puts forward the principles of responding to local health needs in low-resource settings, and refers to other guidelines where they are described in more detail: the scientific and social value of research to promote health (Guideline 1), fair distribution of the burdens and benefits generated by research (Guideline 3), and early engagement with potential participants and communities to ensure that they accept the research and that it is relevant to them (Guideline 7). It further notes an exception to the responsiveness principle: In a demonstration of solidarity during the Ebola outbreak of 2014, phase one studies on investigational Ebola vaccines were carried out in communities not experiencing an Ebola outbreak, and the use of a WHO-supported collaborative platform enabled the rapid development and delivery of effective and safe Ebola vaccines to affected populations [32]. However, lessons learned from Ebola have also confirmed the need to invest in technical and regulatory capacity-building in low-resource settings [33], and to build trust, good governance and collaborative partnerships to combat future infectious disease threats [33] (see also Sect. 3 below).

Guideline 2 further requires that the knowledge and intervention produced by the research be made available to the community, and products be reasonably priced. They note that 'post-trial access plans are of particular concern for research conducted in low-resource settings where governments lack the means or infrastructure to make such products widely available'. The obligation to make available the new interventions is part of a broader obligation to care for participants' health needs (Guideline 6), which also requires that before a study begins, researchers and sponsors make plans for transitioning participants who continue to need treatment after their participation in research to appropriate health services.

Guideline 20 on research in outbreaks requires that the research should be responsive to the health needs or priorities of the disaster victims and affected communities, and that it cannot be conducted outside a disaster situation. It calls for adequate justification when particular populations are targeted or excluded, for example, health workers. Furthermore, researchers must assess the risks and potential individual benefits of experimental interventions realistically, especially those in the early phases of development, and communicate them to potential participants. Well-connected patients should not be further privileged in study enrolment, but it may be acceptable to prioritise certain populations such as front line workers, who would be able to help more patients if experimental interventions are effective.

Guideline 20 also requires researchers to ensure that studies yield scientifically valid results under the challenging and often rapidly evolving conditions of disease outbreaks. Without scientific validity, studies should not be conducted, they may even divert personnel or resources from the disaster response. Research should be planned ahead, with flexible mechanisms to expedite ethical review, such as pre-screening of study protocols or pre-arrangements on data- and sample-sharing. Trial designs other than randomised controlled trials should be explored to increase trial efficiency and access to promising experimental interventions but should be carefully assessed taking into account the promise of the investigational agent, critical background factors (e.g., mortality and infection rates), and measurement of outcomes, among others. Widespread emergency use of investigational products with inadequate data collection about patient outcomes must be avoided, so as not to compromise recruitment of research participants and therefore undermine the conclusion of trials.

2.2.2 Community Engagement

The 2016 CIOMS ethical guidelines newly include a dedicated guideline on community engagement (Guideline 7).

Guideline 2 clarifies that community engagement serves, among other things, to determine the distribution of the study's benefits, and that these can go beyond the benefits associated with study participation; such additional benefits include improving the health infrastructure, training laboratory personnel, or educating the public about research. Guideline 2 says that sponsors and researchers should make every effort to make any intervention or product developed, and knowledge generated, available to the local population as soon as possible, and should assist in building local research capacity (Guideline 8).

In disasters and disease outbreaks, community engagement is essential. The commentary to Guideline 20 calls for early engagement to maintain public trust and conduct studies in a culturally sensitive manner, as disasters often lead to vulnerability and fragile political and social situations. The guideline suggests that creative mechanisms such as social media can be used, and fostering community leadership will often be helpful, but cautions that researchers should be aware of potential conflicts of interest. For example, community leaders might seek to reassert their own authority by providing services to their communities through research.

The spread of fake news has increased dramatically during the COVID-19 pandemic worldwide [34]. In an era of 'infodemics', community engagement is a crucial and increasingly challenging part of an effective research response.

2.2.3 Vulnerable Populations

The 2016 CIOMS guidelines no longer label entire classes of individuals as vulnerable. Rather, certain study participants may be at increased risk of being harmed or wronged because of specific context-dependent characteristics. Researchers and research ethics committees must evaluate these characteristics and devise special context-dependent protections. These include, for example, allowing for no more than minimal risks for research procedures that offer no potential individual benefits for participants, or requiring that the research be carried out only when it targets conditions that affect these groups. Research designed to obtain knowledge relevant to the health needs of the pregnant and breastfeeding woman must be promoted, but special protections are warranted in research involving this group (Guideline 19).

The guidelines emphasise that children and persons who are incapable of giving informed consent must be included in research, provided that appropriate safeguards are in place, unless a valid scientific reason justifies their exclusion. Guideline 20 requires individual informed consent even in a situation of duress unless the conditions for a waiver of informed consent are met (Guidelines 9 and 10), and the difference between research and humanitarian aid must be explained clearly.

2.2.4 Data-Driven Research

With the rapid progress of information technologies, the divide between research and everyday practice is disappearing, making way for learning health care systems, in which data-driven innovation with rapid feedback loops can lead to more beneficial treatments and more information on optimal use [35]. Real-world data are increasingly being used to derive real-world evidence, and this approach has been accelerated in the COVID-19 pandemic, for example, in the USA [36].

The 2016 CIOMS guidelines consider that an individual whose biological materials and/or data are used in research is a study participant, and ethical guidelines that apply to research participants are applicable in this situation. They have adopted concepts of broad informed consent and informed opt-out procedures for research in this area (Guidelines 11 and 12). In broad informed consent, an individual's control over her or his data and biological material is substituted by adequate governance systems that specify—among other items—to which legal entity the material is entrusted, how authorisation from the study participant (the donor) is obtained, and what procedure determines whether unsolicited findings should be disclosed. Proper governance systems are also crucial in the modern information technology environment, where increasing digitalisation is making it challenging to ensure confidentiality: Data are now exposed to many types of security breaches, and they may be de-anonymized by cross-matching against other large data sets [30].

The CIOMS Guideline 12 clarifies that even if an entity does not collect data deliberately but 'mines' data for health-related research (for example, queries in search engines, consumer choices on websites), it must strive for governance structures and mechanisms to obtain authorisation for future use of these data in research as discussed in the Guideline.

Data protection is an evolving and challenging issue. The concept paper for the CIOMS Working Group XIII on real-world data notes that the group has an opportunity to articulate its point of view on ethical issues in data-driven research such as the appropriate use of individualised medical data, privacy, consent, data ownership and financial considerations, and the sharing of insights and results with study populations, as a basis for developing an updated set of ethical guidelines for the use of real-world data and real-world evidence [37].

3 CIOMS Working Group Report on *Clinical Research in Resource-Limited Settings*

3.1 Background and Overview

While most of the preventable morbidity and mortality occurs in LMICs, most clinical research is still being conducted in and for high-income countries (HICs). Also, for a variety of reasons, some research projects in LMICs have not benefited the study participants or the communities involved.

Against this backdrop, in November 2017 CIOMS convened a Working Group of senior scientists from drug regulatory authorities, the pharmaceutical industry, public–private partnerships for product development and academia to develop a consensus report with pragmatic recommendations to promote good quality clinical research in resource-limited settings. All 32 group members had first-hand experience in conducting and/or regulating clinical research in resource-limited settings, eight of them were based in LMICs. The draft report was posted on the CIOMS website for comment from mid-March 2021 for a period of 5 weeks, and Working Group members actively invited comments from their peers working in LMICs. Over 130 comments were received and were addressed before the report was published in June 2021.

The report builds on the 2016 CIOMS *International Ethical Guidelines for Health-Related Research Involving Humans* but is not intended to supersede those guidelines. The CIOMS report on *Clinical Research in Resource-limited Settings* comprises five chapters and seven appendices:

- The opening chapter, 'Backdrop and problem statement', describes how clinical research in resource-limited settings has evolved over time, and calls for more such research, so that communities in those settings can access new interventions to address their specific health needs.
- The chapter on 'Obstacles and enablers' describes the main factors that hinder clinical research in LMICs—including corruption, autocracy, legal uncertainties, regulatory weaknesses that create loopholes for players with undue interests, excessive bureaucracy, limited public funding and the fact that research funders' agendas do not always address the most pressing problems—and points to practical ways how research infrastructure and capacity can be created in resource-limited settings.
- The chapter on 'Guiding principles for clinical research' outlines the history of current good clinical practice (GCP) principles, which have evolved in industrialised countries, and recommends that they should be applied meaningfully to suit the ethical and scientific requirements of the study, with a proportional level of detail that is sufficient to answer the research question. It also calls for more international regulatory collaboration and reliance. The chapter goes on to discuss benefit–risk assessment in emergencies, emphasising the need to maintain solid, scientific and evidence-based principles in regulatory review of studies and new interventions and giving practical recommendations on how to achieve this.

- The chapter titled 'Ethical considerations' highlights five main aspects, which are briefly summarised in the next section.
- The last Chapter, titled 'Scientific considerations', discusses ways to conceptualise and design scientifically valid research, and to ensure responsible data- and information-sharing.

This is followed by five appendices on specific aspects of clinical research in resource-limited settings: (1) Inclusion and protection of special populations, i.e., women and children; (2) digital technologies and electronic health records; (3) disease outbreaks; (4) an example of the debate on standard of care; and (5) the use of pharmacogenetics and personalised medicine. The last two annexes list the Working Group members and meetings, and the persons who provided comments on the draft report.

3.2 Selected Key Aspects

3.2.1 Ethical Considerations

The Working Group report discusses five aspects of ethical considerations when conducting research in low-resource settings. It refers to the CIOMS *International ethical guidelines for health-related research involving humans* [9] where appropriate and provides practical recommendations and examples of how to best to implement these.

Vulnerability

Issues that require special attention in resource-limited settings include the scientific and medical validity of studies, informed consent, compensation for participation in research, indemnity in the event of research-related harm, and caring for participants' health needs during and after the study. As the benefit–risk balance of research may differ between studies, and between sites participating in a multi-site clinical trial, researchers and sponsors should do a tailored analysis for each study and site.

Protecting Research Participants

The consensus report discusses the main practical aspects of addressing the above-mentioned issues in the context of resource-limited settings. Two examples of how the rights of women can be safeguarded in clinical research are provided in an annex to the report.

Avoiding Exploitative Research

In recent years, HIC organisations and companies have been increasingly conducting clinical trials in LMICs. Such partnerships can be highly advantageous for both parties, but they can also pose significant risks of exploitation as a result of the economic inequity. Adherence to the Global Code of Conduct for Research in Resource Poor Settings [38] will support long-term equitable research relationships between partners in lower-income and high-income settings. Research ethics committees (RECs) have a central role in ensuring that the general ethical principles for clinical research are followed, including in public health emergencies. In LMICs several constraints threaten the RECs' ability to facilitate good clinical research efficiently and to function to an acceptable global standard. An informed, unbiased, efficient and effective REC is critical to the research process. Capacity-building, including training for ethical review, should be supported by governments, funders and RECs themselves.

Ethical Review and Capacity-Building

Research ethics committees (RECs) have a central role in ensuring that the general ethical principles for clinical research are followed. The consensus report discusses the responsibilities of RECs, the uses of accelerated review, and the main shortcomings in LMICs that need to be addressed by capacity-building.

Community Engagement

The involvement of relevant communities through a meaningful participatory process is critical for clinical research. Guideline 7 of the 2016 CIOMS ethical guidance requires that both potential study participants and communities be involved [9]. Community engagement is particularly important in resource-limited settings, where the realities of life and the understanding of medical science are often vastly different from those of the researchers or sponsors. The community advisory board is an example of a useful approach. The consensus report includes various references to clinical research where the local context was respected; an example of a study is given in an annex titled 'Special populations'. Community engagement should advance good quality clinical research in resource-limited settings by building trust, managing expectations, facilitating communication of research outcomes to participants, and enabling negotiations for investments in research projects and infrastructure. Formal communication plans should state in advance how a researcher will encourage, moderate and sponsor community engagement. Sponsors have a duty to inform clinical trial participants and their communities about the research being conducted. Doing this in an appropriate, yet realistic manner is particularly important in resource-limited settings in order to build trust and facilitate implementation of research findings.

3.2.2 The COVID-19 Pandemic: Lessons Learned

Conducting clinical research in outbreak settings is always challenging [39]. The problems confronting the conduct of urgent clinical research in low-resource settings are described in an annex to the consensus report, using the examples of the Ebola outbreak in West Africa in 2014–2016 and of the COVID-19 pandemic that unfolded towards the end of the Group's work. Many challenges were common to both outbreaks: Lack of needed resources, equipment and laboratory capacity, heightened logistic problems, challenges in coordinating research, slow regulatory and ethical review processes, and the added stress of working in an outbreak environment. However, the COVID-19 pandemic affected the global community on an unprecedented scale, with a disproportionate and heterogeneous impact in resource-limited settings. Effective vaccines were developed in record time, but a major concern remained how equitable access to these products could be ensured. The Working Group concluded that it was still too early to consider all the lessons learned from COVID-19, but that a conducive environment, collaboration, effective communication and engagement with local communities clearly underlie an effective research response at the international level.

3.3 Recommendations

The CIOMS Working Group agreed on a total of 20 high-level recommendations, which are grouped into three sets for different target audiences (Table 1):

- 'Governments and regulatory authorities', including relevant ministries, e.g., of health or science; authorities in charge of regulating health products, and bodies in charge of scientific and ethical review of research protocols,
- 'Researchers', including those from academic institutions, the health care industry, contract research organisations, and non-commercial entities conducting research in low-resource settings, and
- 'Funders', including organisations such as the Bill & Melinda Gates Foundation or the Wellcome Trust, public–private partnerships such as the Drugs for Neglected Diseases initiative (DNDi), Medicines for Malaria Venture (MMV); and other new actors in the evolving research landscape as described in the CIOMS Working Group report.

The report can foster joint efforts to remove existing barriers and mobilise sustainable investments in good quality clinical research in resource-limited settings. This is needed as part of sustainable development globally.

Table 1 Overview of CIOMS Working Group recommendations on clinical research in resource-limited settings [16]

To governments and regulatory authorities
Research environment (abridged): 1. Invest in a sustainable research environment. 2. Avoid other countries' mistakes in setting up electronic health record systems. 3. Combat inefficiency and corruption. 4. Create incentives for research *Guiding principles* (abridged): 5. Clarify and simplify regulatory requirements *Ethical considerations*: 6. Establish and enforce effective regulations for ethical review; ensure appropriate protection—Which does not mean exclusion—Of vulnerable persons and groups in research 7. Support the establishment of platforms for researchers to engage with patient representatives and communities, e.g., community advisory boards; request and consider formal communication plans as part of applications for clinical studies *Scientific considerations* (abridged): 8. Invest in constructive dialogue with stakeholders ensure that the research findings are implemented in national health systems
To researchers
Research environment (abridged): 9. Understand and respect the local context; aim to build sustainable research capacity in resource-limited settings *Guiding principles* (abridged): 10. Apply the principles of good clinical practice *Ethical considerations* 11. Engage local study participants and communities throughout the research, from an early stage of study design, to ensure that the research adheres to high ethical standards. This will help to generate relevant findings and facilitate their translation into health benefits, thereby justifying the burdens of the study for the local population. Do not divert resources from already overstretched local health care systems 12. Plan in advance how to communicate and engage, throughout all phases of the clinical research, with community stakeholders such as participants, participants' partners and families, community, traditional and religious leaders, community engagement or advisory boards; be transparent about the aims and interests of all parties involved *Scientific considerations* (abridged): 13. Ensure scientifically justified research questions, robust study designs and data collection methods; where relevant, contribute to systematic reviews and meta-analyses. 14. Consider innovative, study designs and technologies. 15. Invest in data integrity, transparency and confidentiality at all stages
To international organisations and funders
Research environment (abridged): 16. Support policies and multi-functional coalitions that facilitate a conducive *1. Research environment* *Guiding principles* (abridged): 17. Support multi-country systems and coalitions for effective oversight of clinical research *Ethical considerations* 18. Prioritise research that answers important questions definitively and is relevant for the specific setting and the health care systems of the communities involved 19. Educate, empower, and support patient organisations and communities to foster an understanding of the value of clinical research *Scientific considerations* (abridged): 20. Make collaboration and data-sharing agreements; avoid research fragmentation; support dissemination of information and results

Note: Only the recommendations linked to 'Ethical considerations' are reproduced in full; the other recommendations are abridged for an easier overview.

4 CIOMS Working Group Report on Patient Involvement

4.1 Background

In recent years it has increasingly been recognised that the unique expertise and perspective of people who live with a serious or long-term disease and of those who care for these patients, can advance the development and use of new medical technology significantly. The CIOMS Working Group XI on Patient Involvement in the development, regulation and safe use of medicines was established in 2018 to produce the first global guidance document on this topic. The group members included patients, patient advocates, regulators, academics, and industry experts, representing perspectives from the European Union, Japan, and the United States of America as well as other regions. Additional input was sought at an open meeting in April 2019 from patient representatives, including from low-resource settings, and at a workshop held in Kampala, Uganda, in August 2019.

The Working Group report was published in September 2022 [40]. It is a pragmatic handbook for involving patients at all stages of the life cycle of medicines. In this context, 'patients' refers to the entire patient community and can include the patient's family, caregivers, patient organisations, and patient representatives in various situations where medicines are discussed. 'Medicines' include any products that are subject to approval by medicines regulatory authorities, e.g., pharmaceuticals, biologicals, vaccines and medicine-device combinations. The report recommends best practices of patient involvement wherever possible, for readers to review and select those which fit with their organisational needs.

4.2 Selected Key Aspects

4.2.1 Ethical Considerations

Patient and public involvement rests mainly on ethical and democratic principles of respecting patients as equal and valuable partners. The introductory section of the Working Group includes a discussion of the ethical considerations for patient involvement. This emphasises that patients' contribution of their preferences, concerns, understandings, and lived experiences of their medical condition can significantly improve clinical care, and discusses the ways in which responsible patient involvement will uphold the ethical principles of respect for persons, beneficence and nonmaleficence (protection of the person's welfare), and justice.

This foreword notes that, regrettably, not all who can benefit from medicines have access to them, because the obligation to provide medicines competes with obligations for meeting other fundamental needs. What constitutes fair distribution is often defined through policy, and not all individuals may consider the solutions to be just. In such cases, patients have historically engaged in advocacy to secure policies that align with their views. However, this can only happen if patients know what research is underway, what medicines exist and who has access to them, or the comparative efficacy and safety of different treatments. Patient involvement will empower likely medicines users to advocate for changes they consider necessary.

The body of the report makes comprehensive best practice recommendations for fair, sustainable and ethical patient engagement activities at all stages of medicines development and use. Some underlying principles include the endeavour to involve a full range of patient views, including from marginalised communities, caregivers, and those often excluded on account of their physical, mental, social, or educational status, including children and the very elderly. Patients should be adequately reimbursed for their time and expenses, and their involvement should be made as convenient as possible. The independence of patients and patient organisations should be maintained. Clear communication will support a long-lasting and trusting relationship, and training should be offered to all stakeholders as needed.

4.2.2 Low- and Middle-Income Countries

While patient involvement is a formal part of regulatory processes, for example, in Europe, the USA, and Japan, it is still in its infancy, or entirely absent, in LMICs. Patient involvement can be encouraged through local research and development initiatives, close collaboration with international institutions and patient organisations, by raising health literacy in the population, and by training health providers to look upon patients as partners in delivery of healthcare.

4.2.3 Pandemic Considerations

The report describes how activist groups organised effectively during the HIV/AIDS pandemic for access to treatment and were able to demonstrate a model for affecting policy. In the first year of the COVID-19 pandemic the CIOMS Working Group XI issued a statement, noting that 'we are all patients or potential patients' and calling for a united fight against the virus, with patient input as an integral part of all ethical, patient consent, scientific and public health processes that were in place prior to the pandemic. The ongoing pandemic has given patients the chance to become involved at all stages of medicine and vaccine development and their use in practice. Some specific concerns have come to light, including a general lack of knowledge on how new medicines and vaccines are developed, and the crucial need to deal with misinformation. COVID–19 has highlighted the scope of patient involvement to improve outcomes by quickly identifying and addressing public concern about vaccination, providing comprehensive information for patients to make an informed decision on vaccination; and making robust preparations for future pandemics.

As patient involvement evolves and expands across different countries and regulatory jurisdictions, much more will be learned. Sharing the lessons widely among diverse audiences—e.g., through professional conferences, social media, and peer-reviewed publications—will advance patient involvement and firmly entrench it in the development, regulation, and safe use of medicines.

5 Conclusion

As the 2016 CIOMS ethical guidelines state, progress towards a world where all can enjoy optimal health and health care is crucially dependent on scientific research supporting innovation, including research involving humans. It has been said that, while addressing the fundamental ethical issues, the guidelines strike an exemplary and reasonable balance between the protection of participants and promotion of research to meet the medical needs of future patients [31].

This balance is constantly changing. New developments are raising new issues that need to be assessed. CIOMS has been revising its ethical guidelines in response to emerging challenges and changing global perspectives (see Sect. 1.2). In addition, it has convened Working Groups on specific topics dedicated to medical product development and safety. This will continue to be the case in the future.

Conflict of Interest The authors have no competing interests to declare that are relevant to the content of this chapter.

References

1. Statutes of the Council for International Organizations of Medical Sciences (CIOMS). 2021. Available at: https://cioms.ch/wp-content/uploads/2021/09/CIOMS-Statutes_2021_English.pdf.
2. CIOMS website. Available at: https://cioms.ch, Accessed 7 July 2022.
3. Biomedical Science and the Dilemma of Human Experimentation; round table organized with the assistance of Unesco and the World Health Organization, Paris, Oct 7, 1967. Fattorusso V, editor. Council for International Organizations of Medical Sciences (CIOMS); 1968.
4. The World Medical Association. The declaration of Helsinki. Ethical principles for medical research involving human subjects. First adopted in 1964, last amended 2013. Available from: https://www.wma.net/policies-post/wma-declaration-of-helsinki-ethical-principles-for-medical-research-involving-human-subjects/.
5. Council for International Organizations of Medical Sciences (CIOMS). Proposed international ethical guidelines for biomedical research involving human subjects. Geneva: CIOMS; 1982. Available at: https://cioms.ch/publications/product/proposed-international-guidelines-for-biomedical-research-involving-human-subjects/. Accessed 30 May 2023.
6. Council for International Organizations of Medical Sciences (CIOMS). International ethical guidelines for biomedical research involving human subjects. Geneva: CIOMS; 1993. Available at: https://cioms.ch/publications/product/international-ethical-guidelines-for-biomedical-research-involving-human-subjects/. Accessed 30 May 2023.
7. Council for International Organizations of Medical Sciences (CIOMS). Ethical review of epidemiological studies. Geneva: CIOMS; 1991. Available at: https://cioms.ch/publica-

tions/product/1991-international-guidelines-for-ethical-review-of-epidemiological-studies/. Accessed 12 July 2022.
8. Council for International Organizations of Medical Sciences (CIOMS). International ethical guidelines for biomedical research involving human subjects. Geneva: CIOMS; 2002. Available at: https://cioms.ch/publications/product/international-ethical-guidelines-for-biomedical-research-involving-human-subjects-2/.
9. Council for International Organizations of Medical Sciences (CIOMS). International ethical guidelines for epidemiological studies. Geneva: CIOMS; 2009. Available at: https://cioms.ch/publications/product/international-ethical-guidelines-for-epidemiological-studies/. Accessed 12 July 2022.
10. Council for International Organizations of Medical Sciences (CIOMS). International ethical guidelines for health-related research involving humans. Geneva: CIOMS; 2016. Available at: https://doi.org/10.56759/rgxl7405. Accessed 6 Sep 2022.
11. Council for International Organizations of Medical Sciences (CIOMS). Principles of medical ethics relevant to the protection of prisoners against torture. Geneva: CIOMS; 1983. Available at: https://cioms.ch/publications/product/1983-principles-of-medical-ethics-relevant-to-the-protection-of-prisoners-against-torture/. Accessed 7 July 2022.
12. Council for International Organizations of Medical Sciences. Genetics, ethics and human values: human genome mapping, genetic screening and gene therapy. J Int Bioethique. 1991;2(2):96–8. PMID: 11659360. Available at: https://cioms.ch/publications/product/genetics-ethics-and-human-values/. Accessed 30 May 2023.
13. Council for International Organizations of Medical Sciences (CIOMS). Biomedical research involving animals: proposed international guiding principles. Geneva: CIOMS; 1984. Hardcopy availably by order from: https://cioms.ch/publications/product/biomedical-research-involving-animals-proposed-international-guiding-principles/.
14. Council for International Organizations of Medical Sciences (CIOMS)/International Council For Laboratory Animal Science (ICLAS). International guiding principles for biomedical research involving animals. 2012. Available at: https://iclas.org/cioms-and-iclas/. Accessed 7 July 2022.
15. Drug development research in resource-limited countries: how to succeed in implementation of good clinical practice guidelines. Draft report of the Joint CIOMS/WHO Working Group. 2005. Available at: https://cioms.ch/wp-content/uploads/2017/05/DrugDevelopRpt14Dec2005.pdf. Accessed 7 July 2022.
16. Council for International Organizations of Medical Sciences (CIOMS). Clinical research in resource-limited settings. A consensus by a CIOMS Working Group. Geneva: CIOMS; 2021. Available at: https://doi.org/10.56759/cyqe7288. Accessed 6 Sept 2022.
17. Council for International Organizations of Medical Sciences (CIOMS). Practical approaches to risk minimisation for medicinal products: report of CIOMS Working Group IX. Geneva: CIOMS; 2014. https://cioms.ch/publications/product/practical-approaches-to-risk-minimisation-for-medicinal-products-report-of-cioms-working-group-ix/. Accessed 7 July 2022.
18. Statement of Council for International Organizations of Medical Sciences (CIOMS) International Expert Working Group XI: patient contribution to the development and safe use of medicines during the Covid-19 pandemic. 2020. Available at: https://cioms.ch/wp-content/uploads/2018/03/CIOMS_WGXI_Statement_7December2020.pdf, Accessed 7 July 2022.
19. Council for International Organizations of Medical Sciences (CIOMS). CIOMS Cumulative Glossary, with a focus on pharmacovigilance. Version 2.1. Geneva; 2023. Available at: https://doi.org/10.56759/ocef1297. Accessed 30 May 2023.
20. CIOMS. Publications [website]. Available at: https://cioms.ch/publications/page/2/?filterby=pharmacovigilance. Accessed 7 July 2022.
21. International Council for Harmonization of Technical Requirements for Pharmaceuticals for Human Use (ICH). 30th anniversary publication. Geneva: ICH; 2021. Available at: https://admin.ich.org/sites/default/files/inline-files/ICH30thAnniversaryPublication_Interactive_2021_1014.pdf. Accessed 12 July 2022.

22. Danan G, Teschke R. Drug-induced liver injury: why is the Roussel Uclaf Causality Assessment Method (RUCAM) still used 25 years after its launch? Drug Saf. 2018;41(8):735–43. https://doi.org/10.1007/s40264-018-0654-2.
23. Council for International Organizations of Medical Sciences (CIOMS). Drug-induced liver injury (DILI): current status and future directions for drug development and the post-market setting. A consensus by a CIOMS Working Group. Geneva: CIOMS; 2020. Available at: https://doi.org/10.56759/ojsg8296. Accessed 6 Sept 2022.
24. Council for International Organizations of Medical Sciences (CIOMS). CIOMS guide to active vaccines safety surveillance. Geneva: CIOMS; 2017. Available at: https://doi.org/10.56759/hnuw8440. Accessed 6 Sept 2022.
25. Council for International Organizations of Medical Sciences (CIOMS). CIOMS guide to vaccine safety communication. Geneva: CIOMS; 2018. Available at: https://doi.org/10.56759/zphi4166. Accessed 6 Sept 2022.
26. World Health Organization. Global vaccine safety blueprint 2.0 (GVSB2.0) 2021–2023. Geneva: WHO; 2021. License: CC BY-NC-SA 3.0 IGO. Available at: https://www.who.int/publications/i/item/9789240036963, Accessed 7 July 2022.
27. World Health Organization. COVID-19 vaccines: safety surveillance manual. Geneva: WHO; 2020. License: CC BY-NC-SA 3.0 IGO. Available at: https://apps.who.int/iris/handle/10665/338400. Accessed 7 July 2022.
28. Medicines assessment during public health emergencies needs good science, best practices and proper communication. Statement of Council for International Organizations of Medical Sciences (CIOMS) International Expert Working Group. 2020. Available at: https://cioms.ch/wp-content/uploads/2020/06/CIOMS_WGXII_Statement.pdf. Accessed 7 July 2022.
29. Universal Declaration on Bioethics and Human Rights. Paris, France: United Nations Educational, Scientific and Cultural Organization (UNESCO); 2005. Available from: https://unesdoc.unesco.org/ark:/48223/pf0000142825.page=80.
30. van Delden JJ, van der Graaf R. Revised CIOMS international ethical guidelines for health-related research involving humans. JAMA. 2017;317(2):135–6. https://doi.org/10.1001/jama.2016.18977.
31. Ehni HJ, Wiesing U. Research ethics for a globalised world: the revised CIOMS international guidelines. Indian J Med Ethics. 2017;2(3):165–8. https://doi.org/10.20529/IJME.2017.046.
32. Akanmori BD, Mukanga D, Bellah A, Traore T, Ward M, Mihigo R. The role of the African Vaccine Regulatory Forum (AVAREF) in the accelerated clinical evaluation of Ebola vaccine candidates during the large West Africa epidemic. J Immunol Sci. 2018;S(011):75–9. https://doi.org/10.29245/2578-3009/2018/si.1111.
33. Katz RA, Salamanca-Buentello F, Silva DS, Upshur RE, Smith MJ. R&D during public health emergencies: the value(s) of trust, governance and collaboration. BMJ Glob Health. 2022;7(3):e007873. https://doi.org/10.1136/bmjgh-2021-007873.
34. Balakrishnan V, Ng WZ, Soo MC, Han GJ, Lee CJ. Infodemic and fake news - a comprehensive overview of its global magnitude during the COVID-19 pandemic in 2021: a scoping review. Int J Disaster Risk Reduct. 2022;78:103144. https://doi.org/10.1016/j.ijdrr.2022.103144.
35. Eichler HG, Bloechl-Daum B, Broich K, Kyrle PA, Oderkirk J, et al. Data rich, information poor: can we use electronic health records to create a learning healthcare system for pharmaceuticals? Clin Pharmacol Ther. 2019;105(4):912–22. https://doi.org/10.1002/cpt.1226.
36. U.S. Food and Drug Administration. COVID-19 Real World Data (RWD) data elements harmonization project [webpage]. Available at: https://www.fda.gov/drugs/coronavirus-covid-19-drugs/covid-19-real-world-data-rwd-data-elements-harmonization-project. Accessed 11 July 2022.
37. Defining intent, and guiding harmonization and ethics standards for real-world data and real-world evidence in regulatory decision-making. CIOMS concept paper. 2020. Available at: https://cioms.ch/wp-content/uploads/2020/03/CIOMS_Concept-Note_-RWD-and-RWE_31Mar2020.pdf. Accessed 11 July 2022.

38. TRUST consortium. Global code of conduct for research in resource-poor settings Available at: https://www.globalcodeofconduct.org/wp-content/uploads/2018/05/Global-Code-of-Conduct-Brochure.pdf. Accessed 6 Mar 2022.
39. Macklin R, Cowan E. Conducting research in disease outbreaks. PLoS Negl Trop Dis. 2009;3(4):e335. https://doi.org/10.1371/journal.pntd.0000335.
40. Council for International Organizations of Medical Sciences (CIOMS). Patient involvement in the development, regulation and safe use of medicines. CIOMS Working Group XI report. Geneva: CIOMS; 2022. Available at: https://doi.org/10.56759/iiew8982. Accessed 6 Sept 2023.

Ethics of Placebo-Controlled Trials: Historical Analysis Including Experiences During the COVID-19 Pandemic

Chieko Kurihara, Dirceu Greco, Ames Dhai, Takeo Saio, and Hiroe Tsubaki

Abstract The ethical justifiability of "placebo-controlled trials in the presence of proven intervention" is an issue that hinges on the principle of justice in bioethics. Since the onset of the HIV/AIDS pandemic in the 1980s, the issue of placebo-controlled trials after the discovery of efficacious drugs for HIV in the mid-1990s has sparked international controversy. This topic needs to be revisited with the experiences accumulated during the COVID-19 pandemic where the international community tried to overcome a specific disease with a call for global solidarity and cooperation.

Based on the core principle of the World Medical Association's Declaration of Helsinki that the rights and interests of the study participants take precedence over scientific goals, it is not appropriate to justify "placebo-controlled trials in the presence of proven intervention," even where an upper limit for the risk of harm is specified. Acceptance conditions for comparative clinical trials should be "clinical equipoise" between the new intervention being tested and an intervention recognized as safe and efficacious on a global perspective. The intervention being tested is not to be compared with an intervention that is economically affordable in the area where the study is to be conducted.

C. Kurihara (✉)
Kanagawa Dental University, Yokosuka, Kanagawa, Japan
e-mail: chieko.kurihara@nifty.ne.jp

D. Greco
Department of Internal Medicine, School of Medicine, Federal University of Minas Gerais, Belo Horizonte, Minas Gerais, Brazil

A. Dhai
School of Clinical Medicine, University of the Witwatersrand, Johannesburg, Gauteng, South Africa

T. Saio
Department of Internal Medicine and Psychiatry, Fuji Toranomon Orthopedic Hospital, Shizuoka, Japan

H. Tsubaki
The Institute of Statistical Mathematics, Tokyo, Japan

C. Kurihara et al. (eds.), *Ethical Innovation for Global Health*,
https://doi.org/10.1007/978-981-99-6163-4_11

Thus ethical principles must be articulated from the perspective of ethical and social values to be achieved and not from the perspective of feasibility. Refining research practices that make this principle achievable is essential to truly overcoming a global pandemic.

This chapter historically traces these controversies, analyzes international normative documents, and clarifies the ethical principle for placebo-controlled trials.

Keywords Placebo-controlled trial · Clinical equipoise · Human dignity · Medical ethics · Exploitation

1 Introduction: Focus of the Issue

The World Medical Association (WMA)'s Declaration of Helsinki (DoH) was first adopted in 1964, with the latest revision in 2013 [1], and the next revision being planned for October 2024. Its article since 1996 addressing placebo-controlled trials where there is a proven intervention aroused international controversies during the HIV/AIDS pandemic of the 1980s and onwards. Similar debates resurfaced in the COVID-19 pandemic since the year 2020. In both pandemics concerns have been raised that participants in low-resource settings have been subjected to placebo-controlled trials, without access to the already proven interventions, while these interventions were already available in wealthier nations. This is evidence of an imbalance in the risk-benefit distribution, against the established ethical principle of "justice" [2–4]. Providing the best proven intervention to those in need is required from the principle of "beneficence" [3, 4]. This should not be a matter to be left to the consent of an individual, within the principle of "respect for persons" [3] (or "respect for autonomy" [4]), because it is a matter of "human dignity" [5–7] which is the bases for "respect for persons" and "the right to health" [8, 9].

In the situations where a proven intervention exists but its safety and efficacy for specific populations (e.g., children, the elderly, pregnant women) have not been established, it is ethically acceptable to conduct a placebo-controlled trial on these populations. However, especially for economic reasons, placebo has been used for a control group when testing a new intervention, rather than the best proven intervention. The biggest problem is that the economic demands are allowed to overwhelm ethical principles [10], allowing utilitarian ethics to take precedence over deontological ethics [11], which mandates respect for the dignity and rights of individuals.

The WMA's DoH and the Council of International Organizations of Medical Sciences (CIOMS)'s International Ethical Guidelines for Health-related Research Involving Humans (last updated in 2016) ("CIOMS Guidelines" hereafter) [12] are well-recognized international standards. Both open up the possibility of allowing placebo in special situations. They have defined the acceptable range of risk of harm for participants in controlled trials using a placebo or an intervention less effective than the best proven intervention (for simplicity we only mention "placebo" where relevant). Yet, their limits for proper use of placebo (choice of words) are different from each other (Table 1). This is an indication that the controversy has not yet been resolved.

This chapter reviews the history of debates over placebo-controlled trials and proposes that global ethical principles be applied in relation to the use of placebo, based on the global experience of justices and injustices during the AIDS and COVID-19 pandemics. The norm of ensuring study participants "post-trial access" to the interventions proven effective in controlled trials has emerged in parallel with the controversy over placebo-controlled trials, but this topic is discussed in the next chapter.

In addition, in this chapter we do not focus on the specific responsibilities of the researcher and sponsor. The premise is that any study that is conducted must be ethical and there are combined responsibilities of all stakeholders in the ethics ecosystem to ensure benefits are maximized, harms are minimized, participants are not exploited, and the research is carried out in a fair and just manner.

2 Foundations for Human Rights and Research Ethics Norms

2.1 International Instruments and Establishment of Research Ethics Principles

Research involving humans makes use of a person to generate scientific knowledge and therefore could violate Kant's maxim [13, 14] that one should not use people only as a means to an end, in the sense of respecting human dignity. The cruel human experimentation by Nazi Germany during World War II was condemned as a "crime against humanity" at the post-war "Doctor's Trial" in the subsequent "Nuremberg trials" by the United States (US). Its judgment of 1947 set out the conditions for permissible human experimentation, called the "Nuremberg Code" [15].

Table 1 Justification conditions of placebo-controlled clinical trials in the Declaration of Helsinki and CIOMS Guidelines (bullet points rewritten)

1. There is no proven intervention[a]; **OR**
2. When there is a proven intervention[a]
(a) Where compelling and scientifically sound methodological reasons **AND**
(b) No additional risks of serious or irreversible harm (DoH)/
Minor increase above minimal risk (CIOMS)
Abuse of this option must be avoided; premise of adherence of all the other principles (DoH)

[a]Instead of "(best) proven intervention" in the DoH, CIOMS use "established effective intervention." Although some different interpretations have been expressed for each of these phrases, this chapter uses terminology used by the DoH and will not focus on this difference

Other cruel human experimentation by the Imperial Japanese Army were condemned by the Khabarovsk Military Tribunal [16], but the accused were exonerated for providing the results of their experiments to the US government [17]. Such political trade-off cannot be justified based on principles of ethics. And in the deontological context, it is seen as a means to an end that suited the US government as this involved an exchange of scientific "know how" that could possibly be used to serve Western ideals. It was also a means to an end for the Japanese Government as it escaped prosecution for its war crimes.

Immediately after World War II, the Constitution of the World Health Organization (WHO) in 1946 recognized people's equal right to health [8] and the Universal Declaration of Human Rights of the United Nations (UN) in 1948 [5] established a Bill of Rights for people based on "human dignity." The Geneva Conventions of the International Committee of the Red Cross Convention (ICRC) was expanded in 1949 as an international wartime law, prohibiting medical experimentation on prisoners of war that are not in their own interest [18]. The International Covenant on Human Rights in 1966 [7], which legislated the UN Declaration of Human Rights [5] as an international treaty, prohibits scientific experimentation without consent. However, the last two have not provided comprehensive norms for protecting human research participants.

Even well after World War II, continuing discussion on ethically questionable research that involves humans led to the WMA's DoH in 1964, and the CIOMS Guidelines in 1982, as well as the US Government Commission's "Belmont Report" in 1979 [3]. These are widely referred to as standard setting in the context of international codes of ethics in research contexts.

In 1996, the International Conference on Harmonisation of Technical Requirements for Registration of Pharmaceuticals for Human Use (ICH)'s Good Clinical Practice (GCP) [19] was agreed as the standard for clinical trials for marketing approval of drugs.

The Council of Europe Convention on Biomedicine and Human Rights in 1997 [20] and the Declaration on Bioethics and Human Rights adopted by the United Nations Educational, Scientific and Cultural Organization (UNESCO) in 2005 [21] clarified the basis for the protection and promotion of human rights in biomedical research. The former is a comprehensive legal instrument but applies only to Member States, while the latter is not legally binding, and its scope is not limited to research.

Here we will review how the topic of placebo-controlled trial has been discussed in the international research ethics guidelines mentioned above (Table 2).

Table 2 Characteristics of international documents for ethics in research and points related to justifiability of placebo-controlled clinical trial

Year	Author/contributor, title (both in **bold**, some abbreviated), characteristics of documents, points related to justifiability of placebo control (**bold**). Verbatim texts are in *Italic*
1947	**International Military Tribunal "Nuremberg Code"** [15]: First internationally recognized condition of permissible human experiments, described in the judgment of doctors' war crime
1949	**ICRC "Geneva Convention"** [18] (the first revision after the WW2): Humanitarian law for the time of war prohibits medical experimentation on prisoners of war that are not in their own interest
1964	**WMA "The Declaration of Helsinki"** [22] (first version): First international agreement of physicians' ethical principle of research involving humans, including following two statements from the first version to date: • "The Declaration of Geneva": ***"The health of my patient will be my first consideration"*** • "The International Code of Medical Ethics": ***"A physician shall act in the patient's best interest when providing medical care"***
1966	**UN "International Covenant on Human Rights"** [7]: International treaty of basic human rights, prohibiting scientific experimentation without consent.
1975	**WMA "The Declaration of Helsinki"** [23] (2nd amendment): Revision to include the following core principles kept to date: • **Rights and interests of individual research subjects take precedence over the goal of research** • **A new method should be compared against *"the best current"* method**; and • **Every patient (study subject) *"should be assured of the best proven"* method**
1979	**US National Commission "Belmont Report"** [3]: Internationally recognized principles of research ethics issued in the United States, which includes: • Clarification of **boundary of "research" aiming at generation of new knowledge and "practice" aiming at individual care**, recognizing "***both often occur together***" • Clarification of principles of "respect for persons," "beneficence," and "justice," the last one promotes **fairness in distributive justice of risks and benefits of research, among disadvantaged and advantaged individuals**
1982, 93	**CIOMS "CIOMS guidelines"** [24, 25] (1982, 1993 versions): Guidelines from the perspective of research in developing countries, following the principles of the DoH
1996	**ICH "Good Clinical Practice (GCP) (E6) Guideline"** [19]: International standard for clinical trials of drugs for regulatory approval, agreed among the EU, US, and Japan, now adopted in the world, defining its principle to be in accordance with the ones in the DoH.
1997	**Council of Europe "Convention on Biomedicine and Human Rights"** [20]: Comprehensible legal instrument for research involving humans by the Council of Europe
1996; 2000	**WMA "The Declaration of Helsinki"** [26, 27] (4th, 5th amendments): The **condition of placebo control** was first defined in 1996 to be **only when there is no proven intervention**, and kept in the 2000 version
2000	**ICH "E10 Guidelines"** [28]: On the choice of control in comparative study, **permitting placebo when there is proven intervention to be no additional risks of serious or irreversible harm**

(continued)

Table 2 (continued)

Year	Author/contributor, title (both in **bold**, some abbreviated), characteristics of documents, points related to justifiability of placebo control (**bold**). Verbatim texts are in *Italic*
2002	**WMA "The Declaration of Helsinki"** [29] (6th amendment): "Note of Clarification" was added, permitting placebo control only because of **"compelling and scientifically sound methodological reasons,"** connecting by "or" with another condition, **"no *additional risk of serious or irreversible harm*"**
2002	**CIOMS "CIOMS guidelines"** [43] (2002 version): Adopts the condition of placebo control when there is an established effective intervention to be **no increase of risk of serious or irreversible harm**
2005	**UNESCO "Universal Declaration of Bioethics and Human Rights"** [21]: Its scope is not limited to research
2008, 2013	**WMA "The Declaration of Helsinki"** [1, 30] (8th, 9th amendments): "Or" in "Note of Clarification" was amended to "and" thus the condition was revised to **"no *additional risk of serious or irreversible harm*"** (Article 33)
2016	**CIOMS "International Ethical Guidelines for Health-related research Involving Humans"** [12] (2016 version): Revised the condition to be "***minor increase above minimal risk***" (Guideline 5)

2.2 *Ethical Principles for Controlled Trials*

2.2.1 Declaration of Helsinki

The DoH is a non-legally binding voluntary code of practice by a professional association of physicians, but it has become the most widely recognized international standard, incorporated into the laws or regulations of many countries and adopted as a condition by journals for acceptance of study reports for publication [31]. It is also widely referred to by non-physicians. The WMA was founded in 1947, just after the Second World War, and the DoH cites the following principles from the 1948 Declaration of Geneva (DoG) [32] (which the WMA regards as a modern version of the Hippocratic Oath [33]) and the 1949 International Code of Medical Ethics (ICoME) [34] since the first adoption in 1964 [22] to date. After the war, the WMA reflected considerably on the inhumane practices of physicians as well as the unethical research during World War II [35]. They attempted to draft ethical principles for research in collaboration with lawyers, but the idea was rejected [36] and the DoH was adopted as a professional code for physicians [35]. Therefore, the responsibilities of physicians in medical practice were clarified as in the following principles, while logically making a distinction between "research" and "practice."

- ***"The health of my patient will be my first consideration."*** (The third sentence of the DoG, 2006 version, cited in Article 3 of the DoH.)
- ***"A physician shall act in the patient's best interest when providing medical care."*** (The 14th sentence of the ICMoE, 2006 version, cited in Article 3 of the DoH.)

- **Rights and interests of individual research subjects take precedence over the goal of research.** (Article 8 of the DoH, negative sentence rephrased into positive sentence.)

Based on this core principle, the 1975 revision of the DoH [23] defined the principle for comparative studies stating that:

- **a new method should be compared against *"the best current"* method** (II-2); and
- **every patient (study subject) *"should be assured of the best proven"* method** (II-3).

Although there was no change in the 1983 version, the 1996 revision added the statement that the above-mentioned principle does not exclude use of placebo where there is no proven method [26]. It was in response to warnings of abuse of placebo-controlled studies [37, 38], and in the meantime concern about the gap between the DoH and the requirement for placebo control by US and European regulators [39]. Shortly afterwards, as mentioned in the Introduction, the debate over the appropriateness of placebo-controlled trials in the presence of proven interventions intensified. The principle since the 1975 version was affirmed in the 2000 revision [27]. In addition, a new norm of ensuring study participants "post-trial access" to the proven intervention in the study was first adopted in this version (this topic will be further discussed in the next chapter). However, succeeding amendments [1, 29, 30] have provided exceptions that contradict the above core principle, and permit placebo when proven intervention exists, conditional to "no additional risks of serious or irreversible harm" (Article 33) (Table 1).

2.2.2 CIOMS Guidelines

CIOMS is a council of academic bodies, not limited to physicians, and includes academia, regulators as well as pharmaceutical companies. CIOMS was established jointly by WHO and UNESCO in 1949 [40] and in cooperation with the WHO builds on the UN's international human rights norms [5–7]. In the late 1970s, CIOMS embarked on a project for the application of the Nuremberg Code [15] and the DoH to research involving humans in developing countries, in contrast to the criminal human experimentation that took place in World War II. It also attempted to prevent exploitation in research driven by industrial countries and conducted in low-resource settings (including collaborative research). The first guidelines were published in 1982 [24], followed by another separate set of guidelines for epidemiological studies in 1991 [41], being revised in 2009 [42]. The guidelines for biomedical research were revised in 1993 [25] and 2002 [43], and the latest 2016 version [12] was combined with the previously separated guidelines for epidemiological studies. The 1993 guidelines recommended that the statement in the DoH 1989 assuring that patients in controlled trials received the best proven interventions be followed. However, as detailed in Sect. 3, it also came to define the acceptability of placebo control when established intervention exists to be "no additional risk of serious or irreversible harm" in 2002 in advance to the DoH, and in the 2016 version this acceptability was more restricted to situations with "minor increase above minimal risk" (Guideline 5).

2.2.3 The Belmont Report

The Belmont Report was presented in 1979 [3] by a US National Commission in response to what had occurred in the Tuskegee Syphilis Study. Although it is a US Government document, it expounds the internationally recognized idea of distinguishing between "research" and "practice," and the three principles of "respect for persons," "beneficence," and "justice." These notions are closely related to the debates on placebo-controlled studies and post-trial access.

The Tuskegee Syphilis Study was conducted in Alabama, US, and lasting from 1932 to 1972. It recruited poor African Americans affected by syphilis, and the researchers did not provide effective treatment even after penicillin became available to the general population [44]. It was only discontinued in 1972 after being publicly exposed in the lay press. A series of investigations into the study was then carried out by a committee established at the behest of the US Congress and this led to the comprehensible reports on questionable studies [45] and establishment of the US National Research Act 1974 [46] (applicable to research receiving federal research funding). The Act mandated study evaluation by an institutional review board (IRB) and required obtaining informed consent from study participants.

The Belmont Report defines "research" as being designed "*to develop or contribute to generalizable knowledge*" and "practice" as "*designed solely to enhance the well-being of an individual patient or client*" but it also states that "*both often occur together.*" It revised four principles by Beauchamp and Childress [4] replacing "respect for autonomy" with "respect for persons" (from which informed consent derives) integrating "non-maleficence" into "beneficence" (from which risk-benefit assessment derives), and maintained "justice" (distributive justice, in terms of selection of participants), and established the three principles of research ethics. Most importantly, in line with the justice principle, it is unethical that disadvantaged people are more exposed to research risks while benefit generated from research flows to the advantaged.

Using Claude Bernard's explanation [47] to clarify the distinction between observation and experimentation and extending it to the Hippocratic maxim "do no harm" to research, "*saying that one should not inure one person regardless of the benefits that might come to others,*" the Report also states that "*learning what will in fact benefit may require exposure to risk.*" Thus, by applying the principle of "beneficence" to "*do no harm*" and "*maximize possible benefits and minimize possible harms,*" it allows for decision-making on justifiable risks involved.

This means that unlike the DoH within the system of physician's professional ethical codes, which prioritize the rights and interests of individuals over the benefits of generating new knowledge, the logical structure of the Report is that "research" is different from "practice" and the risks taken by individuals must be minimized to the extent that research can generate new knowledge. This concept has been the theoretical basis for US opposition to restrictions on placebo-controlled studies since the 1996 DoH, as discussed below.

2.2.4 The International Conference on Harmonisation: Good Clinical Practice (ICH-GCP)

In 1996, the ICH-GCP [19] as the standard for clinical trials for marketing approval of drugs was agreed among the regulatory authorities and pharmaceutical companies from the European Union (EU), US, and Japan. This was later updated [48] and its member/observer countries have expanded into other regions, have included academics, and the full name of the organization changed [49]. It refers to the DoH as a principle for clinical trials and defines the IRB review system and the informed consent of research participants. It also defines strict procedures for obtaining data credible enough for marketing approval. Several detailed guidelines have been issued for individual topics, one of which was the Guideline ("E10 guideline") on the choice of control group in comparative studies [28] published in 2000. This E10 guideline has a similar threshold as the DoH of "no increase of risk of serious or irreversible harm" of placebo control in the presence of proven intervention.

3 Process of the Pragmatic Modification of the Ethical Principles of Controlled Trials

3.1 Controversies in the HIV/AIDS Pandemic

This section traces the controversial process by which DoH principles over a 30-year period since 1975 were modified by demands of utilitarian pragmatism. In the 1980s, the HIV/AIDS pandemic led to massive mortality in Africa, Asia, and Latin America, and HIV also spread in developed countries.

Clinical studies in the US and France (ACTG 076) showing the efficacy of the use of regimens with zidovudine to prevent perinatal transmission were published in 1994 (25.5% placebo vs 8.3% zidovudine at 72 weeks (67.5 RRR)) [50]. Its positive results in significantly reducing the risk of HIV mother to child transmission established a global standard of care. However, because the cost for zidovudine alone was estimated to be 600 times more expensive than the annual medical costs of one person, researchers proposed a shorter regimen hoping that it could also be useful [51], and a number of clinical trials to test this modified regimen were conducted in 15 developing countries, sponsored by the US National Institutes of Health, Center for Disease Control, and French research institutes. But, instead of using the full regimen shown as efficacious in the ACTG 076 study as a comparator, they used placebo control to evaluate the efficacy of the shorter regimen. Reacting to this situation Lurie and Wolf [52] published an article in the *New England Journal of Medicine* (*NEJM*) arguing that it was unethical to conduct placebo-controlled trials in developing countries that could not be conducted in developed countries. Their position was reinforced by Angell [53, 54] also in the *NEJM*. In an editorial she stated that these studies were a cruel reminder of the Tuskegee Study [54].

In the 16 trials conducted in developing countries in 1994 and after, 15 were compared with placebo or non-antiretroviral drugs [52], while one was active controlled [55]. This active controlled study showed that the shorter regimen could be effective compared to placebo in other studies, but not as effective as the full regimen. Subsequently, of the 26 trials results, which were included in the table of evidence in the 2014 recommendation of treatment regimen [56] published by the US Department of Health and Human Services, 5 of the early trials, including 076, were placebo-controlled and the later trials were active drug-controlled. This table of evidence shows that the research question was answered in both the active control and in the placebo-controlled trial; therefore it confirms that it was not acceptable to expose participants to risk in the placebo-controlled trials. It is noteworthy that a placebo-controlled trial conducted in Thailand was stopped after an interim analysis showed that a short-course regimen could reduce infection rates, and pregnant women in the placebo arm trial were offered short-course regimen treatment [57].

The US CDC announced that *"these studies were not designed to address perinatal prevention needs in the U.S. and other industrialized nations. Because HIV-infected pregnant women in the U.S. already have access to the more effective longer treatment regimen, recommendations for perinatal HIV prevention in the U.S. will not change*" [57]. The appendix in the above recommendation stated that "*direct comparison of results from trials of these regimens is not possible because the studies involved diverse patient populations residing in different geographic locations, infected with diverse viral subtypes, and with different infant feeding practices*" [56]. This infers a possible reason for conducting a placebo-controlled study in different regions where background to support the proof is different. Yet, it is not a (this does not provide general) justification for not providing best-proven therapy to the control group, despite the existence of a global standard of care.

Lurie also criticized in 2001 the proposals for company-sponsored trials with Surfaxin for the fatal Respiratory Distress Syndrome (RDS) in neonates [58]. Placebo-controlled trials of new drugs were planned for four South American countries where the efficacious drugs existed but were expensive and not available there, and at the same time trials in Europe using active drug as a control were also planned. Furthermore, Angel pointed out in her 2004 book that the ease of conducting placebo-controlled trials led to the approval of a series of "me too drugs" that did not work [59].

3.2 The Twists and Turns of the Placebo Clause

In 1997 the American Medical Association (AMA), led by Levine, submitted a proposal of comprehensive revision of the DoH to include changing "best proven" to "best available" but it was not accepted [60]. Levine, who contributed to the clarification [61] of the distinction between "practice" and "research" in the Belmont Report, consistently raised objections [62] to the "therapeutic" and "non-therapeutic" research dichotomy in the DoH and argued that placebo-controlled trials should not be subject to specific conditions but should be subject to risk-benefit assessment. Temple [63] and Ellenberg [64] in the US Food and Drug Administration (FDA) argued that placebo-controlled trials were acceptable or needed where proven intervention existed in situations when the disease was mild, such as seasonal rhinitis, or when previous trials of existing drugs did not show adequate "assay sensitivity" ("*the ability of a study to distinguish between active and inactive treatments*"), or otherwise when short-term efficacy did not necessarily guarantee long-term efficacy. They further argued that placebo-controlled trials were ethically acceptable as long as they did not cause harm to the subjects.

In 2000, the ICH issued a guideline on choices of control group [28]. In their opinion, after considering careful study designs (e.g., methodological aspects including sample size estimation of active versus inactive controlled studies) and patient safety measures (e.g., early escape, rescue treatment, and interim analysis by a data safety monitoring board), the use of placebo was allowed when there was a proven drug if there was "no increased risk of irreversible harm." It is worth remembering that this originated from regulatory agencies and industries in the US, Europe, and Japan, which specifies the (high) threshold of harm acceptable for the placebo arm participants.

The revision of the DoH in 2000 maintained the norms of the 1996 version rejecting the AMA's proposal. However, a subsequent small working group convened by WMA decided to add a Note of Clarification in 2002 [29]. This note reversed the 2000 General Assembly decision and made the use of placebo more flexible, contrary to the established rule that limited its use and which has remained since 1975. This Note of Clarification included "*compelling and scientifically sound methodological reasons*" or (instead of "and") no "*additional risk of serious or irreversible harm*" as acceptability for placebo control when there was a proven therapy. Thus, it allows placebo controls only if scientifically justified. Regarding the "or" mentioned above, Human [60] and Kloiber [65], the former and current WMA Secretary Generals, explained that it was agreed by the WMA at the time as it was considered that exploitation would not occur if all other articles of the DoH were adhered to. Tsuboi, who was President of the WMA at the time of 2000 revision, stated that the Japan Medical Association (JMA) did not accept this modification and did not publish a Japanese translation of the note on the JMA website [66]. Tsuboi explained that Japan objected to Levine's proposal because developing countries were not in a position to object because they benefited from the US, thereby expressing the non-Western spirit that ethical reason takes precedence over scientific rationale and pragmatism. This position was also defended at the 2000 General Assembly in Edinburgh by representatives of other medical associations including, but not limited to, South Africa and Brazil. Tsuboi stated that the placebo clause in the 2000 version is a perfect, prima facie norm.

During this period, and concurrently, the 2002 revision of the CIOMS guidelines adopted the principle of "no increased risk of serious or irreversible harm." Subsequently in the 2008 revision of the DoH, the "or" in the above note was amended to "and" and the description in note was placed in the main text, which was maintained in the 2013 revision with small changes to the wording (Table 1). At this point, the high threshold risk for the acceptability of placebo when there was a proven intervention was aligned with the ICH and to the 2002 CIOMS guidelines.

However, in the 2016 revision of the CIOMS guideline, the threshold of risk was modified to "*minor increase above minimal risk.*"

3.3 Declining in Authority: FDA Abandoned the DoH

In 2008, the US FDA changed its rules to no longer use the DoH, but instead to ICH-GCP, for clinical trials conducted abroad and the results of which were included in new drug applications (NDA) in US [67]. Articles in the lay media and leading journals such as *Nature* [68] and *The Lancet* [69] warned that FDA was abandoning the DoH [70, 71]. According to Lurie and Greco [69] certain studies outside the US without an Investigational New Drug application (IND) might not come to the attention of the FDA until a drug company later seeks approval in the US by filing an NDA. At that time FDA regulations required studies submitted in support of NDA to have been done in a manner consistent either with the DoH or any local laws, whichever is more protective for patients. The FDA removed these requirements entirely and mandated only that the submitted studies be consistent with the ICH-GCP. However, the GCP guidelines were developed without involving developing countries and mainly address procedural issues, not overarching ethical ones. The GCP guidelines do not, for example, address placebo use, conflict of interest, the need to publish results, or post-trial availability of successful treatments to study participants or community members—topics included in the DoH. The agency stated that it was concerned with "ensuring quality of data" and that the GCP guidelines are therefore necessary. In the Lurie and Greco opinion the FDA also worried that the DoH could be modified "independent of FDA authority," although the GCP guidelines themselves are not immutable, and the agency does acknowledge that any revisions "could not supersede U.S. laws and regulations." These changes were aimed at expanding the acceptability of placebo-controlled trials that drug companies conduct outside the US.

Later on, Temple, who has been one of the opinion leaders of clinical trial methodology at the US FDA since the 1970s, agreed with the 2008 and 2013 versions of the DoH and explained that it was not prudent to refer to any regulation that the US had no control over, although many of their own principles "*had their origin in the Declaration*" [72].

The DoH and other relevant ethics documents are principle-based guidelines taking into consideration the basic values and norms of ethical research. GCP is standard setting from the technical perspective. These two guidance documents ought to complement each other and function as the adequate standard against which countries can work towards aspiring to improve on these standards. They should be read together and not at the exclusion of each other.

3.4 Declining in Authority: Latin America Rejected the DoH

On the other hand, Brazil has adopted requirements that emanated from the 2000 version of the DoH in their legal documents and do not accept the article on placebo use in the 2008 and 2013 versions [73]. Furthermore, other Latin American countries have also rejected the DoH since 2008 [74]. In 2008, the "Declaration of Cordoba" [75] was agreed on at the meeting of "South American and Caribbean UNESCO Bioethics Network," where 300 bioethics scholars from 12 Latin-American countries participated [76], to propose that all the governments and institutions should not accept the DoH 2008 version and should adopt the UNESCO Declaration on Bioethics and Human Rights. The "Buenos Aires Declaration" in 2008 [77] and the "Declaration of Pachuca" in 2013 [78, 79], both rejecting placebo control when there is a therapeutic method, were agreed on by other Latin American organizations. These declarations were issued responding to an increase of unethical placebo-controlled trials [80] and various other types of abuses [81] in Latin American regions.

The Brazilian National Research Ethics Commission of the National Health Council (CONEP/CNS) has adopted and even made the articles of placebo-controlled trials and of post-trial access in DoH 2000 version more stringent. It is noteworthy to add the position taken by the Brazilian Federal Council of Medicine (CFM) which is included in the Code of Medical Ethics that it is unethical for physicians to participate in any trial that did not use an active comparator [82, 83] and in Brazil the pharmaceutical companies adhered to these requirements [10, 84].

In 2018, a survey conducted revealed a diversity of results on the interpretation and use of placebo in national regulations [85]. Participants included 42 organizations and 103 regulatory authorities. Some countries interpret it strictly, others loosely, and the author concluded that "*The Declaration should be continued to be strengthened to enforce the appreciation of conducting medical research with the highest ethical standard.*"

3.5 Controversies Related to Research for Specific Illnesses

Controversy regarding placebo-controlled trials in the presence of a standard drug has also been evoked in the area of psychiatry [86–88]. The US FDA issued in 1977 psychotropic drug treatment guidelines [89] and, referring to the DoH 1975 version, detailed the situation where placebo-controlled trials were needed in the presence of a standard drug because of placebo response, while discussing various ethical questions and measures to prevent deterioration of patients in the placebo group. In 2001, data from a total of 10,118 cases from previously FDA-approved trials were analyzed, comparing established antipsychotics, new antipsychotics and placebo for suicide, attempted suicide and deterioration, where no significant difference for suicide and suicide attempt was shown [90]. Similar studies found similar results for depression [91–95] and schizophrenia [96], but there was evidence of difference in the symptom reduction rate [90, 91]. This raises the question of which risk threshold between the DoH and CIOMS is most appropriate.

The European Medicines Agency has stated in 2001 that placebo controls are required in both acute and maintenance phase studies. It recommends that studies to prove efficacy in the maintenance phase should include responders in the acute phase and randomize them to treatment and placebo groups (the duration of treatment in the acute phase should be longer than 6 weeks) [97].

In cardiology, while arrhythmic drugs have been recognized as the standard of care, the Cardiac Arrhythmia Suppression Trial (CAST) [98], a placebo-controlled trial with death as an endpoint, is widely known. It involved a 10-month treatment period for patients with asymptomatic or mildly symptomatic ventricular arrhythmias after myocardial infarction. Death due to arrhythmia and nonfatal cardiac arrest was 33/730 (4.5%) in the active drug group; 9/725 (1.2%) on placebo, and the relative risk was 3.6 (CI: 1.7-8.5). The trial was terminated prematurely. This trial is used by those that support the theory that when standard therapy has been proven for surrogate endpoints it may need to be tested against placebo for true endpoints.

There is also an example of a placebo-controlled trial of an angiotensin receptor blocker (ARB) showing 50% mortality reduction in patients who are intolerant of angiotensin-converting–enzyme (ACE) inhibitors [72]. This trial suggests that even when there is a standard intervention there would be some group of individuals not responding to it, indicating the absence of the proven intervention for them. These examples show that there may not be a proven best intervention for some groups or specific situations and support the argument that a placebo is only acceptable when there is no proven effective intervention.

3.6 Three Contentious Issues in the Placebo Debate

The difference of opinions on the ethics of placebo control in the presence of a proven intervention since the 1996 revision to the current 2013 version of the DoH can be categorized mainly into three points: (1) the standard of care, (2) clinical equipoise, and (3) the acceptable risk of harm. Each can be classified in deontological views that prioritize the rights and interests of individuals, and in pragmatic, utilitarian views that prioritize optimization for an individual and a society through risk-benefit assessments (Table 3).

1. **Standards of care** [99]: Deontologists argue that a new intervention should be compared to the intervention proven to be the best in the world as the standard of care. This is regarded as "universalism" to set ethical justifiability of a study as the same anywhere in the world; as compared to utilitarians who argue that a new intervention could be compared to the intervention available in the host community, which is regarded as ethical "pluralism" to allow for differences in justifiability to include the local standard of care.

   Levine argued the latter view is respecting the local cultural values [100]; meanwhile, the universalists argue that international ethical norms, differently from national rules, are essentially applied regardless of local settings, respecting diversity in the region [101].

   The DoH since 1975 describes the duty of physicians to provide patients with the best proven intervention in the world, regardless of the local standard, and new interventions should be compared to this best proven one [102]. In case the best proven intervention may not be the best for a specific population (e.g., not only untested subgroups, but also due to viral variants, or populations not responding to the standard treatment), it is acceptable not to use this best proven one for this specific group. However, not using the best one due to economic reasons cannot be rephrased as ethically justifiable on the basis of the DoH.

   Some other authoritative documents on the ethics of clinical trials in developing countries issued during this period (US Presidential Commission [103], UK

Table 3 Three contentious issues in the placebo debate

	Deontological view		Utilitarian (pragmatic) view
Standard of care (choice of control in comparative study)	Intervention proven to be the best in the world: "universalism"		Intervention available in the host community of the study: "pluralism"
Clinical equipoise	Clinical equipoise is a necessary condition for comparative study		It is deceptive to regard clinical equipoise to be a condition for comparative study
Acceptable risk threshold where there is a proven intervention	Setting a risk threshold is not acceptable (DoH 2000)	"minor increase above minimal risk" (CIOMS 2016)	"no increase of risk of serious or irreversible harm" (CIOMS 2002, DoH 2008~2013, ICH-E10)

Nuffield Council [104], EU Ethics Group [105], UNAIDS [106]) take the position of universal standard but accept local standards as an exception on a conditional basis. Although these commentary documents are different in nature from the DoH, which provides a code of conduct of physicians, consistency is required [60].

2. **Clinical equipoise**: "Clinical equipoise" [107, 108] (or uncertainty [109]) has been traditionally recognized as the premise of ethically acceptable comparative studies and this is consistent with the core principle of the DoH (Article 8, to prioritize patient right and interest to research goal). The opponents have argued that applying such obligations in "practice" to "research" leads to the "therapeutic misconception" of a patient to misunderstand experimental intervention to be treatment; thus clinical equipoise is deceptive [110–112]. This position neglects the statement in the Belmont Report that "*both* (research and practice) *often occur together*" and is a recommendation for appropriate risk-benefit assessment (utilitarian view) rather than adherence to the article 8 of the DoH (deontological view).
3. **Acceptable risk threshold:**

   The range of acceptable risk of harm is an additional discussion point [113]. It seems to be an utilitarian view that physicians could preemptively set risks that they can intentionally inflict on their patients, and should avoid only the risk of serious and irreversible harm. Moreover, such a point of view seems to be incompatible with the deontological view of the article 8 of the DoH. The current risk threshold of the DoH permits intentional risk of harm, such as pain, burden, or discomfort, which is not serious (in pharmaceutical regulations harm without hospitalization is not included in the definition of "serious"), and which may be long lasting but the participant could recover someday.

   CIOMS Guidelines 2016 adopted a different risk threshold from the one in DoH, "*minor increase above minimal risk,*" explaining that: "*Risks of receiving placebo count as minimal when the risk of serious harm is very unlikely and the potential harms associated with more common adverse events are small,*" on the premise that this risk is "*considered acceptable by a reasonable person.*" On the other hand, the DoH threshold for risk does not mention any limits up to "serious and irreversible harm." The WMA seems to support that such risks could be left to the decision-making of study participants who are assumed to be "reasonable human beings," while CIOMS is stricter and does not accept even a moderate risk of harm.

4 Controversies in the Ebola Outbreak

The debate on the ethics of placebo-controlled trials took on a different complexion during the 2014-2015 Ebola outbreak. The pros and cons of placebo use even in the absence of proven interventions were presented, and the theory and method for providing unproven interventions to patients for humanitarian purposes were promoted.

In August 2014, a 12-member expert panel invited by the WHO unanimously concluded, from an ethical perspective, compassionate use for Ebola was conditionally acceptable [114]. Considering this opinion, the WHO in its reports in October [115] and November [116] suggested careful choice of study designs including observational studies associated with compassionate use or RCTs, stating that in any case all patients should receive the best available standard of care. In this case where there is no best proven intervention, we should seek "the best standard of care in the world" if there is one, rather than the "locally available standard of care." At this point in time, there were symptomatic treatments for Ebola worldwide, but no proven intervention, and phase I trials of new drugs were about to begin. In October, a joint statement by 17 authors from both affected or Western countries was issued, stating that it was not practical to conduct randomized controlled trials (RCTs) during the outbreak [117]. It argued that clinical equipoise could not be achieved due to the severity of the disease, and that the political instability in the epidemic area did not allow for RCTs to be conducted. The February 2015 report of the US President's Commission for the Study of Bioethical Issues stated both views for the need for placebo-controlled trials and for the need for collecting data from observational studies [118]. Of note, a framework proposed in WHO's report in October 2014 [112] for safety monitoring and data collection in the use of unapproved medicines called MEURI (monitored emergency use of unregistered and experimental interventions) was updated later when the COVID-19 pandemic needed to be addressed [119].

In fact, one RCT to compare a study drug with a symptomatic treatment-only control, involving 72 participants, that conducted frequent interim analyses using Bayesian analysis and adaptive design, was terminated in January 2016 (a report published on 13 October 2016) [120] after affected countries announced the end of the outbreak. It could not prove the efficacy of the study drug. There were other reports on another RCT comparing active drugs [121], Expanded Access Programs [122], or a vaccine RCT comparing immediate and delayed vaccinated groups [123], as well as a placebo-controlled study looking at antibody responses [124].

With the trend to promote data-driven research utilizing "real-world data," as well as stakeholder engagement, with "GCP Renovation" [125] to update some fundamental guidelines [126, 127], the research community is seeking for ways to allow for other scientifically valid controls instead of placebo-controlled trials. This situation implies that all the stakeholders are seeking for minimization of participant risks, keeping the scientific integrity of the studies, with minimization of bias from observational studies. In such situations stricter assessment of "*compelling and scientifically sound methodological reasons*" and "risk minimization," as demanded in the 2016 CIOMS guidelines [12], must be included in any exceptional justification for placebo use. In this scenario, the current DoH with a substantial risk threshold on the placebo item must be modified.

5 Revisiting Placebo Clause in the COVID-19 Pandemic

5.1 *Best Proven COVID-19 Vaccines*

During the COVID-19 pandemic, a misinterpretation of the placebo clause of the DoH by the WHO Ad Hoc Expert Group on vaccine trials [128] had resulted in resurfaced controversy, similar to the case of HIV/AIDS perinatal transmission prevention trials [129, 130]. One year after the outbreak in December 2019, with a subsequent pandemic declaration by the WHO in March 2020 [131], eventually in December 2020, the interim analysis results of placebo-controlled, multinational, large-scale clinical trials of three European- and US-oriented prophylactic vaccines, which began around July 2020, were issued.

The mRNA vaccine, co-developed by BioNTech and Pfizer achieved an efficacy rate of 95% [132, 133], and another mRNA vaccine co-developed by Moderna and the US National Institute of Allergy and Infectious Diseases achieved 94.1% [134, 135], while adenovirus vector vaccine co-developed by Oxford University and AstraZeneca showed overall 70.4% efficacy (90% in a group receiving some specific regimen) [136]. All of this evidence was from large-scale placebo-controlled global studies. An increasing number of countries granted emergency use authorization. The development of a vaccine with efficacy well above the 50% standard set by the US FDA [137], which normally takes 5-10 years, was achieved in less than a year. Large-scale vaccination on a global scale has proceeded and real-world evidence of efficacy has been demonstrated [138, 139]. By August 2022, about 20 vaccine products had been approved for emergency use, 10 of which had received official approval [140]. It raises the criticism that some of the large-scale placebo-controlled studies starting after 2021 April in Low- and Middle-Income Countries were not acceptable in countries where vaccine was well available [141].

At the end of 2020/beginning of 2021, the situation was such that "best proven interventions" existed for the target population (adult men and women, excluding elders, children and pregnant women) in the areas where clinical trials were conducted and similar areas with high prevalence, severe disease, and mortality rates. There was consensus from the fact that mass vaccination programs were underway in many countries, and that regulators were still of the opinion that the benefits outweighed the risks even after some safety concerns had been raised, and from the views expressed in peer-reviewed journals. Nevertheless, even when there may be scientifically compelling reasons to continue with the placebo-controlled trials, this does not necessarily meet the conditions of ethical justifiability. The results of trials were from interim analyses at six months of a trial planned for two years, and there was a need to assess long-term effects beyond the planned two years. It was also needed to assess efficacy on variants and to assure supplies being produced by the multiple companies. Studies on the population excluded from initial trials were also needed. Some of these topics, e.g., studies on groups excluded in initial trials, on emerging variants, or in low prevalence regions/specific group, may be ruled out from the situation where there is a "best-proven intervention."

At this point, in accordance with prevailing ethical requirements, Pfizer/BioNTech made a change in the study design whereby study participants could switch from the placebo group to the active vaccine group upon request once they were eligible for vaccination in the region [142]. In view of the size of the population to be vaccinated, it was obvious that it would not be possible to complete vaccination of everyone immediately, and since vaccination was initiated in the order of the population most in need in each country, and vaccination was at the level of strong recommendation rather than mandatory, the options adopted by the individual trial design were consistent with the principles on justifiable placebo control as well as post-trial access.

5.2 *Misunderstanding of a WHO Ad Hoc Expert Group*

At the same time, however, the WHO Ad Hoc Expert Group on the Next Steps for Covid-19 Vaccine Evaluation stated that an EUA alone does not constitute a "best proven intervention" as stated in the DoH or an "established effective intervention" in the CIOMS Guidelines, and argued that it is acceptable to conduct placebo-controlled trial in countries with limited or no access to effective vaccines [128]. This assertion evoked the same controversy as the one for the HIV/AIDS perinatal transmission prevention trials discussed above. WHO's statement was similar to the French doctor's statement in April 2020 which the WHO condemned as "racist" [143]. The French doctor proposed that vaccine studies could be conducted in Africa, where there are no masks nor treatments (see 3rd Chap.). It is clear that the utilitarian view of the WHO Expert Panel as stated in December of 2020 represented a "local standard," which differs from the principles of the DoH.

The DoH does not allow placebo-controlled trials on the basis of "limited or no access" in the country where the clinical trial is conducted; the WHO describes principles that differ from those in the DoH feigning that they are in line with it. This is misleading; thus, in order to avoid such errors, it needs to be clarified that the DoH's position on "best proven intervention" means "interventions proven to be best for a specific population in the world."

6 Discussion

We have revisited the ethical principles of placebo-controlled trials, historically, and in light of the global experience of the COVID-19 pandemic. The three controversial points about placebo (standard of care; clinical equipoise; acceptable risk threshold) are now discussed:

6.1 Standard of Care

To avoid misinterpretation, "best proven intervention" should be clearly described as being the best in the world, not accepting a less stringent local standard.

In the AIDS pandemic, people sought to test treatments in low-resource settings where people could not afford the best proven intervention, using placebo, assuming it as "local standard." But the crux of the issue is that the best proven interventions were unaffordable for those who need them due to intellectual property rights issues and lack of technology transfer. This inequitable access to research results has become even more prominent in vaccine allocation during the COVID-19 pandemic. This point will be explored further in the next chapter as a post-trial access issue. If the DoH enforces intolerance to the "local standard principle" it will accelerate efforts to make the best-proven intervention available for those who need it worldwide, rather than justifying ethically questionable placebo-controlled trials. This is what we should learn from the placebo controversy in the COVID-19 pandemic.

6.2 Clinical Equipoise Versus Acceptable Risk Threshold

In the case of justifying an exceptional possibility for using a placebo when an effective comparator exists the DoH should also revisit its risk of harm, removing the high threshold prohibiting only serious or irreversible harm, and perhaps bringing it to line with the 2016 CIOMS Guidelines requirements. The current boundaries in the 2013 DoH are not consistent with the core principle of article 8 that a study participant's rights and interests must be always prioritized in research—indeed a truly deontological approach.

Since 1975 the DoH principles for controlled studies state that a new method should be compared against the best method and study participants should be assured of the best proven method. This infers that comparative studies can be justified only when this "best-proven" method is uncertain. It is prudent that physicians, ethics committees, regulatory authorities, and other relevant stakeholders, including civil society, should agree on the best way forward. In the process of ethics committee review it would be important to include a lay member, and to ensure that the informed consent process is well understood. Risk threshold wording in DoH would be difficult for lay people, even experts, to truly understand. While the history of debates over "clinical equipoise" and "uncertainty" has been complex [144], these notions can be reformulated so that they are understandable and unambiguous. This kind of ethical lexicon must be shared among physicians, patients, and the general public.

In the COVID-19 pandemic, if the current situation prevails where a less virulent variant of the virus spreads, the pandemic could become endemic. However, as SARS-CoV 2 is extremely mutable, new types of vaccines may be needed and they must be tested against current vaccines. It must be clear that the inability to pay can never justify a placebo-controlled trial for those who, for example, did not have access to the first or second dose.

7 Conclusion

From the historical analysis of placebo-controlled trial ethics controversies and considering our global experience with both the AIDS pandemic and COVID-19, refinement of specific DoH principles is required:

1. The meaning of "best proven intervention" must be clarified to be "intervention proven in the world to be best." When efficacy and safety is to be proven on a global population, the possible difficulties of local availability for economic reasons cannot be a reason to discard this principle.
2. Setting of risk threshold for placebo-controlled study when there is a proven intervention should be based on wording similar to the 2016 CIOMS guidelines and "clinical equipoise" or "uncertainty" must be rephrased such that they can be understood and shared with physicians, patients, and the public.

In particular, ethical principles must be articulated not from the view of feasibility but from the view of the promotion and protection of human rights, with the aim being that of achieving the ethical and social values of the research.

The history of pandemics reinforces that the objective to be achieved is to guarantee the best proven intervention for all the people who need it. In addition, the refinement of research practices should make clear the limits for the use of placebo, based on the primary duty to protect research participants. Respect for research participants everywhere will be essential to help overcome global pandemics and other prevalent illnesses.

Acknowledgments The authors appreciate following financial supports for traveling abroad to attend meetings directly related to this manuscript:

- The 23rd Annual Conference of the Japanese Society of Transcultural Psychiatry, 2016, with the presidency of Professor Kazutaka Shimoda, to invite AD to Tochigi, Japan.
- The 40th Annual Scientific Meeting of the Japanese Society of Clinical Pharmacology and Therapeutics (JSCPT), 2019, with the meeting presidency of Professor Kazutaka Shimoda, and the society presidency of Professor Hiroshi Watanabe, to invite DG to Tokyo, Japan.
- The International Federation of Associations of Pharmaceutical Physicians and Pharmaceutical Medicine (IFAPP), with the presidency of Dr. Varvara Baroutsou for travel of CK to attend the World Medical Association's regional conference in Tel Aviv, Israel, 2022, on the Declaration of Helsinki.

Conflict of Interest There is no conflict of interest to be disclosed related to the content of this manuscript.

References

1. World Medical Association. WMA Declaration of Helsinki: ethical principles for medical research involving human subjects. Adopted by the 18th WMA General Assembly, Helsinki, Finland, June 1964, and last amended by the 64th WMA General Assembly, Fortaleza, Brazil, Oct 2013. https://www.wma.net/policies-post/wma-declaration-of-helsinki-ethical-principles-for-medical-research-involving-human-subjects/.
2. Rawls J. A theory of justice. 1971; Belknap Press.
3. The National Commission for the Protection of Human Subjects of Biomedical and Behavioral Research. The Belmont Report: ethical principles and guidelines for the protection of human subjects of research. 18 Apr 1979. https://www.hhs.gov/ohrp/regulations-and-policy/belmont-report/read-the-belmont-report/index.html.
4. Beauchamp TL, Childress JF. Principles of biomedical ethics. Oxford University Press; 1979.
5. United Nations. The universal declaration of human rights. 10 Dec 1948. https://www.un.org/en/about-us/universal-declaration-of-human-rights.
6. United Nations. International covenant on economic, social and cultural rights. Adopted by General Assembly resolution 2200A (XXI) of 16 Dec 1966, entry into force 3 Jan 1976. https://www.ohchr.org/en/instruments-mechanisms/instruments/international-covenant-economic-social-and-cultural-rights.
7. United Nations. International covenant on civil and political rights. Adopted by General Assembly resolution 2200A (XXI) of 16 Dec 1966, entry into force 23 Mar 1976. https://www.ohchr.org/en/instruments-mechanisms/instruments/international-covenant-civil-and-political-rights.
8. World Health Organization. Constitution of the World Health Organization. 22 Jul 1946. https://www.who.int/about/governance/constitution.
9. Office of the United Nations High Commissioner for Human Rights, World Health Organization. The right to health. Fact sheet 31. 1 Jun 2008. https://www.ohchr.org/en/publications/fact-sheets/fact-sheet-no-31-right-health.
10. Greco DB. Revising the Declaration of Helsinki: ethics vs economics or the fallacy of urgency. Can HIV AIDS Policy. Law Rev. 2000;5(4):98–101. https://www.hivlegalnetwork.ca/site/hivaids-policy-and-law-review-54-2000/?lang=en.
11. Michels KB, Rothman KJ. Update on unethical use of placebos in randomised trials. Bioethics. 2003;17(2):188–204. https://doi.org/10.1111/1467-8519.00332. PMID: 12812185.
12. Council for International Organizations of Medical Sciences. International ethical guidelines for health-related research involving humans. 2016. https://cioms.ch/publications/product/international-ethical-guidelines-for-health-related-research-involving-humans/.
13. Kant I. Groundwork of the metaphysics of morals. 1785.
14. van der Graaf R, van Delden JJ. On using people merely as a means in clinical research. Bioethics. 2012;26(2):76–83. https://doi.org/10.1111/j.1467-8519.2010.01820.x. Epub 2010 Apr 5. PMID: 20377571.
15. International Military Tribunal. Trials of war criminals before the Nuremberg Military Tribunals under Control Council law no. 10 Nuernberg, October 1946–April 1949. vol. 2, p. 181–2. (The Nuremberg Code.)
16. International Committee of the Red Cross. Convention (III) relative to the treatment of prisoners of war. Geneva, 12 Aug 1949. Commentary of 2020 Article 13: humane treatment of prisoners. 2020. https://ihl-databases.icrc.org/applic/ihl/ihl.nsf/Comment.xsp?action=openDocument&documentId=3DEA78B5A19414AFC1258585004344BD#73.
17. Kimura R. Contemporary Japan, history of medical ethics, South and East Asia. In: Reich WT, editor. Encyclopedia of bioethics, revised ed. New York: Simon & Schuster Macmillan; 1995. p. 1496–505. http://www.bioethics.jp/bios-japanall.html.
18. International Committee of the Red Cross. Convention (III) relative to the Treatment of Prisoners of War. Geneva; 1949. https://ihl-databases.icrc.org/applic/ihl/ihl.nsf/INTRO/375?OpenDocument.

19. International Conference on Harmonisation of Technical Requirements for Registration of Pharmaceuticals for Human Use. ICH harmonised tripartite guideline: guideline for Good Clinical Practice E6(R1). 10 June 1996.
20. Council of Europe. Convention for the protection of human rights and dignity of the human being with regard to the application of biology and medicine: convention on human rights and biomedicine. (European Treaty Series No. 164). 1997. https://www.coe.int/en/web/conventions/full-list?module=treaty-detail&treatynum=164.
21. United Nations Educational, Scientific and Cultural Organization. Universal declaration on bioethics and human rights. 19 Oct 2005. https://www.unesco.org/en/legal-affairs/universal-declaration-bioethics-and-human-rights?hub=66535.
22. World Medical Association. Declaration of Helsinki: recommendations guiding doctors in clinical research. 1964. Rescinded. https://www.wma.net/wp-content/uploads/2018/07/DoH-Jun1964.pdf.
23. World Medical Association. Declaration of Helsinki: recommendations guiding medical doctors in biomedical research involving human subjects. 1975. Rescinded. https://www.wma.net/wp-content/uploads/2018/07/DoH-Oct1975.pdf.
24. Council for International Organizations of Medical Sciences. Proposed international ethical guidelines for biomedical research involving human subjects. 1982.
25. Council for International Organizations of Medical Sciences. International ethical guidelines for biomedical research involving human subjects. 1993.
26. World Medical Association. Declaration of Helsinki: recommendations guiding physicians in biomedical research involving human subjects. 1996. Rescinded. https://www.wma.net/wp-content/uploads/2018/07/DoH-Oct1996.pdf.
27. World Medical Association. Declaration of Helsinki: ethical principles for medical research involving human subjects. 2000. Rescinded. https://www.wma.net/wp-content/uploads/2018/07/DoH-Oct2000.pdf.
28. International Conference on Harmonisation of Technical Requirements for Registration of Pharmaceuticals for Human Use. ICH harmonised tripartite guideline: choice of control group and related issues in clinical trials E10. 2000. https://database.ich.org/sites/default/files/E10_Guideline.pdf.
29. World Medical Association. Declaration of Helsinki: ethical principles for medical research involving human subjects. 2004. Rescinded. https://www.wma.net/wp-content/uploads/2018/07/DoH-Oct2004.pdf.
30. World Medical Association. Declaration of Helsinki: ethical principles for medical research involving human subjects. 2008. Rescinded. https://www.wma.net/wp-content/uploads/2018/07/DoH-Oct2008.pdf
31. International Committee of Medical Journal Editors. Recommendations for the conduct, reporting, editing, and publication of scholarly work in medical journals. Updated May 2023. https://www.icmje.org/recommendations/.
32. World Medical Association. WMA Declaration of Geneva. Adopted in Sept 1948, last amended in Oct 2017. https://www.wma.net/policies-post/wma-declaration-of-geneva/.
33. Rheinsberg Z, Parsa-Parsi R, Kloiber O, Wiesing U. Medical oath: use and relevance of the Declaration of Geneva. A survey of member organizations of the World Medical Association (WMA). Med Health Care Philos. 2018;21(2):189–196. https://doi.org/10.1007/s11019-017-9794-x. PMID: 28785834.
34. World Medical Association. WMA international code of medical ethics. Adopted in Oct 1949, last amended in Oct 2022. https://www.wma.net/policies-post/wma-international-code-of-medical-ethics/.
35. World Medical Association. History: the story of the WMA. https://www.wma.net/who-we-are/history/.
36. Schmidt U, Frewer A, Sprumount D, editors. Ethical research: the Declaration of Helsinki, and the past, present and future of human experimentation. Oxford University Press; 2020.

37. Rothman KJ, Michels KB. The continuing unethical use of placebo controls. N Engl J Med. 1994;331(6):394–8.
38. Aspinall RL, Goodman NW. Denial of effective treatment and poor quality of clinical information in placebo controlled trials of ondansetron for postoperative nausea and vomiting: a review of published trials. BMJ. 1995;311(7009):844–6.
39. Collier J. Confusion over use of placebos in clinical trials. Better guidelines needed. BMJ. 1995;311:821.
40. Council for International Organizations of Medical Sciences. About. https://cioms.ch/about/.
41. Council for International Organizations of Medical Sciences. International guidelines for ethical review of epidemiological studies. 1991.
42. Council for International Organizations of Medical Sciences. International guidelines for epidemiological studies. 2009.
43. Council for International Organizations of Medical Sciences. International ethical guidelines for biomedical research involving human subjects. 2002.
44. Reverby SM. 'Normal exposure' and inoculation syphilis: PHS 'Tuskegee' doctors in Guatemala, 1946-1948 and at Sing Sing prison, Ossining, New York, 1953–1954. 2010.
45. U.S. Department of Health & Human Services, Office for Human Research Protections. Related historical documents from the National Commission for the Protection of Human Subjects of Biomedical and Behavioral Research. https://www.hhs.gov/ohrp/regulations-and-policy/belmont-report/access-other-reports-by-the-national-commission/index.html.
46. National Research Service Award Act of 1974, Public Law 93-348, 88 Stat. 342. 1974.
47. Bernard C. An introduction to the study of experimental medicine. 1865.
48. International Council for Harmonisation of Technical Requirements Pharmaceuticals for Human Use (ICH). ICH harmonised guideline: integrated addendum to ICH E6(R1): guideline for good clinical practice E6(R2). 2016. https://database.ich.org/sites/default/files/E6_R2_Addendum.pdf.
49. International Council for Harmonisation of Technical Requirements for Pharmaceuticals for Human Use. Members & observers. https://www.ich.org/page/members-observers.
50. Connor EM, Sperling RS, Gelber R, Kiselev P, Scott G, O'Sullivan MJ, VanDyke R, Bey M, Shearer W, Jacobson RL, et al. Reduction of maternal-infant transmission of human immunodeficiency virus type 1 with zidovudine treatment. Pediatric AIDS Clinical Trials Group Protocol 076 Study Group. N Engl J Med. 1994;331(18):1173–80. https://doi.org/10.1056/NEJM199411033311801. PMID: 7935654.
51. Varmus H, Satcher D. Ethical complexities of conducting research in developing countries. N Eng J Med. 1997;337(14):1003–5.
52. Lurie P, Wolfe SM. Unethical trials of interventions to reduce perinatal transmission of the human immunodeficiency virus in developing countries. N Engl J Med. 1997;337(12):853–6. https://doi.org/10.1056/NEJM199709183371212. PMID: 9295246.
53. Angell M. Investigators' responsibilities for human subjects in developing countries. N Engl J Med. 2000;342(13):967–9. https://doi.org/10.1056/NEJM200003303421309. PMID: 10738056.
54. Angell M. The ethics of clinical research in the third world. N Engl J Med. 1997;337:847–9.
55. Lallemant M, Jourdain G, Le Coeur S, Kim S, Koetsawang S, Comeau AM, Phoolcharoen W, Essex M, McIntosh K, Vithayasai V. A trial of shortened zidovudine regimens to prevent mother-to-child transmission of human immunodeficiency virus type 1. Perinatal HIV Prevention Trial (Thailand) Investigators. N Engl J Med. 2000;343(14):982–91.
56. Panel on Treatment of HIV-Infected Pregnant Women and Prevention of Perinatal Transmission. Recommendations for use of antiretroviral drugs in pregnant HIV-1-infected women for maternal health and interventions to reduce perinatal HIV transmission in the United States. 2014.
57. Centers for Disease Control and Prevention. Short-course regimen of AZT proven effective in reducing perinatal HIV transmission: offers hope for reducing mother-to-child HIV transmis-

sion in developing world. Feb 10, 1998 (archived document). https://www.cdc.gov/media/pressrel/r980210.htm.

58. Lurie P, Wolfe SM. Commentary 9.2: the developing world as the "Answer" to the dreams of pharmaceutical companies: the Surfaxin story. In: Lavery JV, Grady C, Wahl ER, Emanuel EJ, editors. Ethical issues in international biomedical research: a casebook. Oxford University Press; 2007. p. 159–70.
59. Angell M. The truth about the drug companies: how they deceive us and what to do about it. New York: Random House; 2004.
60. Human D, Kurihara C, Mitsuishi T. Interview. interview with Delon human, secretary general of the world medical association-discussion on the 2000 revision to the declaration of Helsinki and onwards. Clin Eval. 2002, 29:307–13.
61. Levine RJ. The boundaries between biomedical or behavioral research and the accepted and routine practice of medicine. In: The National Commission for the Protection of Human Subjects of Biomedical and Behavioral Research. The Belmont Report: ethical principles and guidelines for the protection of human subjects of research. Appendix vol. 1. DHEW Publication No. (OS) 78-0013. 1975. p. 1-1-1-44. https://repository.library.georgetown.edu/bitstream/handle/10822/779133/ohrp_appendix_belmont_report_vol_1.pdf#page=1.
62. Levine RJ. The need to revise the Declaration of Helsinki. N Engl J Med. 1999;341(7):531–4.
63. Temple R, Ellenberg SS. Placebo-controlled trials and active-control trials in the evaluation of new treatments. Part 1: ethical and scientific issues. Ann Intern Med. 2000;133(6):455–63.
64. Ellenberg SS, Temple R. Placebo-controlled trials and active-control trials in the evaluation of new treatments. Part 2: practical issues and specific cases. Ann Intern Med. 2000;133(6):464–70.
65. Kloiber O, Kurihara C, Saio T, Interview. Discussion toward the 50th anniversary of the Declaration of Helsinki: interview with Dr. Otmar Kloiber, Secretary General, World Medical Association. Clin Eval. 2013;41(2):351–72. http://cont.o.oo7.jp/41_2/p351-72eng.pdf.
66. Tsuboi E, Kurihara C, Interview. The 2000 revision of the Declaration of Helsinki and its implications in medical ethics: interview with the WMA Immediate Past President, Eitaka Tsuboi. Clin Eval. 2002;30(1):99–107. Japanese. http://cont.o.oo7.jp/30_1/p99-107.pdf.
67. Department of Health and Human Services Food and Drug Administration. 21 CFR Part 312 Human subject protection: foreign clinical studies not conducted under an investigational new drug application. Federal Register. 2008;73(82):22800–16. https://www.govinfo.gov/content/pkg/FR-2008-04-28/pdf/E8-9200.pdf.
68. Trials on trial. Nature. 2008;453(7194):427–8. https://doi.org/10.1038/453427b.
69. Lurie P, Greco DB. US exceptionalism comes to research ethics. Lancet. 2005;365(9465):1117–9.
70. Anderson M. FDA abandons Declaration of Helsinki for international clinical trials. Social Medicine Portal. 2008. https://www.socialmedicine.org/2008/06/01/ethics/fda-abandons-declaration-of-helsinki-for-international-clinical-trials/.
71. Integrity in Science. FDA scraps Helsinki Declaration on protecting human subjects. 2008. https://web.archive.org/web/20081022213154/http://www.cspinet.org/integrity/watch/200805051.html#2.
72. Temple R, Kurihara C, Interview. Interview with Dr. Robert Temple on drug evaluation policy of FDA: ethics, science of placebo-control and comparative effectiveness studies. Clin Eval. 2014;42(2):539–51. http://cont.o.oo7.jp/42_2/p539-51eng.pdf.
73. Greco D, Shimoda K, Watanabe H, Organizers. The past, present, and future of ethics of international health research: research as a stepping-stone to universal public health care access. Clin Eval. 2020;48(1):207–31. http://cont.o.oo7.jp/48_1/w29-w53.pdf.
74. Homedes N, Ugalde A, editors. Clinical trials in Latin America: where ethics and business clash. Springer; 2014.
75. Red Latinoamericana y Del Caribe de Bioetica – Redbioetica/UNESCO. Carta de Córdoba sobre ética en investigaciones con seres humanos. Córdoba, Argentina. 2008. https://www.

psi.uba.ar/academica/carrerasdegrado/psicologia/sitios_catedras/obligatorias/723_etica2/material/casuistica/carta_de_cordoba_08.pdf.
76. Solbakk JH. National ethics advisory bodies and committees in Norway: history, lessons learnt, and common challenges ahead. National Bioethics Committees in Action. United Nations Educational, Scientific and Cultural Organization; 2010. https://unesdoc.unesco.org/ark:/48223/pf0000189548.
77. The Buenos Aires Declaration on Ethics and Clinical Trials. Buenos Aires, May 13, 2008. In: Homedes N, Ugalde A, editors. Clinical trials in Latin America: where ethics and business clash. Springer; 2014.
78. Confederation Medica latinoamericana Y del Caribe (Latin American and Caribbean Medical Confederation). Declaration of Pachuca; 2013.
79. Greco DB. Ethical dilemmas, research, and access to treatment. In: International encyclopedia of the social & behavioral science, vol. 8, 2nd ed. 2015. p. 31–6.
80. Schlemper BR Jr. The placebo and the Declaration of Pachuca: dead letters? Rev. bioet. (Impr.). 2014;22(3):461–9.
81. Homedes N, Ugalde A. Outsourcing clinical trials to Latin America: causes and impact. In: Rivera-Lopez E, Hevia M, editors. Controversies in Latin American bioethics. Springer; 2019.
82. Brazilian National Health Council (CNS), Brazilian National Research Ethics Commission (CONEP/CNS/MS). Guidance manual: frequent pending issues in clinical research protocols. Version 1.0. 2015. https://abracro.org.br/images/legislacao/guidance_manual_frequent_pending_issues_in_clinical_research_protocols_CONEP_ V1_2015.pdf.
83. Brazilian Research Ethics Commission Resolution 466/2012, which succeeded Resolution 404/2008.
84. Greco D, Kimura R, Victoria Perottino M, Saio T, Kurihara C. Ethics of international collaborative research: perspectives from Brazil: part 1 selected notes on Paulo Freire: part 2 access, compulsory license, case study. Clin Eval. 2020;48(1):273–301. http://cont.o.oo7.jp/48_1/w95-w123.pdf.
85. Skierka AS, Michels KB. Ethical principles and placebo-controlled trials - interpretation and implementation of the Declaration of Helsinki's placebo paragraph in medical research. BMC Med Ethics. 2018;19(1):24. https://doi.org/10.1186/s12910-018-0262-9.
86. Miller FG. Placebo-controlled trials in psychiatric research: an ethical perspective. Biol Psychiatry. 2000;47:707–16.
87. Carpenter WT, Appelbaum PS, Levine RJ. The Declaration of Helsinki and clinical trials: a focus on placebo-controlled trials in schizophrenia. Am J Psychiatry. 2003;160(2):356–62.
88. Charney DS, Nemeroff CB, Lewis L, et al. Consensus Development Panel. National Depressive and Manic-Depressive Association consensus statement on the use of placebo in clinical trials of mood disorders. Arch Gen Psychiatry. 2002;59(3):262–70.
89. Wittenborn JR. Guidelines for clinical trials of psychotropic drugs. 1977.
90. Khan A, Khan SR, Leventhal RM, Brown WA. Symptom reduction and suicide risk among patients treated with placebo in antipsychotic clinical trials: an analysis of the food and drug administration database. Am J Psychiatry. 2001;158(9):1449–54.
91. Khan A, Warner HA, Brown WA. Symptom reduction and suicide risk in patients treated with placebo in antidepressant clinical trials. Arch Gen Psychiatry. 2000;57:311–7.
92. Khan A, Khan S, Kolts R, Brown W. Suicide rates in clinical trials of SSRIs, other antidepressants, and placebo: analysis of FDA reports. Am J Psychiatry. 2003;160:790–2.
93. Storosum JG, Zwieten van BJ, Brink van den W, Gersons BPR, Broekmans AW. Suicide risk in placebo-controlled studies of major depression. Am J Psychiatry. 2001;158:1271–5.
94. Gunnell D, Seperia J, Ashby D. Selective serotonin reuptake inhibitors (SSRIs) and suicide in adults: meta-analysis of drug company data from placebo controlled, randomised controlled trials submitted to the MHRA's safety review. BMJ. 2005;330(7488):385–93.
95. Seperia J, Ashby D, Gunnell D. Suicidal behavior and SSRIs: updated meta-analysis. BMJ. 2006;332(7555):1453.

96. Laughren TP. The scientific and ethical basis for placebo-controlled in depression and schizophrenia: an FDA perspective. Eur Psychiatry. 2001;16(7):418–23.
97. European Commission, Directorate-General III Industry. Clinical guidelines: medical products for treatment of depression. The rules governing medicinal products in the European union. Volume III, part 2. Eudra/G/96/010.
98. Cardiac Arrhythmia Suppression Trial (CAST) Investigators. Preliminary report: effect of encainide and flecainide on mortality in a randomized trial of arrhythmia suppression after myocardial infarction. N Engl J Med. 1989;321(6):406–12. https://doi.org/10.1056/NEJM198908103210629. PMID: 2473403.
99. Lie RK, Emanuel E, Grady C, Wendler D. The standard of care debate: the Declaration of Helsinki versus the international consensus opinion. J Med Ethics. 2004;30(2):190–3. https://doi.org/10.1136/jme.2003.006031. PMID: 15082816; PMCID: PMC1733825.
100. Levine RJ, Tsutani K, Sakanoue M, et al. Ethics in clinical trial in the days of globalization of drug development. Clin Eval. 1999;29(3):341–80. Japanese. http://cont.o.oo7.jp/26_3/p341-80.pdf.
101. Macklin R. Against Relativism: cultural diversity and the search for ethical universals in medicine. Oxford University Press; 1999.
102. Kloiber O, Greco D, Watanabe H, Imamura K, Yamamoto Y, Matsuyama K, Saio T, Kurihara C. International collaborative research and new trends of research ethics: follow-up session. Clin Eval. 2020;48(1):233–65. http://cont.o.oo7.jp/48_1/w55-w87.pdf.
103. National Bioethics Advisory Commission. Report on ethical and policy issues in international research: clinical trials in developing countries. April 2001.
104. Nuffield Council on Bioethics. Report on the ethics of research to healthcare in developing countries. 2002.
105. European Group on Ethics in Science and New Technologies to the European Commission. Opinion Nr 17 on ethical aspects of clinical research in developing countries. 4 February 2003.
106. Joint United Nations Programme on HIV/AIDS (UNAIDS). Ethical considerations in biomedical HIV prevention trials [Additional guidance point added in 2012]: UNAIDS/WHO guidance document. 2012.
107. Freedman B. Equipoise and the ethics of clinical research. NEJM. 1987;317:141–5.
108. Rothman KJ. Placebo mania. Br Med J. 1996;313:3–4.
109. Sackett DL. CMAJ 2000;162:1311–4, CMAJ 2001;164:1831–2.
110. Miller FG, Brody H. Clinical equipoise and the incoherence of research ethics. J Med Philos. 2007;32(2):151–65. https://doi.org/10.1080/03605310701255750. PMID: 17454420.
111. Miller FG, Brody H. What makes placebo-controlled trials unethical? Am J Bioeth. 2002;2(2):3–9. https://doi.org/10.1162/152651602317533523. PMID: 12189059.
112. Miller FG, Brody H. A critique of clinical equipoise. Therapeutic misconception in the ethics of clinical trials. Hastings Cent Rep. 2003;33(3):19–28. PMID: 12854452.
113. Emanuel EJ, Miller FG. The ethics of placebo-controlled trials--a middle ground. N Engl J Med. 2001;345(12):915–9. https://doi.org/10.1056/NEJM200109203451211. PMID: 11565527.
114. World Health Organization. Ethical considerations for use of unregistered interventions for Ebola viral disease: report of an advisory panel to WHO. WHO/HIS/KER/GHE/14.1. 2014. https://apps.who.int/iris/bitstream/handle/10665/130997/WHO_HIS_KER_GHE_14.1_eng.pdf?sequence=1.
115. World Health Organization. Ethical issues related to study design for trials on therapeutics for Ebola virus disease. Report of the WHO Ethics Working Group meeting, 20–21 October. Geneva: https://apps.who.int/iris/handle/10665/137509.
116. World Health Organization. Interim guidance: potential Ebora therapies and vaccines. 5 November 2014. https://apps.who.int/iris/bitstream/handle/10665/137590/WHO_EVD_HIS_EMP_14.1_eng.pdf?sequence=1&isAllowed=y.
117. Adebamowo C, Bah-Sow O, Binka F, Bruzzone R, Caplan A, Delfraissy JF, Heymann D, Horby P, Kaleebu P, Tamfum JJ, Olliaro P, Piot P, Tejan-Cole A, Tomori O, Toure A, Torreele

E, Whitehead J. Randomised controlled trials for Ebola: practical and ethical issues. Lancet. 2014;384(9952):1423–4. https://doi.org/10.1016/S0140-6736(14)61734-7. Epub 2014 Oct 13. https://www.ncbi.nlm.nih.gov/pmc/articles/PMC4392883/pdf/nihms677442.pdf.

118. Presidential Commission for the Study of Bioethical Issues. Ethics and ebola: Public health planning and response. February 2015.

119. World Health Organization. Emergency use of unproven clinical interventions outside clinical trials: ethical considerations. 12 Apr 2022. https://www.who.int/publications/i/item/9789240041745.

120. PREVAIL II Writing Group, Multi-National PREVAIL II Study Team, Davey RT Jr, Dodd L, Proschan MA, Neaton J, Neuhaus Nordwall J, Koopmeiners JS, Beigel J, Tierney J, Lane HC, Fauci AS, Massaquoi MBF, Sahr F, Malvy D. A randomized, controlled trial of ZMapp for Ebola virus infection. N Engl J Med. 2016;375(15):1448–56. https://doi.org/10.1056/NEJMoa1604330. PMID: 27732819; PMCID: PMC5086427.

121. Mulangu S, Dodd LE, Davey RT Jr, Tshiani Mbaya O, Proschan M, Mukadi D, Lusakibanza Manzo M, Nzolo D, Tshomba Oloma A, Ibanda A, Ali R, Coulibaly S, Levine AC, Grais R, Diaz J, Lane HC, Muyembe-Tamfum JJ, PALM Writing Group, Sivahera B, Camara M, Kojan R, Walker R, Dighero-Kemp B, Cao H, Mukumbayi P, Mbala-Kingebeni P, Ahuka S, Albert S, Bonnett T, Crozier I, Duvenhage M, Proffitt C, Teitelbaum M, Moench T, Aboulhab J, Barrett K, Cahill K, Cone K, Eckes R, Hensley L, Herpin B, Higgs E, Ledgerwood J, Pierson J, Smolskis M, Sow Y, Tierney J, Sivapalasingam S, Holman W, Gettinger N, Vallée D, Nordwall J, PALM Consortium Study Team. A randomized, controlled trial of Ebola virus disease therapeutics. N Engl J Med. 2019;381(24):2293–303. https://doi.org/10.1056/NEJMoa1910993. Epub 2019 Nov 27. PMID: 31774950.

122. Simpson S, Chakrabarti A, Robinson D, Chirgwin K, Lumpkin M. Navigating facilitated regulatory pathways during a disease X pandemic. NPJ Vaccines. 2020;5:101. https://doi.org/10.1038/s41541-020-00249-5. PMID: 33110630; PMCID: PMC7584587.

123. Samai M, Seward JF, Goldstein ST, Mahon BE, Lisk DR, Widdowson MA, Jalloh MI, Schrag SJ, Idriss A, Carter RJ, Dawson P, Kargbo SAS, Leigh B, Bawoh M, Legardy-Williams J, Deen G, Carr W, Callis A, Lindblad R, Russell JBW, Petrie CR, Fombah AE, Kargbo B, McDonald W, Jarrett OD, Walker RE, Gargiullo P, Bash-Taqi D, Gibson L, Fofanah AB, Schuchat A. STRIVE Study Team. The Sierra Leone trial to introduce a vaccine against Ebola: an evaluation of rVSVΔG-ZEBOV-GP vaccine tolerability and safety during the West Africa Ebola outbreak. J Infect Dis. 2018;217(suppl_1):S6–S15. https://doi.org/10.1093/infdis/jiy020. Erratum in: J Infect Dis. 2018;218(3):508–507. PMID: 29788345; PMCID: PMC5961340.

124. Kennedy SB, Bolay F, Kieh M, Grandits G, Badio M, Ballou R, Eckes R, Feinberg M, Follmann D, Grund B, Gupta S, Hensley L, Higgs E, Janosko K, Johnson M, Kateh F, Logue J, Marchand J, Monath T, Nason M, Nyenswah T, Roman F, Stavale E, Wolfson J, Neaton JD, Lane HC, PREVAIL I Study Group. Phase 2 placebo-controlled trial of two vaccines to prevent Ebola in Liberia. N Engl J Med. 2017;377(15):1438–47. https://doi.org/10.1056/NEJMoa1614067. PMID: 29020589; PMCID: PMC5705229.

125. International Council for Harmonisation of Technical Requirements for Pharmaceuticals for Human Use. ICH reflection on "GCP Renovation": modernization of ICH E8 and subsequent renovation of ICH E6. January 2017. https://admin.ich.org/sites/default/files/2019-04/ICH_Reflection_paper_GCP_Renovation_Jan_2017_Final.pdf.

126. International Council for Harmonisation of Technical Requirements for Pharmaceuticals for Human Use. ICH-E6 good clinical practice (GCP): explanatory note. 19 April 2021. https://database.ich.org/sites/default/files/ICH_E6-R3_GCP-Principles_Draft_2021_0419.pdf.

127. International Council for Harmonisation of Technical Requirements for Pharmaceuticals for Human Use. ICH Harmonised Guideline: general considerations for clinical studies-E8(R1). 6 Oct 2021. https://database.ich.org/sites/default/files/ICH_E8-R1_Guideline_Step4_2021_1006.pdf

128. WHO Ad Hoc Expert Group on the Next Steps for Covid-19 Vaccine Evaluation, Krause PR, Fleming TR, Longini IM, Peto R, Beral V, Bhargava B, Cravioto A, Cramer JP, Ellenberg SS, Figueroa JP, Halloran E, Henao-Restrepo AM, Ryan MJ, Levine MM, Nason M, Nohynek HM, Plotkin S, Rees H, Singh JA, Swaminathan S. Placebo-controlled trials of Covid-19 vaccines - why we still need them. N Engl J Med. 2021;384(2):e2. https://doi.org/10.1056/NEJMp2033538. Epub 2020 Dec 2. PMID: 33264543.
129. Greco DB. Ethical limits to placebo use and access to Covid-19 vaccines as a human right. Indian J Med Ethics. 2021;VI(2):1–14. https://doi.org/10.20529/IJME.2021.027. PMID: 33908350.
130. Macklin R. Double standards redux. Indian J Med Ethics. 2021;VI(2):1–7. https://doi.org/10.20529/IJME.2021.021. PMID: 33908355.
131. World Health Organization. WHO Director-General's opening remarks at the media briefing on COVID-19 - 11 March 2020. 2020 [Cited 2023 Mar 13]. https://www.who.int/director-general/speeches/detail/who-director-general-s-opening-remarks-at-the-media-briefing-on-covid-19%2D%2D-11-march-2020.
132. Polack FP, Thomas SJ, Kitchin N, Absalon J, Gurtman A, Lockhart S, Perez JL, Pérez Marc G, Moreira ED, Zerbini C, Bailey R, Swanson KA, Roychoudhury S, Koury K, Li P, Kalina WV, Cooper D, Frenck RW Jr, Hammitt LL, Türeci Ö, Nell H, Schaefer A, Ünal S, Tresnan DB, Mather S, Dormitzer PR, Şahin U, Jansen KU, Gruber WC, C4591001 Clinical Trial Group. Safety and efficacy of the BNT162b2 mRNA Covid-19 vaccine. N Engl J Med. 2020; https://doi.org/10.1056/NEJMoa2034577. Epub ahead of print. PMID: 33301246.
133. Food and Drug Administration. Letter of authorization. 2021. https://www.fda.gov/media/144412/download.
134. Baden LR, El Sahly HM, Essink B, Kotloff K, Frey S, Novak R, Diemert D, Spector SA, Rouphael N, Creech CB, McGettigan J, Kehtan S, Segall N, Solis J, Brosz A, Fierro C, Schwartz H, Neuzil K, Corey L, Gilbert P, Janes H, Follmann D, Marovich M, Mascola J, Polakowski L, Ledgerwood J, Graham BS, Bennett H, Pajon R, Knightly C, Leav B, Deng W, Zhou H, Han S, Ivarsson M, Miller J, Zaks T, COVE Study Group. Efficacy and Safety of the mRNA-1273 SARS-CoV-2 Vaccine. N Engl J Med. 2020; https://doi.org/10.1056/NEJMoa2035389. Epub ahead of print. PMID: 33378609.
135. Food and Drug Administration. Letter of authorization. 2023. https://www.fda.gov/media/144636/download.
136. Voysey M, Clemens SAC, Madhi SA, Weckx LY, Folegatti PM, Aley PK, Angus B, Baillie VL, Barnabas SL, Bhorat QE, Bibi S, Briner C, Cicconi P, Collins AM, Colin-Jones R, Cutland CL, Darton TC, Dheda K, Duncan CJA, Emary KRW, Ewer KJ, Fairlie L, Faust SN, Feng S, Ferreira DM, Finn A, Goodman AL, Green CM, Green CA, Heath PT, Hill C, Hill H, Hirsch I, Hodgson SHC, Izu A, Jackson S, Jenkin D, Joe CCD, Kerridge S, Koen A, Kwatra G, Lazarus R, Lawrie AM, Lelliott A, Libri V, Lillie PJ, Mallory R, Mendes AVA, Milan EP, Minassian AM, McGregor A, Morrison H, Mujadidi YF, Nana A, O'Reilly PJ, Padayachee SD, Pittella A, Plested E, Pollock KM, Ramasamy MN, Rhead S, Schwarzbold AV, Singh N, Smith A, Song R, Snape MD, Sprinz E, Sutherland RK, Tarrant R, Thomson EC, Török ME, Toshner M, Turner DPJ, Vekemans J, Villafana TL, Watson MEE, Williams CJ, Douglas AD, Hill AVS, Lambe T, Gilbert SC, Pollard AJ, Oxford COVID Vaccine Trial Group. Safety and efficacy of the ChAdOx1 nCoV-19 vaccine (AZD1222) against SARS-CoV-2: an interim analysis of four randomised controlled trials in Brazil, South Africa, and the UK. Lancet. 2020;397(10269):99–111. https://doi.org/10.1016/S0140-6736(20)32661-1. Epub ahead of print. PMID: 33306989; PMCID: PMC7723445.
137. US Dept of Health and Human Services, Food and Drug Administration, Center for Biologics Evaluation and Research. Guidance for Industry: development and licensure of vaccines to prevent COVID-19. Jun 2020.
138. Dagan N, Barda N, Kepten E, Miron O, Perchik S, Katz MA, Hernán MA, Lipsitch M, Reis B, Balicer RD. BNT162b2 mRNA Covid-19 vaccine in a nationwide mass vaccination

setting. N Engl J Med. 2021;384:1412–23. https://doi.org/10.1056/NEJMoa2101765. Epub ahead of print. PMID: 33626250; PMCID: PMC7944975.

139. Thompson MG, Burgess JL, Naleway AL, Tyner HL, Yoon SK, Meece J, Olsho LEW, Caban-Martinez AJ, Fowlkes A, Lutrick K, Kuntz JL, Dunnigan K, Odean MJ, Hegmann KT, Stefanski E, Edwards LJ, Schaefer-Solle N, Grant L, Ellingson K, Groom HC, Zunie T, Thiese MS, Ivacic L, Wesley MG, Lamberte JM, Sun X, Smith ME, Phillips AL, Groover KD, Yoo YM, Gerald J, Brown RT, Herring MK, Joseph G, Beitel S, Morrill TC, Mak J, Rivers P, Harris KM, Hunt DR, Arvay ML, Kutty P, Fry AM, Gaglani M. Interim estimates of vaccine effectiveness of BNT162b2 and mRNA-1273 COVID-19 vaccines in preventing SARS-CoV-2 infection among health care personnel, first responders, and other essential and frontline workers - eight U.S. locations, December 2020-March 2021. MMWR Morb Mortal Wkly Rep. 2021;70(13):495–500. https://doi.org/10.15585/mmwr.mm7013e3. PMID: 33793460.
140. New York Times. Vaccine tracker. Last updated Aug 31, 2022. https://www.nytimes.com/interactive/2020/science/coronavirus-vaccine-tracker.html.
141. Dal-Ré R. Placebo control group in COVID-19 vaccine trials: context and timing matters. Eur J Clin Pharmacol. 2022;78(3):523–6. https://doi.org/10.1007/s00228-021-03259-x. Epub 2021 Nov 27. PMID: 34837495; PMCID: PMC8626740.
142. U.S. National Library of Medicine, Clinical Trials.gov archive. History of changes for study: NCT04368728 [cited 2021 Apr 20]. Available from: https://clinicaltrials.gov/ct2/history/NCT04368728?V_22=View#StudyPageTop.
143. BBC News. Coronavirus: Africa will not be testing ground for vaccine, says WHO. 2020. https://www.bbc.com/news/world-africa-52192184.
144. van der Graaf R, van Delden JJ. Equipoise should be amended, not abandoned. Clin Trials. 2011;8(4):408–16. https://doi.org/10.1177/1740774511409600. Epub 2011 Jul 11. PMID: 21746767.

Post-Trial Access: Historical Analysis Considering the Experience of COVID-19 Pandemic

Chieko Kurihara, Dirceu Greco, and Ames Dhai

Abstract The ethical obligation to ensure clinical trial participants "post-trial access" to an intervention proven effective in the trial emerged from the controversy at the end of the 1990s on clinical trials to overcome the HIV/AIDS pandemic. This ethical principle was drawn upon out of concern that sponsor companies would conduct placebo-controlled trials in resource poor countries that could not be conducted in wealthier parts of the world, and that the drugs proven to be effective would not become available in the host communities. This situation violates justice principle in bioethics. Considering the COVID-19 pandemic, we need to revisit this issue to overcome inequities in the distribution of vaccines developed through global collaboration of various stakeholders.

The post-trial access clause in the World Medical Association's Declaration of Helsinki was first introduced in the 2000 revision and appears to be downgraded in subsequent revisions from an ethical obligation to an item to be described in the research protocol and informed consent process.

Reviewing the past history and our global experience of COVID-19 pandemic, we argue that access to interventions proven to be effective is not only a right of study participants, but also for those in need of that intervention globally. Pursuing the latter must be recognized as the ethical responsibility of those who are engaged in research involving humans, to achieve access to health for all.

C. Kurihara (✉)
Kanagawa Dental University, Yokosuka, Kanagawa, Japan
e-mail: chieko.kurihara@nifty.ne.jp

D. Greco
Department of Internal Medicine, School of Medicine, Federal University of Minas Gerais, Belo Horizonte, Minas Gerais, Brazil

A. Dhai
School of Clinical Medicine, University of the Witwatersrand, Johannesburg, Gauteng, South Africa
e-mail: Ames.Dhai@wits.ac.za

C. Kurihara et al. (eds.), *Ethical Innovation for Global Health*,
https://doi.org/10.1007/978-981-99-6163-4_12

Keywords Post-trial access · Right to health · Intellectual property rights · Benefit sharing · Global health

1 Introduction: Focus of the Issue

During the expansion of global efforts in 1990s to overcome HIV/AIDS pandemic since 1980s [1], the placebo-controlled trials on HIV mother-to-child transmission sponsored by the United States (US) and French institutions, conducted in developing countries after the establishment of effective intervention in developed countries [2], resulted in international controversies [3–5]. Ongoing concerns included that pharmaceutical companies would conduct placebo-controlled trials in resource poor communities that could not be conducted in the wealthier world, and that the drugs proven to be effective would not become available in host communities [6–8], hence raising issues related to ethics dumping. This situation violates justice principle [9, 10] in bioethics. In response to these controversies, ethical obligation of assuring "post-trial access" of study participants to the intervention proven to be effective in the trial was first introduced in the 2000 version (Article 30) [11] of the World Medical Association (WMA)'s Declaration of Helsinki (DoH) to avoid exploitation of trial participants in vulnerable communities.

However, many objections have been raised, including pressure especially from the pharmaceutical industry [12], that argue it is impossible to guarantee "post-trial access" immediately at the completion of the study, mainly because it takes time to make proven interventions available to the trial participants without regulatory authorization prior to introduction into the public health system in the host community [8]. Accepting the feasibility arguments rather than the ethical reasons, the DoH post-trial access clause has been downgraded in subsequent revisions from an ethical obligation of the researcher to an item to be described in the research protocol and informed consent process (Table 1). This topic was also discussed in the Council of International Organizations of Medical Sciences (CIOMS)'s International Ethical Guidelines for Health-related Research Involving Humans ("CIOMS Guidelines" hereafter) [13]. Its 2002 version provided ideas of "reasonable availability" (Guideline 10) and "capacity building" (Guideline 20) of the host community, considering the difficulty to achieve "post-trial access" [8]. But its 2016 revision provided strengthened ideas stressing "social value" of research (Guideline 1) and community engagement strategies (Guideline 7) to achieve post-trial access, as well as the principle of continuous caring for individual research participants (Guideline 6) [13] (Table 1).

During the COVID-19 pandemic, we faced injustice and inequality in the distribution of vaccines [14, 15] which were developed through global collaboration among various stakeholders such as government, industry, researchers, and citizens. Noteworthy, companies sponsoring vaccine trials changed their protocol when efficacy was proven by interim analysis and ensured study participants access to the vaccine [16]. However, this did not mean that vaccines became available for those

Table 1 Downgraded provisions of "post-trial access" in the Declaration of Helsinki and related principles in CIOMS guidelines

Declaration of Helsinki (2000 to 2013 versions. Key descriptions of the principles are in **bold, underlined**; verbatim texts are in *Italic*.)
2000 version [11]**:** "*At the conclusion of the study,* ***every patient entered into the study should be assured of access to the best proven prophylactic, diagnostic and therapeutic methods identified by the study*"**. (Article 30)
2004 Note of clarification [23]**:** "*…it is necessary during the study planning process to identify post-trial access …* ***or access to other appropriate care****. Post-trial access arrangements or other care must be described in the study protocol so the ethical review committee may consider such arrangements during its review.*"
2008 version [24]**:** "*At the conclusion of the study, patients entered into the study are* ***entitled to be informed*** *about the outcome of the study* ***and to share any benefits that result from it****, for example,* ***access to interventions identified as beneficial in the study or to other appropriate care or benefits.***" (Article 33)
2013 version [25]**:** (described in bullet points by authors): • "*In advance of a clinical trial,* ***sponsors, researchers and host country governments should make provisions*** *for post-trial access ….*" • *This information* ***must also be disclosed to participants during the informed consent process*** in addition to above) (Article 34)
CIOMS guidelines (2002 [8], 2016 [13] versions. Related principles.)
• Reasonable availability (Guideline 10); capacity building (Guideline 20) (2002) (These ideas seemed to be proposed considering the difficulty to achieve "post-trial access".) • Social value (Guideline 1); Community engagement (Guideline 7); caring for participant's health needs (Guideline 6) (2016) (These ideas were proposed to strengthen for achievement of post-trial access" as well as continuous caring for individual research participants.)

who most needed it in the world. As we review our global experience of the pandemic, we need to revisit this norm of "post-trial access." Our argument is that access to interventions proven to be effective is not only a right of study participants, but also should be available for those in need of that intervention globally. This is because health is a human right [17, 18] derived from "human dignity" [19–21]. The guarantee of this right is a goal that must be achieved without leaving anyone behind, as suggested in the United Nation (UN)'s Sustainable Development Goals (SDGs) [22]. Thus, guaranteeing post-trial access must be recognized as an ethical responsibility of those who are engaged in research involving humans.

This chapter historically traces the debate over post-trial access, which is related to the debate on placebo-controlled trials discussed in the previous chapter (Chap. "Ethics of Placebo-Controlled Trials: Historical Analysis Including Experiences During the COVID-19 Pandemic"), and identifies ethical principles not only to confront more serious pandemics that may occur in the future, but also to achieve equitable access to global health against any types of disease, to ensure the fundamental rights to health for those who need it most.

2 Post-Trial Access Debate in Response to the HIV/AIDS Pandemic

2.1 Downgrade in the DoH and Alternative Ideas

The norm of post-trial access was first established in the 2000 revision [11] of the DoH. However, due to operational difficulties in its implementation and immense pressure from the pharmaceutical industry, the WMA in 2004 included a Note of Clarification [23] that made "*post-trial access arrangements or other care*" subject to review by an ethics committee, and that left room for the researcher to only provide "*other care*" without necessarily ensuring proven effective intervention. This weakened the justice position of the DoH. The 2008 revision [24] stated that the study participants are "*entitled to share any benefits*" resulting from the study, however, this benefit may be "*access to interventions identified as beneficial*" but also may be "*other appropriate care of benefits.*" Subsequently, the improvement in 2013 [25] (current version, to be updated in 2024) is the recommendation that "*Prior to a clinical trial, sponsors, investigators and host country governments should make arrangements for post-study access,*" which may allow research ethics committees to review the scope of this provision. However, this addition does not really guarantee post-study access and is weaker than the 2000 version where it stated that "*At the conclusion of the study, all patients enrolled in the study must be satisfied that access to the best proven prophylactic, diagnostic and therapeutic methods identified by the study.*"(Table 1).

Some notions, such as "fair benefit," "fair additional benefit" have been proposed to replace "post-trial access" in the DoH further undermining the rights of participants. As a similar issue, "ancillary care" [26] (question of how far to take responsibility for care of non-target diseases when conducting a trial) were discussed in relation to the post-trial access.

Emanuel and other US bioethicists involving researchers from developing countries advocated that the "how" is more important than the "what" in order to avoid exploitation, and that there should be a shift towards the concept of "fair benefits" [27] (not ensuring "what" is proven to be the best). This issue was further discussed at the WMA, but according to Mungherera, who represented the Uganda Medical Association and was President of the WMA (2013–2014), requiring people in low-resource settings to make fairness decisions in trials does not protect them from exploitation. Under her leadership, in the final stage of the DoH 2013 revision, the "fair additional benefits" was removed from the final draft, due to concerns that it would lead to "dependency syndrome" (local community may be induced by financial benefit provided by external researchers rather than long-term benefit of healthcare improvement intended by local researchers or government) in developing countries [28].

2.2 Strengthened Alternatives in CIOMS Guidelines

The CIOMS 2002 guidelines, on the other hand [8], affirm that it is an ethical obligation of sponsors to ensure the availability of beneficial intervention or product developed from research for the population or community in question, but this availability was stated as "reasonable availability." Also, their guidelines recommend "capacity building" of the host community. Both seemed to be substitutes for the "post-trial access" which sometimes seemed difficult to achieve right away. Such considerations were not included in the DoH but were expanded in the 2016 revision [13] of the CIOMS Guidelines.

In 2004, Emanuel proposed "seven requirements" [29] replacing the three principles of the "Belmont Report" (respect for persons, beneficence, justice) with procedural descriptions, and later in 2008 proposed "eight principles" adding "social value" and "collaborative partnership," for collaborative study with developing countries [30, 31].

This logical framework was also specified in the 2016 CIOMS Guideline 1, with a clear statement that the human rights of individuals must be guaranteed. Furthermore, Guideline 7 recommends "community engagement" from the early stage of product development for returning research results to the society to achieve social value. In addition to the issue of access for the host community, Guideline 6 addresses the health needs of individual research participants, such as interventions that demonstrate significant benefit at the completion of study, as well as ancillary care during the study. This is more protective in relation to post-trial access but has many limitations regarding the duration of such a right.

2.3 Exploring International Consensus

Such difference adopted in the DoH and CIOMS Guidelines reflect the difference of characteristics of both documents. The DoH represents ethical principles within the physician-patient relationship, which are translated into norms and standards in the range of a research protocol whereas the CIOMS Guidelines represent a multi-stakeholder agreement, and addresses all health-related research involving humans, and consider external issues outside or after the completion of research [32]. Thus, in what can be considered with two of the main international guidelines, there is no clear consensus on ethical requirements for "post-trial access."

Another important international declaration, the UNESCO Declaration on Bioethics and Human Rights (2005) [33] in its Article 15 "Sharing of benefits" sets out a "benefit sharing" principle. However, access to developed products described in item (c) "*provision of new diagnostic and therapeutic modalities or products stemming from research*" is just one of seven options defining how benefits should be shared, including item (g) "*other forms of benefit consistent with the principles set out in this Declaration.*" This benefit sharing principle emerged from the discussions towards the Convention on Biological Diversity (CBD, 1992) [34] to avoid exploitation of biological resources in developing countries and to ensure that the benefits from research and development are shared fairly with developing countries, which are owners of these resources. This principle of access and benefit sharing (ABS) is affirmed by Article 15 on Access to Genetic Resources, together with the related Article 16 on Access to and Transfer of Technology, of the CBD and the Nagoya Protocol (2011) [35], thereby allowing implementation of this norm.

2.4 TRIPS Doha Declaration

The issue of access to proven interventions is closely linked to the issue of intellectual property rights in health technology: during the HIV/AIDS pandemic the Doha Declaration in 2001 [36] was adopted to prioritize response to public health crisis over the protection of intellectual property rights in the system of the World Trade Organization (WTO)'s Agreement on Trade-Related Aspects of Intellectual Property Rights (TRIPS Agreement). This opened a possibility for exportation of medicinal products produced executing "compulsory licensing" (a mechanism whereby a State authorizes a non-patent holder to use patent-protected technology without the patent holder's permission), which has traditionally been a part of each country's patent system. In 1998, the South African Pharmaceutical Manufacturing Association and 40 pharmaceutical companies sued against South African government claiming that legal amendment to address a public health crisis through compulsory licensing is against the TRIPS Agreement and South African constitution [37]. Eventually in 2002 the companies withdrew their legal action because of the difficulties of claiming violation against TRIPS, because the related texts in South African law were found to be based on a draft produced by the World Intellectual Property Organization Committee of Experts [38].

The Doha Declaration on the TRIPS Agreement and Public Health was adopted by the WTO Ministerial Conference of 2001 in Doha on November 14, 2001. WTO agreed "*that the TRIPS Agreement does not and should not prevent members from taking measures to protect public health. Accordingly, while reiterating our commitment to the TRIPS Agreement, we affirm that the Agreement can and should be interpreted and implemented in a manner supportive of WTO members' right to protect public health and, in particular, to promote access to medicines for all*" [36]. It provides justification to compulsory licensing and enables developing countries with manufacturing capacity to produce patented drugs and export them to countries with a public health crisis but no manufacturing capacity.

3 Post-Trial Access Challenge During the COVID-19 Pandemic

3.1 COVAX to Overcome "Vaccine Nationalism"

The success of the rapid development of COVID-19 vaccines announced in December 2020 [39–43], 1 year after the onset of the outbreak, can be attributed to significant governmental funding and streamlined review by regulators, as exemplified by "Operation Warp Speed" in the US [44], as well as collaboration with study participants pursuing developments to be rapid and as early as possible. Placebo-controlled trials had been conducted in higher prevalence areas but benefit of the study results had not necessarily been distributed by prioritizing these high prevalence areas. Countries with purchasing power secured their share through bilateral negotiations, criticized as "vaccine nationalism" [45] (see Chap. "Equitable Access to COVID-19 Vaccines, Vaccine Research, and Vaccine Apartheid on the African Continent: Challenges and Recommendations"), and proceeded with vaccination programs prioritizing populations more vulnerable to the virus within a country. For prominent example, Japan, without participating in the global placebo-controlled trials to prove the efficacy of the two mRNA vaccines, signed contracts with the sponsor companies for the purchase of these vaccines to cover their entire population. Later, it was revealed that the government secured 7 times more than the entire population and 60 million doses were discarded (see Chap. "Therapeutic Misconception as the Basis for Vaccine Nationalism of Japan: A Historical Reflection and Perspectives for Global Public Health"). There have also been moves to purchase, sell, and donate vaccines for political influence, which has been termed "vaccine diplomacy" [46]. All these factors, including the TRIPS Agreement, have contributed negatively to accessing of vaccines and other interventions by populations from low-income countries even where their communities have been involved in and participated in the COVID-19 vaccine clinical trials.

The "COVID-19 Vaccine Global Access" (COVAX) initiative has juxtaposed the above trends, with a policy of supplying the poorest countries and for vaccinating the groups most in need. This project is co-led by the Coalition for Epidemic Preparedness Innovations (CEPI), Gavi and the World Health Organization (WHO), alongside key delivery partner United Nations Children's Fund (UNICEF). It formed research and development/joint purchasing mechanism (COVAX facility) selecting candidate vaccines with contributions from high- and middle-income countries, and purchases successfully developed vaccines. It combines official development assistance (ODA) and a framework (COVAX AMC) to provide vaccines to low-income countries based on contributions from private donors.

It formulated an aspirational, challenging framework, but WHO reports in July 2022 stated that *"In the first year of rollouts, COVID-19 vaccines are estimated to have saved 19.8 million lives. Through unprecedently large and rapid rollouts worldwide, over 12 billion doses have been administered globally, in nearly every country in the world, resulting in countries reaching 60% of their populations on average. Yet only 28% of older populations and 37% of health care workers in low-income countries have been vaccinated with their primary series. 27 of WHO's Member States have not yet started a booster or additional dose program, 11 of which are low-income countries."* [47, 48] Unfortunately, COVAX did not realize its laudable objectives. One of the main reasons was that countries which had signed into COVAX concluded bilateral agreements with manufactures in parallel, thereby undermining what COVAX had set out to attain.

3.2 The Proposal for a TRIPS Waiver to Patents

To overcome this situation, a new mechanism is needed to assure the right to health recognized by the international community in the post-war period, which rejects an idea to regard health as a commercial commodity [49]. One milestone, though only a part of the desired new mechanism, was the TRIPS waiver initially proposed by South Africa and India.

In October 2020, these two countries proposed a waiver from the application of TRIPS Agreement to COVID-19-related intellectual property rights [50], but it was only in June 2022 that a final agreement was reached at the 12th WTO Ministerial Conference, allowing a limited waiver just for the COVID-19 vaccine [51]. South Africa, India, and the US, which had already expressed in May 2021 support for the proposal, as well as the European Commission received this limited waiver positively. Médecins Sans Frontières (Doctors without Borders), on the other hand expressed disappointment that it was limited to the COVID-19 vaccine and would not reach all countries [52]. The Pharmaceutical Manufacturers Associations expressed criticism and disappointment stating that patent protection has supported safe, effective, and fastest product development, as well as supply at global scale [53–56]. This view makes no sense because fast development was achieved by substantial investment of public money and greatly streamlined regulatory procedures as well as specifically collaborative review practices. Additionally, safety, efficacy, quality, and supply are not controlled by the patent system but by pharmaceutical regulations, and by other quality control measures through life cycle management of the drug. At the time of the 2022 December Council meeting, discussion continued as to whether or not to expand the range of waiver [57]. Of note, however, several middle-income countries already manufacture a host of important vaccines (See Chapters "Equitable Access to COVID-19 Vaccines, Vaccine Research, and Vaccine Apartheid on the African Continent: Challenges and Recommendations" and "Medicines Development for Global Health: Learning from COVID-19 Vaccines R&D"). Responding to the voices for extension of the waiver to achieve equitable and universal access to COVID-19 diagnostics and therapeutics [58], it was decided in December 2022 to continue discussions [59].

4 Discussion

While the ideals envisioned by initiatives such as COVAX and by the TRIPS waiver were never fully implemented, the global community learned what could be done to achieve global health for all by overcoming this unprecedented pandemic. UNESCO [14, 15] and the United Nations Independent Expert on human rights [60] have suggested that this initiative and the TRIPS waiver could be a step towards global vaccine equity and solidarity, with the aim of fully respecting and realizing fundamental equal right to health [17–21].

Considering the global experience of the HIV/AIDS and COVID-19 pandemics in the context of research involving humans, we should revisit the "post-trial access" clause in the DoH, having regard to this WMA declaration being valued as the standard-bearer for international research ethics, and that it could be the driving force at achieving adequate and equalitarian global health through the results of health research. Given that "post-trial access" is an aspirational norm, its implementation has been deemed to be difficult. However, faced with the reality of inequitable vaccine distribution during the COVID-19 pandemic, we found it necessary to specify some approaches towards making the already existing methods of turning the idealistic norm to reality. Such approaches could be described as globally common principles of ethics in research involving humans. Towards this end, we will discuss as below in terms of (1) the Latin American experience; (2) expanding stakeholder participation; and (3) efforts for post-trial access for global health.

4.1 Latin American Experience

Of note, responses among countries to the changes on post-trial access in the different versions of the DoH were varied, with some weighing down heavily against the weakened WMA position. Examples have been seen in Brazil in its legal documents [61–64], and in the majority of Latin American countries in declarations agreed in Buenos Aires [65] and Cordova both in 2008 [66], as well as in Pachuca in 2013 [67, 68], among relevant stakeholders [69]. Consistent with commitment to post-trial access and limited use of placebo they had rejected the post-2008 DoH, which incorporated in main text the 2002 Note of Clarification on placebo item, as discussed in the previous chapter (Chap. "Ethics of Placebo-Controlled Trials: Historical Analysis Including Experiences During the COVID-19 Pandemic"), and also the 2004 Note of Clarification where post-trial access was downgraded. The 2008 Brazilian research ethics resolution (404/2008) succeeded by 2012 resolution (466/2012) [63] made the guarantee of post-trial access essential condition when conducting clinical trials there. It was demonstrated that the sponsoring companies had no difficulty in complying with this requirement, remembering that the costs for this are small compared to post-approval marketing expenses [70].

It is also worth mentioning that the Brazilian Federal Constitution of 1988 [71] establishes that "health is everyone's right and the State's duty," and an example of this right is clear in the fight against the AIDS pandemic where all those in need have access to antiretrovirals, through the Unified Health System, without any out-of-pocket payments [64]. Thus, access was not limited to the post-study, demonstrating that the ethical principles related to the right to health reflect the need for global access to the benefits of the research.

4.2 Expanding Stakeholder Participation

The DoH's statement on post-trial access is a mix of obligations among the sponsor, researchers, and host country government. It does not include participation in this process of non-governmental organizations or patients eligible for the trial, and civil societies. According to the current DoH, the assessment of the post-trial arrangements is left only to the research ethics committee and the potential trial participants, and it explicitly refers only to prior arrangements and not to what should be done after the completion of the trial. While the DoH is a document by the WMA directed to physicians, it could not exclude participatory contributions from relevant stakeholders. However, the current version requires only limited involvement of the stakeholders. The key stakeholder is the participant/patient population, especially disadvantaged groups that are vulnerable to exploitation. Hence it will be necessary to bring in the voices of patient advocates, civil society organizations, and other similar groups in order to achieve post-trial access for all. This is important as it moves away from paying lip service to ethics in action in particular for the vulnerable and disadvantaged groups.

In addition to an individual researcher and the whole team responsibility in ensuring post-trial access, the overall responsibility lies with the sponsor to guarantee this access. Furthermore, interventions that have been successfully developed must be available and affordable for those who need it worldwide, in national and global health systems according to the magnitude of the needs. This comprehensive plan should be developed in full consultation among the sponsor, the host government, civil society and the international community and institutions, such as WHO, and it must be before the initiation of study planning, with responsibilities by all being shared equitably.

It should be also noted that the obligation of physicians to provide best proven care to individual study participants after the completion or discontinuation of study [72] must be clearly recognized, as a part of the norm of "post-trial access." Furthermore, this obligation is a part of obligations of physicians, as a whole, of "caring for each individual participant's health needs." This was clarified by CIOMS Guideline 6 [13] and is also implicit in the current DoH [72].

4.3 *Efforts for Post-Trial Access for Global Health*

The experience of post-trial access during COVID-19 and other global pandemics, including the AIDS pandemic are seminal examples [73, 74], underscoring that ethics must be applied to all research involving humans and must be used to improve world health at large. Pivotal trial results for new drug application are a crucial part of the long-term global accumulation of data. These drugs are possible achievements as the result of this accumulation of data from research participants that exist on a global scale. If there are people in different parts of the world who need interventions that have been proven to be the best, they might be involved in some way in generating the data on which the evidence is based. It will become more obvious in the emerging landscape where data-driven research is further promoted. Investigators conducting research on humans have a responsibility to contribute to ensuring the right of access to the best proven interventions for those who need them, regardless of geographical location.

Post-trial access to proven interventions should not only be seen as a right for trial participants, but could be also considered as a right of access for those in need of that intervention worldwide. It must be recognized that pursuing this normative trajectory is the ethical responsibility of all those planning and conducting research involving humans.

5 Conclusion

Our conclusion regarding the norm of "Post-trial access" is that it is not just a matter of evaluation by research ethics committees and candidate participants.

First, researchers have a responsibility to provide the best possible care, including the best proven interventions, to individual research participants at the time of completion or discontinuation of the study, in line with obligation of caring for participants' health needs.

Second, they have an ethical responsibility to participate in activities for realization of the effective access of target populations to developed interventions, including not only study participants but also a local community hosting the trial, as well as to the people in the world who need it, to achieve adequate and equitable global health. This would be a true application of the principle of justice and solidarity, achieved through equitable participation of all the relevant stakeholders.

Ethical principles must be described in such a way that they are oriented towards a goal to be shared and achieved globally. Refining research practice to guarantee post-trial access is a mandatory prerequisite for overcoming the pandemic we experienced and others that may occur in the future. Furthermore, this norm must be applied to any type of disease, and for those who most need it.

For this, it is necessary to ensure that health is not a commodity but one of the human rights agreed upon in international declarations and covenants.

Acknowledgments Authors appreciate following financial supports for traveling abroad to attend meetings directly related to this manuscript:

- The 23rd Annual Conference of the Japanese Society of Transcultural Psychiatry, 2016, with the presidency of Professor Kazutaka Shimoda, to invite AD to Tochigi, Japan.
- The 40th Annual Scientific Meeting of the Japanese Society of Clinical Pharmacology and Therapeutics (JSCPT), 2019, with the meeting presidency of Professor Kazutaka Shimoda, and the society presidency of Professor Hiroshi Watanabe, to invite DG to Tokyo, Japan.
- The International Federation of Associations of Pharmaceutical Physicians and Pharmaceutical Medicine (IFAPP), with the presidency of Dr. Varvara Baroutsou for travel of CK to attend the World Medical Association's regional conference in Tel Aviv, Israel, 2022, on the Declaration of Helsinki.

Conflict of Interest There is no conflict of interest to be disclosed related to the content of this manuscript.

References

1. Beyrer C. A pandemic anniversary: 40 years of HIV/AIDS. Lancet. 2021;397(10290):2142–3. Epub 2021 Jun 1. https://doi.org/10.1016/S0140-6736(21)01167-3.
2. Connor EM, Sperling RS, Gelber R, Kiselev P, Scott G, O'Sullivan MJ, VanDyke R, Bey M, Shearer W, Jacobson RL, et al. Reduction of maternal-infant transmission of human immunodeficiency virus type 1 with zidovudine treatment. Pediatric AIDS Clinical Trials Group protocol 076 study group. N Engl J Med. 1994;331(18):1173–80. https://doi.org/10.1056/NEJM199411033311801.
3. Lurie P, Wolfe SM. Unethical trials of interventions to reduce perinatal transmission of the human immunodeficiency virus in developing countries. N Engl J Med. 1997;337(12):853–6. https://doi.org/10.1056/NEJM199709183371212.
4. Angell M. The ethics of clinical research in the third world. N Engl J Med. 1997;337:847–9.
5. Varmus H, Satcher D. Ethical complexities of conducting research in developing countries. N Engl J Med. 1997;337(14):1003–5.
6. Angell M. Investigators' responsibilities for human subjects in developing countries. N Engl J Med. 2000;342(13):967–9. https://doi.org/10.1056/NEJM200003303421309.
7. National Bioethics Advisory Commission. Report on ethical and policy issues in international research: clinical trials in developing countries. April 2001.
8. Council for International Organizations of Medical Sciences. International ethical guidelines for biomedical research involving human subjects. 2002.
9. Beauchamp TL, Childress JF. Principles of biomedical ethics. Oxford: Oxford University Press; 1997.
10. The National Commission for the Protection of Human Subjects of Biomedical and Behavioral Research. The Belmont Report: ethical principles and guidelines for the protection of human subjects of research. April 18, 1979. https://www.hhs.gov/ohrp/regulations-and-policy/belmont-report/read-the-belmont-report/index.html.
11. World Medical Association. Declaration of Helsinki: Ethical principles for medical research involving human subjects. 2000. Rescinded. https://www.wma.net/wp-content/uploads/2018/07/DoH-Oct2000.pdf.
12. Greco DB. Special editorial: the proposed changes for paragraph 30 of the declaration of Helsinki 2000 will reduce requirements relative to healthcare access for clinical trial volunteers. Rev Bras Epidemiol. 2003;6(4):284. https://doi.org/10.1590/S1415-790X2003000400002 English version: https://www.scielo.br/j/rbepid/a/kYxSTQDqHc4dWhYW3fnxD5K/?lang=en

13. Council for International Organizations of Medical Sciences. International ethical guidelines for health-related research involving humans. 2016. https://cioms.ch/publications/product/international-ethical-guidelines-for-health-related-research-involving-humans/.
14. Joint Statement by the UNESCO International Bioethics Committee (IBC) and the UNESCO world commission on the ethics of scientific knowledge and technology (COMEST). UNESCO's ethics commissions' call for global vaccines equity and solidarity joint statement. SHS/BIO/IBC-COMEST/COVID-19 vaccines. Paris. 24 Feb 2021.
15. Joint statement of UNESCO ethics commission ensuring equal access for all to vaccines and therapeutic development to confront COVIS-19; joint statement of the UNESCO international bioethics committee (IBC) and UNESCO world Commission for the Ethics of science and technology (COMEST). SHS/IBC-COMEST/COVID-19 Vaccines IP Paris. 21 Sept 2021. https://unesdoc.unesco.org/ark:/48223/pf0000379042.
16. U.S. National Library of Medicine, Clinical Trials.gov archive. History of Changes for Study: NCT 04368728. https://clinicaltrials.gov/ct2/history/NCT04368728?V22=View#StudyPageTop.
17. World Health Organization. Constitution of the World Health Organization. 22 Jul 1946. https://www.who.int/about/governance/constitution.
18. Office of the United Nations High Commissioner for human rights, World Health Organization. The right to health. Fact sheet 31. 1 Jun 2008. https://www.ohchr.org/en/publications/fact-sheets/fact-sheet-no-31-right-health.
19. United Nations. The universal declaration of human rights. 10 Dec 1948. https://www.un.org/en/about-us/universal-declaration-of-human-rights.
20. United Nations. International Covenant on Economic, Social and Cultural Rights. Adopted by General Assembly resolution 2200A (XXI) of 16 December 1966, entry into force 3 January 1976. https://www.ohchr.org/en/instruments-mechanisms/instruments/international-covenant-economic-social-and-cultural-rights.
21. United Nations. International Covenant on Civil and Political Rights. Adopted by General Assembly resolution 2200A (XXI) of 16 December 1966, entry into force 23 March 1976. https://www.ohchr.org/en/instruments-mechanisms/instruments/international-covenant-civil-and-political-rights.
22. United Nations. Resolution adopted by the General Assembly on 25 September 2015—Transforming our world: the 2030 Agenda for sustainable development. 21 Oct 2015. https://www.un.org/en/development/desa/population/migration/generalassembly/docs/globalcompact/A_RES_70_1_E.pdf.
23. World Medical Association. Declaration of Helsinki: ethical principles for medical research involving human subjects. 2004. Rescinded. https://www.wma.net/wp-content/uploads/2018/07/DoH-Oct2004.pdf.
24. World Medical Association. Declaration of Helsinki: ethical principles for medical research involving human subjects. 2008. Rescinded. https://www.wma.net/wp-content/uploads/2018/07/DoH-Oct2008.pdf.
25. World Medical Association. Declaration of Helsinki: ethical principles for medical research involving human subjects. 2013. https://www.wma.net/policies-post/wma-declaration-of-helsinki-ethical-principles-for-medical-research-involving-human-subjects/.
26. Belsky L, Richardson HS. Medical researchers' ancillary clinical care responsibilities. BMJ. 2004;328(7454):1494–6. PMID: 15205296; PMCID: PMC428526. https://doi.org/10.1136/bmj.328.7454.1494.
27. Participants in the 2001 Conference on Ethical Aspects of Research in Developing Countries. Moral standards for research in developing countries: From "reasonable availability" to "fair benefits". Hastings Cent Rep. 2004;34(3):17–27.
28. Mungherera M, Kloiber O, Doppelfeld E, Kumar A, Jorge MR, Kurihara C, Saio T. Interview. The WMA Council Session in Tokyo, 2014: globalized medical ethics and research ethics: -Interview with Dr. Margaret Mungherera, Dr. Otmar Kloiber, Dr. Ajay Kumar, Prof. Dr. Elmar

Doppelfeld, Dr. Miguel R. Jorge. Clin Eval. 2014;42(2):553–90. http://cont.o.oo7.jp/42_2/p553-90eng.pdf.
29. Emanuel EJ, Wendler D, Grady C. What makes clinical research ethical? JAMA. 2000;283(20):2701–11.
30. Emanuel EJ, Wendler D, Killen J, Grady C. What makes clinical research in developing countries ethical? The benchmarks of ethical research. J Infect Dis. 2004;189(5):930–7.
31. Emanuel EJ, Wendler D, Grady C. An ethical framework for biomedical research. In: Emanuel EJ, et al., editors. The Oxford textbook of clinical research ethics. New York: Oxford University Press; 2008. p. 123–35.
32. Kurihara C. Chapter 9. A new horizon in health research ethics: a view from the perspective of radiological protection. In: Zölzer F, Meskens G, editors. Research ethics for environmental health. 1st ed. London: Routledge; 2021.
33. United Nations Educational, Scientific and Cultural Organization. Universal Declaration on Bioethics and Human Rights. 19 Oct 2005. https://en.unesco.org/about-us/legal-affairs/universal-declaration-bioethics-and-human-rights.
34. United Nations. Convention on biological diversity. 1992. https://www.cbd.int/doc/legal/cbd-en.pdf.
35. Secretariat of the Convention on Biological Diversity, United Nations. Nagoya Protocol on Access to Genetic Resources and the Fair and Equitable Sharing of Benefits Arising from their Utilization to the Convention on Biological Diversity : text and annex / Secretariat of the Convention on Biological Diversity. 2011. https://www.cbd.int/abs/doc/protocol/nagoya-protocol-en.pdf. Accessed from https://www.cbd.int/abs/.
36. World Trade Organization. Doha Declaration on the TRIPS Agreement and Public Health. Adopted by the WTO Ministerial Conference of 2001 in Doha on November 14, 2001. https://www.wto.org/english/thewto_e/minist_e/min01_e/mindecl_trips_e.htm.
37. World trade rules and cheaper drugs. Lancet. 2001;357:243.
38. 'tHoen EFM. The global politics of pharmaceutical monopoly power: drug patents, access, innovation and the application of the WTO Doha declaration on TRIPS and public health. AMB. 2009. https://msfaccess.org/sites/default/files/MSF_assets/Access/Docs/ACCESS_book_GlobalPolitics_tHoen_ENG_2009.pdf.
39. Polack FP, Thomas SJ, Kitchin N, Absalon J, Gurtman A, Lockhart S, Perez JL, Pérez Marc G, Moreira ED, Zerbini C, Bailey R, Swanson KA, Roychoudhury S, Koury K, Li P, Kalina WV, Cooper D, Frenck RW Jr, Hammitt LL, Türeci Ö, Nell H, Schaefer A, Ünal S, Tresnan DB, Mather S, Dormitzer PR, Şahin U, Jansen KU, Gruber WC. C4591001 clinical trial group. Safety and efficacy of the BNT162b2 mRNA Covid-19 vaccine. N Engl J Med. 2020;383:2603. Epub ahead of print. https://doi.org/10.1056/NEJMoa2034577.
40. Food and Drug Administration. Letter of Authorization. 10 May 2021. https://www.fda.gov/media/144412/download.
41. Baden LR, El Sahly HM, Essink B, Kotloff K, Frey S, Novak R, Diemert D, Spector SA, Rouphael N, Creech CB, McGettigan J, Kehtan S, Segall N, Solis J, Brosz A, Fierro C, Schwartz H, Neuzil K, Corey L, Gilbert P, Janes H, Follmann D, Marovich M, Mascola J, Polakowski L, Ledgerwood J, Graham BS, Bennett H, Pajon R, Knightly C, Leav B, Deng W, Zhou H, Han S, Ivarsson M, Miller J, Zaks T, COVE Study Group. Efficacy and safety of the mRNA-1273 SARS-CoV-2 vaccine. N Engl J Med. 2020;384:403. Epub ahead of print. https://doi.org/10.1056/NEJMoa2035389.
42. Food and Drug Administration. Letter of Authorization. 18 Apr 2023. https://www.fda.gov/media/144636/download.
43. Voysey M, Clemens SAC, Madhi SA, Weckx LY, Folegatti PM, Aley PK, Angus B, Baillie VL, Barnabas SL, Bhorat QE, Bibi S, Briner C, Cicconi P, Collins AM, Colin-Jones R, Cutland CL, Darton TC, Dheda K, Duncan CJA, Emary KRW, Ewer KJ, Fairlie L, Faust SN, Feng S, Ferreira DM, Finn A, Goodman AL, Green CM, Green CA, Heath PT, Hill C, Hill H, Hirsch I, Hodgson SHC, Izu A, Jackson S, Jenkin D, Joe CCD, Kerridge S, Koen A, Kwatra G, Lazarus R, Lawrie AM, Lelliott A, Libri V, Lillie PJ, Mallory R, Mendes AVA, Milan EP,

Minassian AM, McGregor A, Morrison H, Mujadidi YF, Nana A, O'Reilly PJ, Padayachee SD, Pittella A, Plested E, Pollock KM, Ramasamy MN, Rhead S, Schwarzbold AV, Singh N, Smith A, Song R, Snape MD, Sprinz E, Sutherland RK, Tarrant R, Thomson EC, Török ME, Toshner M, Turner DPJ, Vekemans J, Villafana TL, Watson MEE, Williams CJ, Douglas AD, Hill AVS, Lambe T, Gilbert SC, Pollard AJ, Oxford COVID Vaccine Trial Group. Safety and efficacy of the ChAdOx1 nCoV-19 vaccine (AZD1222) against SARS-CoV-2: an interim analysis of four randomised controlled trials in Brazil, South Africa, and the UK. Lancet. 2020;397(10269):99–111. Epub ahead of print. PMID: 33306989; PMCID: PMC7723445. https://doi.org/10.1016/S0140-6736(20)32661-1.

44. US Dept of Defense. Trump administration announces framework and leadership for 'Operation Warp Speed'. 2020. https://www.defense.gov/News/Releases/Release/Article/2310750/trump-administration-announces-framework-and-leadership-for-operation-warp-speed/. Accessed 14 Feb 2023.
45. Duke Global Health Innovation Center. New study shows rich country shopping spree for COVID-19 vaccines could mean fewer vaccinations for billions in low-income countries. 2 Nov 2020. https://dukeghic.org/wp-content/uploads/sites/20/2020/11/COVID19-Vax-Press-Release__28Oct2020-1.pdf.
46. Su Z, McDonnell D, Li X, Bennett B, Šegalo S, Abbas J, Cheshmehzangi A, Xiang YT. COVID-19 vaccine donations-vaccine empathy or vaccine diplomacy? A narrative literature review. Vaccines (Basel). 2021;9(9):1024. PMID: 34579261; PMCID: PMC8470866. https://doi.org/10.3390/vaccines9091024.
47. World Health Organization. WHO releases global COVID-19 vaccination strategy update to reach unprotected. 22 Jul 2022. https://www.who.int/news/item/22-07-2022-who-releases-global-covid-19-vaccination-strategy-update-to-reach-unprotected. Accessed 14 Feb 2023.
48. World Health Organization. Global COVID-19 vaccination strategy in a changing world: July 2022 update. 2022.
49. Proceedings: COVID-19 and bioethics Part 3: Pandemic and research ethics-democracy, placebo and post-trial access. Clin Eval. 2021; 49(sup 38). http://cont.o.oo7.jp/49sup38/49sup38contents_e.html.
50. Council for Trade-Related Aspects of Intellectual Property Rights. Waiver from certain provisions of the TRIPS Agreement for the prevention, containment and treatment of COVID-19: Communication from India and South Africa. 2 Oct 2020, IP/C/W/669. https://docs.wto.org/dol2fe/Pages/SS/directdoc.aspx?filename=q:/IP/C/W669.pdf&Open=True.
51. World Trade Organization, Ministerial Conference Twelfth Session Geneva, 12–15 June 2022. Ministerial Decision on the TRIPS Agreement. Adopted on 17 June 2022.22 June 2022; WT/MIN(22)/30 WT/L/1141. https://docs.wto.org/dol2fe/Pages/SS/directdoc.aspx?filename=q:/WT/MIN22/30.pdf&Open=True.
52. Médecins Sans Frontières. Disappointing outcome after nearly two years of discussion at the WTO. 17 Jun 2022. https://www.msf.org.za/news-and-resources/press-release/inability-agree-real-pandemic-intellectual-property-waiver-wto. Accessed 14 Feb 2023.
53. The International Federation of Pharmaceutical Manufacturers and Associations. Pharmaceutical industry expresses deep disappointment with decision on waiving intellectual property rights adopted at the World Trade Organization Ministerial Conference. 17 Jun 2022. https://ifpma.org/news/pharmaceutical-industry-expresses-deep-disappointment-with-decision-on-waiving-intellectual-property-rights-adopted-at-the-world-trade-organization-ministerial-conference/. Accessed 14 Feb 2023.
54. The European Federation of Pharmaceutical Industries and Associations. EFPIA and Vaccines Europe are deeply disappointed about the decision taken by World Trade Organization (WTO) at the Ministerial Conference (MC12) to endorse a TRIPS waiver for COVID-19 vaccines. 17 Jun 2022. https://efpia.eu/news-events/the-efpia-view/statements-press-releases/efpia-is-deeply-disappointed-about-the-decision-taken-by-world-trade-organization-wto-at-the-ministerial-conference-mc12-to-endorse-a-trips-waiver-for-covid-19-vaccines/. Accessed 14 Feb 2023.

55. Pharmaceutical Research and Manufacturers of America. PhRMA Statement on the TRIPS Waiver Agreement. 17 Jun 2022. https://phrma.org/resource-center/Topics/Trade/PhRMA-Statement-on-the-TRIPS-Waiver-Agreement. Accessed 14 Feb 2023.
56. Japan Pharmaceutical Manufacturers Association. On the TRIPS Waiver agreement at the 12th World Trade Organization Council. 20 Jun 2022. Japanese https://www.jpma.or.jp/news_room/release/news2022/220620.html. Accessed 14 Feb 2023.
57. World Trade Organization, Council for Trade-Related Aspects of Intellectual Property Rights. Paragraph 8 of the Ministerial Decision on the TRIPS Agreement Adopted on 17 June 2022: Report to the General Council; 16 Dec 2022, IP/C/95. https://dmscdn.vuelio.co.uk/publicitem/62777dec-6118-46f0-8a4d-bdd209d9aac9.
58. World Trade Organization, Decision text on extension of the 17 June 2022 ministerial decision to COVID-19 therapeutics and diagnostics: communication from the delegations of Argentina, Bangladesh, the Plurinational state of Bolivia, Egypt, India, Indonesia, Pakinstan, South Africa and the Bolivarian Republic of Venezuela on behalf of the co-sponsors of the IP/C/W/669/REV.1 proposal; 6 Dec 2022, IP/C/W/694. https://docs.wto.org/dol2fe/Pages/SS/directdoc.aspx?filename=q:/IP/C/W694.pdf&Open=True.
59. World Trade Organization, Council for Trade-Related Aspects of Intellectual Property Rights. Paragraph 8 of the Ministerial Decision on the TRIPS Agreement Adopted on 17 June 2022: Report to the General Council; 16 Dec 2022, IP/C/95. https://docs.wto.org/dol2fe/Pages/SS/directdoc.aspx?filename=q:/IP/C/95.pdf&Open=True.
60. United Nations. A/77/173: Global vaccine solidarity and human rights in the context of the coronavirus disease (COVID-19) pandemic. 15 Jun 2022. https://www.ohchr.org/en/documents/thematic-reports/a77173-global-vaccine-solidarity-and-human-rights-context-coronavirus. Accessed 14 Feb 2023.
61. Greco DB. Thirty years of confronting the Aids epidemic in Brazil, 1985-2015. Cien Saude Colet. 2016;21(5):1553–64. English, Portuguese. http://www.scielo.br/pdf/csc/v21n5/en_1413-8123-csc-21-05-1553.pdf. https://doi.org/10.1590/1413-81232015215.04402016.
62. Brazilian National Health Council (CNS), Brazilian National Research Ethics Committee (CONEP/CNS/MS). Guidance manual: frequent pending issues in clinical research protocols. Version 1.0. 2015.
63. Brazilian Research Ethics Commission Resolution 466/2012, which succeeded Resolution 404/2008.
64. Greco D, Shimoda K, Watanabe H, Organizer. The past, present, and future of ethics of international Health Research: research as a stepping-stone to universal public health care access. Clin Eval. 2020;48(1):207–31. http://cont.o.oo7.jp/48_1/w29-w53.pdf.
65. The Buenos Aires Declaration on Ethics and Clinical Trials. Buenos Aires, May 13, 2008. In: Homedes N, Ugalde A, editors. Clinical trials in Latin America: where ethics and business clash. Cham: Springer International; 2014.
66. Red Latinoamericana y Del Caribe de Bioetica—Redbioetica/UNESCO. Carta de Co'rdoba sobre e'tica en investigaciones con seres humanos. Córdoba, Argentina. 2008. https://www.psi.uba.ar/academica/carrerasdegrado/psicologia/sitios_catedras/obligatorias/723_etica2/material/casuistica/carta_de_cordoba_08.pdf.
67. Confederation Medica latinoamericana Y del Caribe (Latin American and Caribbean medical confederation). Declaration of Pachuca 2013.
68. Greco DB. Ethical dilemmas, research, and access to treatment. In: International Encyclopedia of the Social & Behavioral Science, vol. 8. 2nd ed; 2015. p. 31–6.
69. Homedes N, Ugalde A, editors. Clinical trials in Latin America: where ethics and business clash. Cham: Springer International; 2014.
70. Greco D, Kimura R, Victoria Perottino M, Saio T, Kurihara C. Ethics of international collaborative research: perspectives from Brazil: Part 1 selected notes on Paulo Freire: Part 2 access, compulsory license, case study. Clin Eval. 2020;48(1):W95–W123. http://cont.o.oo7.jp/48_1/w95-w123.pdf.

71. Constitution of the Federative Republic of Brazil. 1988. https://www.oas.org/es/sla/ddi/docs/acceso_informacion_base_dc_leyes_pais_b_1_en.pdf.
72. Kloiber O, Greco D, Watanabe H, Imamura K, Yamamoto Y, Matsuyama K, Saio T, Kurihara C. International collaborative research and new trends of research ethics: follow-up session. Clin Eval. 2020;48(1):233–65. http://cont.o.oo7.jp/48_1/w55-w87.pdf.
73. Brandt AM. How AIDS invented global health. N Engl J Med. 2013;368(23):2149–52. https://doi.org/10.1056/nejmp1305297.
74. Piot P, Quinn T. Response to the AIDS pandemic—a Global Health model. N Engl J Med. 2013;368:2210–8. https://doi.org/10.1056/nejmra1201533.

Our "WMA Declaration of Helsinki": Opinions and Proposals from Patient and Public for Research Ethics

Chieko Kurihara, Keiko Inoue, Hiroto Kai, Katsura Suzuki, Haruko Saeki, Yoshikazu Funabashi, Noriko Kishi, Akemi Kuge, Toshie Murakami, Yoshiko Saito, Eiko Uchida, Naoki Tsutsumi, and Kyoko Imamura

C. Kurihara (✉)
Kanagawa Dental University, Yokosuka, Kanagawa, Japan
e-mail: chieko.kurihara@nifty.ne.jp

K. Inoue
Association for Medical Malpractice Victims, Tokyo, Japan

H. Kai
Faculty of Education, Aichi University of Education (Science Course, Major in High School Education, Faculty of Education), Kariya, Aichi, Japan

K. Suzuki
Person with Parkinson, Kawasaki, Kanagawa, Japan

H. Saeki
My Informed Consent, Tokyo, Japan

Y. Funabashi
Cancer Patient and Survivor, Tokyo, Japan

N. Kishi
CMT Association Japan, Tokyo, Japan

A. Kuge
Hyogo Association of Certified Social Workers, Kobe, Hyogo, Japan

T. Murakami
Sagamihara Kyodo Hospital Cancer Patients' Association "Hukkiso", Sagamihara, Kanagawa, Japan

Y. Saito
Breast Cancer Survivor, Tokyo, Japan

E. Uchida
Breast Cancer Patient Support Group Bougainvllea/Patients–Voice Council, Tachikawa, Japan

N. Tsutsumi · K. Imamura
Japanese Institute for Public Engagement, Yokohama, Japan

C. Kurihara et al. (eds.), *Ethical Innovation for Global Health*,
https://doi.org/10.1007/978-981-99-6163-4_13

Abstract The Declaration of Helsinki (DoH) was first issued in 1964 by the World Medical Association (WMA), addressed to physicians, and was amended nine times with the latest version being adopted in 2013. While it has been incorporated into research regulations in many countries and is well known to most researchers, most patients and the public see its title for the first time when they are requested to participate in medical research. We therefore formed a group composed mainly of patients and the public together with experts having perspectives of patients and the public. Our activity was intended to enhance our understanding of the DoH and to have it explained in our own language. In this way, patients and the public would be able to better comprehend its scope and contents.

This chapter is resulting from about 2 years of our monthly web-meetings, during COVID-19 pandemic. We found some discussion points not included in the current version of the DoH, such as the value of research aimed at Sustainable Development Goals (SDGs), patient and public involvement, multidisciplinary collaboration, shared decision-making founded on informed consent, patient-oriented research, diversity and fairness of research ethics committees, assuring the rights of those considered to be "vulnerable populations," broad informed consent, dynamic consent, and social contract.

We hope that this chapter will contribute to the future revision of the DoH, as well as stimulate discussion for the international research ethics norms being under development, and that can be agreed to uniformly by all relevant stakeholders.

Keywords Patient and public involvement · Research ethics · Declaration of Helsinki · Human rights · Informed consent · Sustainable development goals (SDGs)

1 Introduction

1.1 Background

This chapter describes the opinions and proposals agreed to among the authors, members of a Japanese working group composed of patients, public, and experts of medicine or research ethics, resulting from discussions on the World Medical Association (WMA)'s Declaration of Helsinki (DoH) [1], having equal perspectives from patients and/or public.

The DoH was first issued in 1964 and was revised nine times with the latest version being adopted in 2013 (and planned to be revised in 2024). It has been developed by and addressed to physicians, as ethical principles of medical research involving humans. It is now incorporated into the medical research regulations in many countries, and physicians and other professionals understand it as a canon of governance when conducting research and submitting papers for publication. However, patients and the general public are unlikely to know about it and will see the title for the first time in an informed consent document when they are requested to participate in the medical research. Even most researchers have not read it in detail, although they may recognize the title and know that the government regulations directly applied to their study are developed in accordance with the DoH [2].

1.2 Methodological Aspects (Fig. 1)

Because of the above reason, a series of projects, based on having monthly web-meeting were started in November 2020, and have been undertaken by a voluntary working group, in partnership with the Japanese Institute for Public Engagement (Ji4pe) [3], an institute established in 2020 for promoting patient and public involvement (PPI) in health research and development. Among the 13 authors of this chapter (Box 1 of Fig. 1), ten are patients, patient families, or lay public (hereafter "non-experts"), including one pharmaceutical company research staff and one social worker. One is an expert in research ethics. Two are experts in medicine, one of which is the representative, another is a board member of Ji4pe. Nine are members of Ji4pe, some became members during the process. Two of the authors are not members of Ji4pe but are participating in this and other projects, including the educational program of Ji4pe. Hereafter, "we" in this chapter represents the authors of this composition. This chapter was generated through the following process.

1. First, a project started to read and understand the contents of the DoH. A similar structure to that of authors but with broader members have participated in this project (Box 1 of Fig. 1, "Other collaborators"). Although experts sometimes responded to the questions or opinions of non-experts, all the members of this project continued the discussion from the perspective of the patient and/or public. An expert with graphic recording skills was included in this part of the project. This resulted in our discussions and thoughts uniquely illustrated in graphics (Fig. 2), which have been published continuously in journal issues (Box 2 of Fig. 1) [4–6]. In addition, some of the non-expert members, held web-based bioethics seminars twice (first, inviting a bioethics professor, second, presentations by two patients), which were also introduced in journal issues (Box 2 of Fig. 1) [7–9].
2. During the above process, some non-expert members initiated the development of a document titled "Our Declaration of Helsinki" to explain the DoH in our own words using the language of patients and the public, addressed to a larger number of patients and the public who are now or may become research participants. This was partly because the expression in the current DoH is difficult for patients and general public to understand. Opinions were raised that the texts should be in plain language, e.g., avoiding "double-negative" phrases. In this process, the members found that there are some points that are not included in the current version of the DoH but are necessary to be shared among patients and the public. The project of developing this document (to be titled "Our WMA

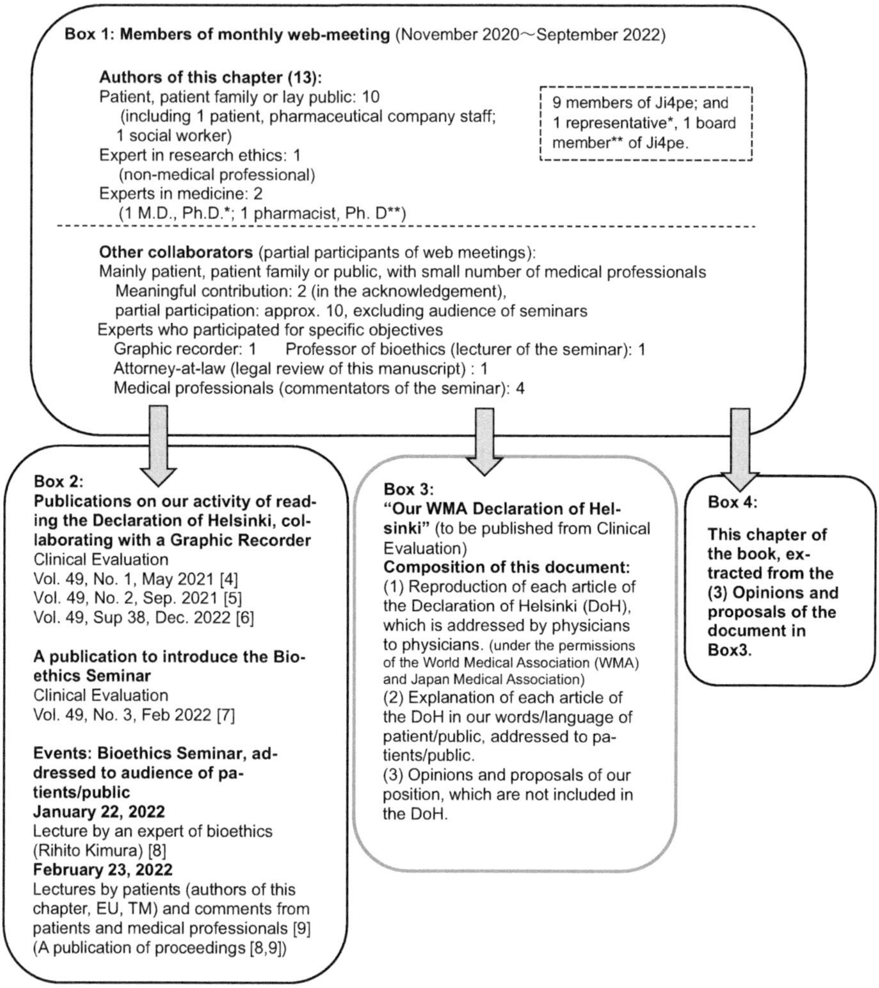

Fig. 1 The organization of the working group, products of the project

Declaration of Helsinki") is continuing separately from and simultaneously with the project of developing a manuscript of this chapter. "Our WMA Declaration of Helsinki" for future publication is composed of (1) Reproduction of each article of the DoH (under permissions of WMA and Japan Medical Association); (2) an Explanation of each article of the DoH in our words/language, to address to patients/public; (3) Opinions and proposals from the perspectives of patients/public which are not included in the DoH. This document will be published from the coming issue of Clinical Evaluation (Box 3 of Fig. 1).

3. While continuing the above process, some members, who are the authors of this chapter, decided to develop a manuscript, by extracting the important points from the "(3) Opinions and proposals" of the above document, to be shared with

Fig. 2 The first work of a series of graphic recordings of our discussions on the Declaration of Helsinki. Graphic Recording by Kanna Yoshikawa. http://cont.o.oo7.jp/grareco.html. English versions of Graphic Recording, which were originally published in Japanese, were provided for this chapter, including the others translated and published from the above website. All these were originally published in Japanese series of articles [4–6] and full texts are available from the website

wider international readers for improvement of international ethical principles of research involving humans. This becomes to be a manuscript of this chapter (Box 4 of Fig. 1), as summarized in Table 1.

The above process coincided with the COVID-19 pandemic. Patients and general public have experienced or observed a stringent process to discuss medical systems and activities, including clinical trials of antiviral products or vaccines as well as data-driven observational research, and the system of product authorization for emergency use. People have witnessed global and domestic inequity of resource allocation, together with a crisis of dwindling resources for management of diseases other than COVID-19. Under these circumstances, the DoH emerged for us as a realistic topic to be discussed seriously, along with the growing need to improve health literacy (knowledge and skill to improve personal or community health) [10] of general public.

Table 1 A summary of authors' opinions and proposals on the points which are not included in the current version of the Declaration of Helsinki and should be discussed for future revision

• Early evaluation of the research and development plan considering the impact on study target populations, society, global situations, and future generations
– To avoid discrimination or stigmatization against the study's target populations
– Aiming at Sustainable Development Goals (SDGs)
– Evaluating the impact on future generations, the environment, and society
• Strengthening patient/public perspectives in the ethical principle beyond the scope of self-autonomous norms by and addressed to physicians
– Using plain language, e.g., avoiding "double-negative" phrases
– The scope of the Declaration needs respecting embryos, fetuses, and the dead when being studied
– Promotion of patient and public involvement in each process of research
– Multidisciplinary collaboration, including patient peer supporters
– Shared decision-making based on informed consent
– Diversity and fairness of Research Ethics Committee membership
– Protecting dignity and rights of an individual incapable of consent and no available relative
– To develop international research ethics norm agreed among all the relevant stakeholders

2 International Norms and Health Systems Related to the DoH

2.1 *International Norms*

During the process of working group discussion to understand the contents of the DoH, the participants found it necessary to share a common understanding of the international norms particularly important and related to the DoH. Some of these are addressed by the lecturers of the Bioethics Seminar [8] organized by non-expert members. Some non-expert members proposed that viewpoints such as Sustainable Development Goals (SDGs) should be also recognized as the goal of health research, considering the most vulnerable populations. Here, these international norms are introduced to the readers of this chapter, including patients and the public, for a better understanding of the foundation of the principles of research ethics.

2.1.1 World Health Organization Constitution (WHO) (Adopted in 1946) [11]

The Constitution of the WHO, adopted shortly after World War II, states: "*Health is a state of complete physical, mental and social well-being and not merely the absence of disease or infirmity. The enjoyment of the highest attainable standard of health is one of the fundamental rights of every human being without distinction of race, religion, political belief, economic or social condition.*"

Article 25 of the Constitution of Japan [12] relates to this right, and many democracies around the world clearly state this right. However, the principle declared by the WHO "*highest attainable standard of health*" has been interpreted in the Constitution of Japan as "*the minimum standards of wholesome and cultured living*." In addition, questions have appeared to ask whether health cannot be achieved if all of these are not realized. In such context, the WHO Constitution provides the ideal status and goal, as well as an opportunity to discuss in depth the meaning and implication of "health."

We agreed that all the research participants, including victims or refugees affected by war, conflict, or natural disaster, must be entitled to this right, and health research must be performed with the objective of equal assurance of this right to health among the people in the world.

2.1.2 United Nations Declaration of Human Rights (1948) [13] and the United Nations International Covenants on Human Rights (1966) [14, 15]

The United Nations Declaration of Human Rights [13] also adopted just after World War II, was declared by member states that pledged not to repeat the war, and it emphasized the dignity of the individual, freedom, and equality and that no one should be discriminated against, under the reign of law. In democracies, the rights stated in this declaration have been related to the constitutional guarantee of rights. The Declaration recognizes fundamental rights such as freedom of thought and belief, life and safety, mobility and career choice, and privacy. It also prohibits slavery, torture, punishment, and detention without trial. Everyone has a responsibility to a society in which these rights are guaranteed, and everyone is also required to respect the freedoms of others and maintain public order. In Japan, Article 13 of the Constitution requires respect for individuals and the right to pursue happiness, and other articles guarantee the fundamental rights of citizens.

The United Nations International Covenants on Human Rights [14, 15] are treaties that makes the contents of the UN Declaration [13] legally binding. Although Japan has ratified it there are still various issues to be addressed. There is a paragraph in Article 7 that "*no one shall be subjected without his free consent to medical or scientific experimentation*" [15]. We believe that Japan should legislate, based on this fundamental bill of rights, comprehensive rules for the equal protection of research participants, irrespective of the scope of various research regulations. In South Africa, this article of the International Covenant is stipulated in the Constitution [16, 17]. In contrast in Japan, this Article has been mentioned only in Article 1 of the Supplementary Resolution by the House of Councilors when the "Clinical Research Act" [18] was enacted in 2017, as an issue for further consideration. In response, protecting the rights of research participants was defined as the basic principle of the enforcement regulation of this Act. However, as this Act only covers medicinal product research, equitable protection of participants of surgical research has become an outstanding issue, because of the difficulties of legally binding force on the discretion of physicians. To avoid such compromise, we need a more stringent and comprehensible international convention to call for a legal system of protection of human participants and research integrity.

2.1.3 Act on the Protection of Personal Information

Although there is no international treaty by the United Nations for the protection of personal information, the recommendations by the Organisation for Economic Co-operation and Development (OECD) (1980) [19] have suggested a common rule for many democracies to guarantee the protection of the individual's right to privacy. In particular, the Regulations by the European Union (EU) (General Data Protection Regulation: GDPR, 2016 [20]) have a great impact internationally because of the system of "adequacy decision," which provides procedural deregulation for cross-border transfer of personal data from EU to the countries which have personal data protection systems equivalent to those in the EU. This "adequacy decision" was granted to the general framework of the Japanese Act on the Protection of Personal Information, excluding the part on exceptional rules for scientific research. Thus, the Japanese Act was revised and implemented in April 2022 seeking for full "adequacy decision" including scientific research.

We have a strong opinion that research ethics principles must be the assurance of protection of this privacy right considering the recent landscape of research utilizing a large amount of data of patients and the public.

2.1.4 Sustainable Development Goals (SDGs) (2015) [21]

The world is now making great strides toward the Sustainable Development Goals (SDGs), which are the 17 goals adopted by the United Nations in 2015. The goals begin with "*end poverty*," "*end hunger*" and "*ensure healthy lives and promote well-being for all at all ages,*" pledging that no one must be left behind for the sustainable development of society, where diversity and inclusiveness must be taken into consideration. This cannot be achieved without strong determination. As a public good, access to healthcare must be ensured regardless of the difference between rich and poor. The COVID-19 pandemic has taught us that in order to achieve these goals, both globally and domestically, we need robust collaborations. UNESCO [22, 23] claims to make COVID-19 vaccines global common, public goods, and the world's top journals [24, 25] are proposing to change the current situation of the patent monopoly. We wish to create a new sustainable, patient-public-centric landscape of medicine and, as a part of it, research ethics that incorporates the notion of the public good.

The goal of the research is to generate scientific knowledge [26]. However, beyond that, it must be aimed at goals common among the international society. It is necessary to develop a research protocol and project design that leads to goals beyond the interests of individual persons or organizations, such as research performance achievements or commercial interests. During our discussion, it was clearly stated by some of the authors who are non-experts that the essential motivation of the patient and public who participate in medical research as study subjects or as members/partners in planning or conducting research is the spirit of altruism. We recognize an opinion expressed by non-expert members that the study results are "gifts" to future patients and the public.

2.2 Other Related Documents of the WMA and the Core Principles of the DoH

The WMA has issued a substantial number of declarations or statements in its post-World War II history [27]. The DoH is one of the most recognized and influential documents among them. We have confirmed that patients and the public should be informed of the scope of the DoH limited to research and that there are other WMA documents covering patient care in the context of service delivery. The documents in Table 2 are particularly important in relation to the responsibilities of physicians and the rights of patients, more generally in "medical care."

Among these documents, the Declaration of Geneva [28] and the International Code of Medical Ethics (ICoME) [29], both adopted shortly after World War II, were based on the reflection of the inhumanity of experimentation on human being and the denial of the right to life of some patients endorsed by doctors during the war (by Nazi Germany [33], Japan Imperial Army [34], judged in post-war trials). It is the firm determination of doctors around the world never to repeat such inhumanities. The DoH sets forth the principles for "research," but its core principle, Article 8 states that the goal of research must "*never take precedence over the rights and interests of individual research subjects*." This means that even if research is carried out for the sake of future patients, the best interests of each patient who participates in research must be prioritized. This core principle is established by quotation in Article 3 of the above two statements (Declaration of Geneva [28] and ICoME [29]).

These principles of the WMA are based on the notion that physicians should serve their patients in such a manner that patients trust them. On the other hand, it is not possible for physicians alone to decide on the "best interests" of the patient. For this reason, in contrast to the previous paternalistic way of thinking that the physician is assumed to take all responsibility and make decisions, the importance of "shared decision-making (SDM)" [35] has been recently attracting attention. This approach facilitates decision-making, which is achieved through discussions among physicians, other professionals, and patient groups with the focus being autonomous choices of a patient.

Table 2 WMA's documents particularly important in relation to the responsibilities of physicians and the rights of patients

• **Declaration of Geneva** (1948, last amended in 2017) [28]
The Declaration of Geneva defines that "*The health and well-being of my patient will be my first consideration*". It is considered to be a modern version of the Oath of Hippocrates, who is regarded as the medical saint of ancient Greece. This Declaration is recommended by the WMA to be pledged by doctors around the world. The above sentence is quoted in the DoH, without the word well-being as it was introduced in the recent amendment. In addition, the term "*patient autonomy*" was introduced in the latest revision. Also, another new idea was introduced to state that physicians should attend to their "*own health, well-being and abilities in order to provide care of the highest standard*"
• **International Code of Medical Ethics** (1949, last amended in 2022) [29]
The Code declares that "*A physician shall act in the patient's best interests when providing care*". It sets out the fundamental rules of professional ethics for physicians. This is also quoted in the Declaration of Helsinki
• **The Declaration of Lisbon for the Rights of Patients** (1981, last revised 2015) [30]
The Declaration of Lisbon stipulates the right of patients to the medical care of good quality, freedom of choice, self-determination, information, confidentiality, health education, and dignity
• **Declaration of Patient Safety** (2002, revised in 2012) [31]
The Declaration of Patient Safety states that "*Physicians strive to provide the highest quality health and medical care to patients*" and "*must ensure the patient safety is always considered during medical decision-making*", and that since individuals are rarely solely responsible for errors, it is necessary to foster a confidential reporting culture and to adopt a systematic, organizational approach
• **The Declaration of Taipei on health databases and biobanks** (2002, revised in 2016) [32]
The Declaration of Taipei lays down the governance of health databases and biobanks, collection, storage, and use of identifiable data and biological material (blood, surgical residual tissue, genome, etc.) beyond the individual care of patients. It also defines the requirements of valid consent of individuals who provide their data or materials to such banks and is positioned as a declaration that complements the DoH

2.3 Medical Research and Public Insurance

We have to recognize that the ethics of medical research are closely related to the public insurance system. There are international disparities in the insurance coverage of medical care, and as a result, the level of standard of care varies from country to country.

While research is conducted in multinational collaboration based on internationally common ethical principles, and drug development is also being promoted through global initiatives, international disparities are considered a challenge. The relationship between medical research and the public insurance system must also be considered from an international viewpoint, considering sustainability.

WHO plays an important role in promoting universal health coverage [36], and the WMA is also promoting it in collaboration with the WHO [37], but this initiative differs from those of the DoH.

3 Scope of the DoH

3.1 Scope of the Protection

Since the revision in 2000, the DoH has covered not only living human participants but also individual-identifiable biological materials and information [38]. The DoH does not clarify whether it covers prenatal fertilized embryos, fetuses, and the deceased. Moreover, national jurisdictions differ regarding the scope of the legal protection, and also how this is covered in research regulations. Even if the details are left to the rules of each country, we believe that an international consensus of principles of research ethics is needed for respecting embryos, fetuses, and the dead when being studied.

3.2 Scope of Obliged Persons

The DoH is addressed to physicians but it encourages others to adopt it as well. Since the DoH is a professional code of self-regulation of physicians, it is essentially impossible to have binding power over research conducted by other professions. However, even if a non-physician researcher conducts research, research participants must be equally protected. The level of the binding power of regulation should not depend on the occupation of the researcher but depend on the level of risk. The establishment of internationally accepted research ethics norms that transcend the boundary of professional associations is urgently needed, a continuing challenge.

We believe that even non-physicians, including us (patient or public) undertaking research, should follow the principles of the DoH. Although patients or the public do not conduct medical interventions, recently many of them have been involved in health research, which deals with personal information with resultant privacy risks. Regardless of the agenda of the researcher, the personal rights and human dignity of study participants must be protected and research integrity must be assured.

4 Visions for Future Generations

4.1 Environmental and Social Impact

Article 11 of the DoH states that "*Medical research should be conducted in a manner that minimises possible harm to the environment.*" This statement is related to the argument that global companies should be responsible for the disposal of drugs and materials especially when clinical trials are conducted in developing countries

[39]. In particular, in research that uses genetically modified organisms such as gene therapy, it must be recognized that researchers must follow legal regulations that are common to the world, according to the international treaty [40].

Considerations of the impact on the environment lead to deliberations about the impact on future generations [41]. In recent years, not only chemically synthesized pharmaceuticals but also products using various advanced technologies such as biologics, nanotechnology, digital devices, artificial intelligence, etc., have been developed. We believe that it is necessary to explore methodologies to evaluate the impact on the global environment, social and economic life, as well as cultural or spiritual aspects of human life when such products have spread in the future in the global market. For this reason, it is important to involve patient and public and listen to their concerns from the early stages of development.

4.2 Responsibility After the Completion of Research

In the process of revisions of the DoH, the topic of "post-trial access" has been discussed as an ethical principle that developed products should be made available to the study participants and the community where this study has been conducted. It is necessary to address not only the North-South issue on a global scale but also the issue of disparity within each region. In order to solve this issue, the Guidelines for Health-related Research by the Council for International Organizations of Medical Sciences (CIOMS) [42], developed in line with the DoH and make it applicable anywhere including low-resource settings, emphasize the importance of engagement of related communities in decision-making from the earliest stages of research and development of the products. Recently, it has been recommended worldwide to facilitate "patient and public involvement" from the early stages of development. During the COVID-19 pandemic, it is argued that people from all over the world, including study participants, have collaborated for vaccine development, therefore vaccines must be made equally available in the world [43].

We believe that it is important to involve patients, the public, and related communities in discussions, from the early stage of product development, considering not only the availability of the product but also to avoid negative impacts on the study target population or related communities such as discrimination or stigmatization [44]. In case of the application of new technologies which manipulate genetic information such as genome editing, consequences for future generations should be also discussed. These technologies can adversely impact the environment, directly, or as a result of their effects on the ecosystem.

5 Multidisciplinary Collaboration and Patient and Public Involvement in Research

5.1 *Multidisciplinary Collaboration*

Article 12 of the DoH states that research "*must be conducted only by individuals with the appropriate ethics and scientific education, training and qualifications. Research on patients or healthy volunteers requires the supervision of a competent and appropriately qualified physician or other health care professional.*"

As medical care becomes more sophisticated, there are issues to be discussed repeatedly among all the relevant stakeholders, including, but not limited to the way to describe in the research ethics principles the responsibilities of physicians, the rights and responsibilities of patients, and responsible collaboration among the multidisciplinary professionals, as well as local communities and civil society.

The DoH is a document whereby physicians address physicians about physicians' responsibilities, but it is now impossible for physicians alone to make decisions in research. Hence, Article 12 states the supervisory responsibilities of professionals not limited to physicians. However, we feel that there are other Articles that need reconsideration, as they emphasize physicians' decision-making, and it is clear that physicians alone cannot make decisions (e.g., Articles 18, 37 as mentioned below).

5.2 *Involvement, Collaboration, and Leadership of Patient and Public*

Recently, "patient-centered medicine" [45] has been promoted, not only integrating interdisciplinary expertise but also incorporating the views and narratives of patients and the public [46]. Many attempts have been made to integrate objective, scientific knowledge and subjective information and to share the experience and knowledge of patients and their families as social resources [47].

Also, in the context of research and development, patient and public involvement has been promoted around the world [48–51]. In addition to participating as a research subject as well as in the research ethics committees, it has been recommended to involve them, from an early stage, in the design of the study protocol or in the development of the clinical program. This is especially important since it has been incorporated into the international standards for medicines development [52]. Moreover, methodologies and systems of incorporating the reporting from patients of adverse events [53, 54] or treatment outcomes [55, 56] into drug evaluation have already been enhanced and facilitated.

In recent years, patients and the public have come to participate in research projects as members of multidisciplinary research teams. In medical care, a patient, a family member or a person who used to be a patient comes to be designated as a

"Peer supporter" [57] who supports patients or their family based on his/her own experience, and a such person with knowledge and experience as a lay person comes to be a member of a research team.

In addition, when the level of engagement becomes more advanced, patients and the public may lead a research project or take on the role of a sponsor. For this to be meaningful, they have to receive education and training so that they acquire the relevant specialized knowledge required for the conduct of those specific studies.

There is also a concept called "open science" or "citizen science." This explains the activities of conducting science by not only personnel of research institutes or pharmaceutical companies, but also by the general public. One example is that when the general public tackles environmental problems they may voluntarily start their research activities gathering data around them. Currently, research institutions, some pharmaceutical companies, and some governments are promoting open science policies, collaborating with people outside the established institutions, although only a small portion of their knowledge has been fairly shared with other stakeholders in need.

In addition, people with mental disorders have also developed an alternative way of research, called "Tojisha-Kenkyu" (also called "emancipatory research" [58]). It started as self-help, emancipating activities of users of mental health services, based on their own experience-based expertise or thought, different from medical research that presupposes medical expertise. This methodology has been spreading to other disease areas. Such research activities have been initiated by patients or the public, in a paradigm (framework of value, theory, and way of thinking) different from conventional scientific research. It makes observations sometimes on themselves, expressing and disseminating their findings. Some people are seeking ways to improve the quality of research and empowerment activities by integrating this kind of knowledge with the way of disseminating scientific and objective information.

In the cases described above, even though the DoH is addressed to physicians, we, as a group of patient and public, have to learn and follow the principles of the DoH, to protect the rights and safety of the study participants. Also, in the future revision of the DoH, appropriate promotion of patient and public involvement, including the cases of conducting research in collaboration with patient and public should be taken into consideration.

5.3 Physician's Responsibilities and Limitations in Risk Management

Article 18 states that "*A physician shall not engage in research involving human subjects unless he or she is confident that the risk has been properly assessed and that the risk can be adequately managed. If the risk is judged to be greater than the potential benefit, or if there is clear confirmation of results, the physician must decide whether to continue, change or immediately stop the study.*" This indicates that it is essential for a physician to be involved in the decision to continue or discontinue the research. The medical expertise of the physician is required for such a decision when an assessment of medical intervention is a target of research. However, in many cases, this decision is not something that a physician alone can make. In some cases, it may have to be based on a recommendation from an organization independent from the conduct of research like a Data Safety Monitoring Board, or otherwise, applicable law may require the decision of the head of the institution to which the researcher belongs to or sometimes of the regulatory authority.

Article 37 also defines the following conditions for the conduct of unproven interventions in clinical practice, outside the research program:

- "*In the treatment of an individual patient, where proven interventions do not exist or other known interventions have been ineffective*";
- "*After seeking expert advice*";
- "*With informed consent from the patient or a legally authorised representative*";
- "*In the physician's judgment it offers hope of saving life, re-establishing health or alleviating suffering*";
- "*This intervention should subsequently be made the object of research, designed to evaluate its safety and efficacy*";
- "*In all cases, new information must be recorded and, where appropriate, made publicly available.*"

(Divided into bullet points by authors.)

Since its first version, the DoH has been based on the idea that physicians can use unproven treatments at their discretion in order to save patients [59]. This could be acceptable as far as this treatment is performed following the ethical code of conduct of a physician, as described in other statements of the WMA. However, in recent years, such "discretionary power" of physicians has been recognized to be problematic, and in many cases, it is no longer possible for physicians alone to make decisions to use unproven interventions. Many countries have regulatory bodies for the use of unproven interventions and hence permission from a regulatory authority is usually required. In Japan, in particular, medical care has been placed in "a sanctuary" and legal control has been avoided in various aspects. We argue that unproven intervention should be used in a research program as suggested in Article 37 of the DoH. Nevertheless, there are many cases where unproven treatment is used at the discretion of the physician outside the research program, and in such cases, it is necessary to explain and communicate adequately which part of the intervention is out of the standard care, and only after the patient is adequately informed and agrees with the procedure, should the physician proceed.

It is desired that unproven interventions are used within a well-designed, publicly controlled regime. As suggested in the concept of MEURI (monitored emergency use of unregistered and experimental intervention), proposed by the WHO responding to the Ebola [60] and COVID-19 [61] pandemics, it is necessary to develop a mechanism to evaluate safety and efficacy, monitoring accumulated data of the use of such interventions, not limited to the cases of emergency use, but for wider cases of the use of the unauthorized intervention.

6 Research Ethics Committee

6.1 Diversity of Membership of the Research Ethics Committee

The DoH stipulates the need for transparency and independence of the Research Ethics Committee (Article 23), but does not mention the diversity of membership, which is pivotal for a democratic decision-making mechanism. The laws or guidelines of each country, including Japan, require the participation of a member representing lay public, not specialized in medical science. This seems to be because the DoH was originally developed as a self-governance regulation for physicians with peer-review systems among physicians only [2]. We believe that the Declaration should clearly state the need for diversity of membership and in particular, the participation of a member representing patients or the public.

In addition, although national regulations require the participation of members representing the public, it is questionable whether they really represent the general public, whether they can express their opinions, and whether the system is designed to encourage the participation of patients or the public in the discussion of the Research Ethics Committee. It is also necessary for patients and the public to be informed about research ethics for their responsible participation in the Committee. In addition, it is also necessary for the committee head to create an environment where lay members can speak without hesitation. More recently, some patients and members of the public are voluntarily educating themselves in research ethics. They are also learning the relevant regulations necessary for responsible performance as a member of the research ethics committee. Furthermore, some groups of patients or the public are formulating training programs for themselves [62]. It is desirable for an open, fair, and equitable recruitment system where highly motivated and trained lay persons can apply to become a member.

This is a challenge not limited to committee members representing the public. For members in any position, a system should be established in which members who are highly motivated and who have acquired the necessary knowledge through training are fairly recruited. There is also a problem with the system that allows researchers to apply to committees so that they are more likely to get approval. In order to ensure the independence and fairness of the committee, we desire, as already adopted in some European countries [63–65], the establishment of a fair system for recruiting members and a system where researchers cannot freely select and apply to a committee that tends to give a favorable opinion.

6.2 Support System of the Committee

It is often said that stringent regulations impede the progress of research, however, this problem is rather due to the insufficiency of supporting staff. The same is true of the secretariat system of the Research Ethics Committee. The limitations of relying solely on research ethics committee review have been noted for decades [66]. In order to keep the secretariats with specialized knowledge and experience, it is necessary for the committee to secure sufficient financial resources.

7 Informed Consent

7.1 Right to Self-Determination and Shared Decision-Making

Article 25 stipulates that informed consent must be given by a person who is going to be studied, not by any others, as far as this person is capable of giving consent. This means respecting human dignity and assuring the right to self-determination as defined in the Lisbon Declaration [30]. It is a prerequisite that informed consent must be freely given, after receiving sufficient explanation and with full understanding, without undue influence. Potential participants also need to be ensured that they can participate or at any time withdraw, without disadvantage or prejudice. Any information that could impact the willingness of continuing participation must be provided to study participants during the process of research.

In such environment with the increasing sophistication of technologies such as genomic medicine, elucidation of the pathophysiology of intractable diseases, along with advancement of information technology, in order to assure adherence to this principle of informed consent, democracy in the process of medical research is a necessary premise. Additionally, in order to bridge the knowledge gap between the experts and the general public, it is important to improve mutual communication skills. For that purpose, it is also necessary for patients and the public to learn for improvement of their health literacy.

The DoH states in Articles 28 and 29 that for a potential research subject incapable of giving informed consent, a physician must seek it from the legally authorized representative, and seek the assent of this research subject, additionally. We argue that it should be defined in the DoH that this legally authorized representative is the advocate of the potential research subject, representing the rights and interests of this person.

Article19 states that vulnerable people "*may have an increased likelihood of being wronged or of incurring additional harm.*" In addition to assuring protection, it is necessary to promote the self-advocacy of these people to exercise their right to express their opinions and make autonomous decisions. We have to respect diversity and assure equity, leading to "solidarity" to support their "emancipation" [67, 68].

On the premise of assuring the right to autonomous decisions of the study participants, it is also necessary to promote the idea of "shared decision-making (SDM)" [35], where the research team and advocates will collaboratively consider and decide with patients, and continuously reconsider this decision at each process of participating in a study. This can be a critical factor to promote informed consent. In such a situation, the patient or public also have to take responsibility, and not only just trust research team.

7.2 *Research Involving an Individual Incapable of Giving Consent, and Having No Relative*

Article 30 describes research involving a person incapable of giving consent where the research has to be conducted without delay, not being able to obtain consent from a legal representative, e.g., research involving an unconscious patient who needs emergency care. In this case, specific reasons for involving such patient in the study have to be stated in the research protocol and this reason has to be approved by a research ethics committee. Moreover, consent must be obtained at a later stage, but as soon as possible, from this study subject or a representative. Such requirement is also defined in research regulations in many countries.

However, the DoH does not mention the situation where the candidate participant is incapable of giving consent and there are no relatives, nor any legally authorized representatives. In such cases the best option for these persons has to be sought. Each country defines within its legal system some specific areas where a public organization provides decision or permission for an individual (e.g., a principle of "parens patriae," a state acts to give protection of an individual who lacks a guardian). The international research community has not had a sufficient discussion regarding research involving such participants. Recently, there is an increasing interest in the ethics of research involving homeless people or those in situations of natural/man-made disasters, and war or conflict settings. The international community has been discussing research with the purpose of improvement of well-being of such population and also issues of the rights of such populations to access clinical trials of promising drugs. This is a difficult issue but it must be discussed from the view point of patients and the public.

7.3 *Declaration of Taipei (DoT) and Alternative Expressions of Consent*

The DoT [32] on health databases and biobanks was adopted as a document to complement the DoH. Article 32 of the DoH defines informed consent for the use of identifiable human material or data. It allows exceptional situations impossible or impracticable to obtain consent, where research may be conducted only after consideration and approval of the Research Ethics Committee.

The DoT defines the "valid" consent of the individual from whom data or biological material (e.g., blood, surgical residual tissue, genomes, etc.) are collected and stored in health databases or biobanks. It also defines the principles of "governance" (an organizational mechanism to assure research ethics) of health databases or biobanks. The DoH seems to be launched for research conducted within the relationship between a doctor and a patient. However, the research landscape has changed drastically to promote the use of substantial number of human data and/or materials. Within this background context, the DoT was adopted to complement the DoH. In such situations, alternative ways of consent have been proposed as follows. These concepts should be discussed in future revisions of the DoH and the DoT.

7.3.1 Broad Informed Consent

We should recognize the recent trend of research where researchers assume secondary data use or future data sharing beyond the original consents of participants in primary research [69]. Sharing data from individual study participants in clinical trials for the purpose of verification of the results of the trial and for meta-analysis (methodology to integrate data from the number of studies, comprehensively, aiming at improved analytical power of safety and efficacy of intervention) is recommended by the International Committee of Medical Journal Editors (ICMJE, academic organization of editors of world prestigious journals to develop international policies or standards for medical publication) to be an ethical obligation of the researcher to respond to the altruism of study participants [70]. CIOMS Ethical Guidelines [42] also recommend this data sharing on the premise of privacy protection, and that an individual person is the only one who can decide on his/her data to be shared. It is also necessary to consider that in situations where patient-oriented research (research proposed, planned, and conducted by patients), patients themselves may be the main actors of data sharing, which means that data are shared among researchers who are patients [71].

Now data sharing based on conventional carte blanche "blanket consent" [72] comes to be ethically unacceptable. The CIOMS Ethical Guidelines [42] proposed the concept of "broad informed consent" to obtain consent after explaining as much as possible how individual data or materials will be used in various forms in the future. We should promote this procedure of "broad informed consent" during the process of prospective research, instead of relying just on "opt-out" procedure (to stop the use of data/material only when a person concerned expresses withdrawal) in retrospective research. This would be indispensable in the current legislation in various countries regarding personal information protection. The DoT already stipulated the idea of voluntary "valid consent" based on the explanation including governance as the ethical organizational mechanism of health databases or biobanks. The implication of "broad informed consent" in CIOMS Ethical Guidelines is the same, and we wish to share these concepts widely.

7.3.2 Dynamic Consent

In such a situation of promoting multiple data use, we wish to endorse a concept of "dynamic consent" [73], which assures the various rights of study participants to withdraw their consent, according to the information of new proposals of secondary research, as well as the rights to access to study information, including "rights to know" and "right not to know" [74] about the incidental findings (causes or symptoms incidentally found, without intension in the study plan) and/or research results. In some cases, up to the way of participation or disease characteristics, study participants may not wish to know about the research results with uncertain clinical implications. In other cases, some participants may wish to know the achievement of the research to which they have contributed. Or otherwise, in rare cases, participants may seek a way to share intellectual property rights specifically due to their contribution.

Recent regulations worldwide of clinical trials, responding to the statement by ICMJE [75], require registration of progress status and results of each trial in public databases. Study participants are informed how they can access such information. In some of the genetic analysis projects, researchers prepare a system using an electronic digital tool to enable participants access to study information according to the progress of research as well as information of secondary use of their data or biological materials, and to make decisions to continue or withdraw their participation considering the changing situations. This type of consent given according to the continuing changes of situation is called "dynamic consent," and is supported by some groups of patients and the public in Europe [51].

7.3.3 Social Contract and Consensus in Society

New legal frameworks have been developed in each country so that completely anonymized/de-identified data can be widely used without the consent of each individual. However, even if there is only minimal risk of privacy violations, we, the patient and public involved in the activity are uncomfortable providing our data for multiple uses, particularly if our data are to be used beyond our knowledge. In such a case, researchers should provide explanations to society, repeatedly, and ensure that society as a whole gains an understanding of what the research enterprise entails. We hope that researchers increase not only transparency and openness, but also actively fulfill their accountability and discuss pros and cons based on information that everyone can access. The use of personal data without the consent of the individual concerned must be carried out with minimal risk and only in specially approved circumstances, based on such a democratic process and on the basis of a contract with society.

8 Discussion: Disagreements and Limitations

Most of the authors of this chapter are patient or public, including several experts in medical areas and one expert in research ethics, who participated in this project having perspectives of patient and/or public. Non-expert members have various positions, e.g., leading a patient group, engaged in research teams, or working for victims of medical malpractices. For this reason, some varieties of opinions have been expressed, and opinions described here may be those of "lay experts" and are not representing those of patients and the public in general. While there was no remarkable disagreement, the following limitations and variety of positions should be addressed.

First, because the DoH is a norm of self-regulation of physicians addressed to physicians, it seems to be difficult to make proposals beyond this nature. However, authors who are patient or public argued much about their opinions beyond the nature of the Declaration, e.g., the goal of the research should be aimed at SDGs, the necessity of evaluation of research considering the impact on future generations beyond the protection of individual research participants, or clarification of the responsibilities of non-physician or patient/public who are researchers. These proposals are from the intention for improvement of research ethics principles, collaborating with physicians. A member who is a physician criticized the paternalistic aspects of the DoH in particular, but such criticisms do not represent the opinions of many of Japanese clinicians, and these criticisms may have come from the expertise of mechanism of research and development projects. While the authors agreed with the position of the DoH as an established international research ethics principle, there have not yet been consensus about the need of another international norm developed by members not limited to physicians, and which organization should lead it.

Second, it was difficult to reach a consensus on how to ensure the quality of researchers who are patient or public conducting observational research involving humans, not for evaluating a medical intervention, but for improvement of the medical environment or promoting self-help activities, and how to ensure the quality of such research.

Third, regarding the most controversial article of the DoH on a placebo-controlled study where the best-proven interventions exist, it was difficult for us to clarify opinions agreed among the authors, although some of the authors who are patient or public participated in the international webinar to discuss this topic [43] and also participated in a discussion session in Japanese using Graphic Recording [6], which resulted in Japanese publications. The understanding of patients and the public seemed to be limited to the ethical problem and scientific necessity of placebo-controlled trials when there is no established therapy. It seemed difficult for patient and public to reach to consensus with sufficient understanding of the meaning of setting conditions on the justifiability of placebo-controlled trials when there is an established intervention somewhere in the world. This may suggest the difficulty of true informed consent to such studies.

9 Conclusion

This chapter was born out of repeated discussions with the aim of learning about the DoH from the standpoint of patients and the public wishing to convey the contents of the DoH to people in the same position, in the language of patients and the public. During the period of COVID-19 pandemic, our group, patients with various diseases, their families, and friends engaged in civic activities met together and interacted through the discussion. Generally, patients and their families are exerting themselves to tackle their own diseases and related societal issues. Our group has learned about the DoH and related ethical issues including the need to take on the role of advocates, and the importance of having broader views for patients and the public in the global society. We found it necessary to bear in mind the situation of patients and the public who do not have adequate access to medicine and education. The DoH goes beyond research ethics, emphasizing the importance of protecting human dignity and human rights, as well as the justice that our world has to achieve. The DoH is described as a "living document" [76] and it is expected to change as the need arises. We are grateful that we are in such environment to be able to continue these discussions, and we wish to keep deliberation on the challenging issues facing the world.

During this learning process, we have clarified important points of research ethics focused on patient/public-oriented perspectives, in order to promote valuable research aiming at the achievement of internationally shared SDGs. We believe that we could record a valuable trajectory of discussions to provide a new landscape for research ethics.

We hope that this chapter will contribute to the future revision of the DoH and the development of international principles of research ethics in which not only physicians but also related diverse stakeholders can participate.

Acknowledgment We would like to express our appreciation to Ms. Kanna Yoshikawa for her contribution of providing English-translated versions of Graphic Recording which was originally published in Japanese [4–6], and Ms. Mariko Mimura, Attorney-at-Law, Nishimura & Asahi, for her legal review on this manuscript, Prof. Rihito Kimura who provided an important lecture on the roots of bioethics and SDGs at the web-based seminar organized mainly by patient and public of our group. We appreciate all of the participants of web-based meetings and seminars, especially Ms. Hiroko Takahashi, Ms. Noriko Iwaya, who provided important input during development of this manuscript. In addition, we appreciate Prof. Dirceu Greco and Prof. Ames Dhai, editors of this book, who provided substantially important suggestions.

Author Contributions This chapter originated from a working group to understand the contents of the DoH, which was started in November 2020 by people from the standpoint of patient or public. HS was the first to conceive to write the contents of the DoH in the language of patient and public. KI, HK, KS, and CK organized activities as group leaders. CK answered questions and provided necessary knowledge as research ethics experts, and KI and NT as medical and pharmaceutical experts, but most of the ideas and contents of this manuscript are derived from the opinions and perspectives of authors positioned as patient or public.

CK, KI, HK, KS, YF, AK, TM, YS, EU, NT and KI are members of the Japanese Institute for Public Engagement (Ji4pe).

Conflict of Interest There is no conflict of interest to be declared related to this manuscript.

References

1. World Medical Association. WMA Declaration of Helsinki: Ethical principles for medical research involving human subjects. Adopted by the 18th WMA General Assembly, Helsinki, Finland, June 1964, and last amended by the 64th WMA General Assembly, Fortaleza, Brazil, October 2013. https://www.wma.net/policies-post/wma-declaration-of-helsinki-ethical-principles-for-medical-research-involving-human-subjects/.
2. Kloiber O, Greco D, Watanabe H, Imamura K, Yamamoto Y, Matsuyama K, Saio T, Kurihara C. International collaborative research and new trends of research ethics: follow-up session. Clin Eval. 2020;48(1):W55–87. http://cont.o.oo7.jp/48_1/w55-w87.pdf.
3. Japanese Institute for Public Engagement. https://ji4pe.tokyo/index_en.html. Accessed 31 Jul 2022.
4. Yoshikawa K, Saeki H, Tsutsumi N, Imamura K. A group of Patient Public Involvement (PPI) shared their feelings after reading "DECLARATION OF HELSINKI": the atmosphere of the meeting is vividly reported with Graphic Recording. Clin Eval. 2020;49(1):109–15. Japanese. http://cont.o.oo7.jp/49_1/p109-15.pdf.
5. Yoshikawa K, Saeki H, Tsutsumi N, Imamura K, Kai H, Inoue K, Suzuki K, Kurihara C. A group of patient public involvement (PPI) shared their feelings after reading "DECLARATION OF HELSINKI": part 2 the launch of developing "MY/OUR DECLARATION OF HELSINKI". Clin Eval. 2021;49(2):305–14. Japanese. http://cont.o.oo7.jp/49_2/p305-14.pdf.
6. Yoshikawa K, Inoue K, Kai H, Suzuki K, Murakami T, Tsutsumi N, Imamura K, Kurihara C. A group of patient public involvement (PPI) shared their feelings after reading "DECLARATION OF HELSINKI": part 3 graphic recording of the discussion on placebo-controlled trial and post-trial access. Clin Eval. 2021;49(Sup 38):217–26. Japanese. http://cont.o.oo7.jp/49sup38/p217-26.pdf.
7. Kurihara C, Uchida E, Murakami T, Kai H, Suzuki K, Inoue K. Bioethics Seminar organized by and for patient and public: Civic Action to link the lives for our future. Clin Eval. 2022;49(3):531–8. Japanese. http://cont.o.oo7.jp/49_3/p531-8.pdf.
8. Kimura R. The roots of global development of bioethics and SDGs to bridge the rights of patients and public to the future generation. 22 Jan 2022. [Proceedings: Kimura R. The 1st Bioethics Seminar Development of bioethics and SDGs: the roots of the global development of bioethics and future-oriented Sustainable Development Goals (SDGs). Clin Eval. 2022;50(3):299-324. Japanese.]
9. Uchida E, Murakami T. Patient-oriented bioethics: considering about ethics based on the patient's experience of themselves of users of medical care. 23 Feb 2022. [Proceedings: Uchida E, Murakami T. The 2nd bioethics seminar patient-oriented bioethics: in consideration of ethics based on patient-centric experiences and users of medical services. Clin Eval 2022;50(3):325-60. Japanese.]
10. World Health Organization. Health promotion glossary. WHO/HPR/HEP/98. 1. 1998. https://iris.who.int/bitstream/handle/10665/64546/WHO_HPR_HEP_98.1.pdf?sequence=1.
11. World Health Organization. Constitution of the World Health Organization. 22 Jul 1946. https://www.who.int/about/governance/constitution.
12. Prime Minister of Japan and His Cabinet. The Constitution of Japan. Promulgated on November 3, 1946, Came into effect on May 3, 1947. https://japan.kantei.go.jp/constitution_and_government_of_japan/constitution_e.html. Accessed 31 Jul 2022.
13. United Nations. The Universal Declaration of Human Rights. 10 Dec 1948. https://www.un.org/en/about-us/universal-declaration-of-human-rights.

14. United Nations. International Covenant on Economic, Social and Cultural Rights. Adopted by General Assembly resolution 2200A (XXI) of 16 December 1966, entry into force 3 January 1976. https://www.ohchr.org/en/instruments-mechanisms/instruments/international-covenant-economic-social-and-cultural-rights.
15. United Nations. International Covenant on Civil and Political Rights. Adopted by General Assembly resolution 2200A (XXI) of 16 December 1966, entry into force 23 March 1976. https://www.ohchr.org/en/instruments-mechanisms/instruments/international-covenant-civil-and-political-rights
16. South African Government. The Constitution of the Republic of South Africa. As adopted on 8 May 1996 and amended on 11 October 1996 by the Constitutional Assembly. https://www.gov.za/documents/constitution/constitution-republic-south-africa-1996-1.
17. Dhai A. Ethics and regulations of clinical research in South Africa. Clin Eval. 2017;45(3):W15–35. http://cont.o.oo7.jp/45_3/w15-w35.pdf.
18. Clinical Trials Act. Act No. 16 of April 14, 2017. https://www.mhlw.go.jp/file/06-Seisakujouhou-10800000-Iseikyoku/0000213334.pdf.
19. Organisation for Economic Co-operation and Development. Recommendation of the council concerning guidelines governing the protection of privacy and transborder flows of personal data. 1980. https://legalinstruments.oecd.org/en/instruments/OECD-LEGAL-0188.
20. Regulation (EU) 2016/679 of the European Parliament and of the Council of 27 April 2016 on the protection of natural persons with regard to the processing of personal data and on the free movement of such data, and repealing Directive 95/46/EC (General Data Protection Regulation) (2016). Official Journal of the European Union. 4.5.2016: L 119/1-88.
21. United Nations. Resolution adopted by the General Assembly on 25 September 2015 - Transforming our world: the 2030 Agenda for Sustainable Development. 21 Oct 2015. https://www.un.org/en/development/desa/population/migration/generalassembly/docs/globalcompact/A_RES_70_1_E.pdf.
22. Joint Statement by the UNESCO International Bioethics Committee (IBC) and the UNESCO World Commission on the Ethics of Scientific Knowledge and Technology (COMEST). UNESCO's ethics commissions' call for global vaccines equity and solidarity joint statement. SHS/BIO/IBC-COMEST/COVID-19 Vaccines. Paris, 24 Feb 2021. https://unesdoc.unesco.org/ark:/48223/pf0000375608.
23. Joint Statement of UNESCO Ethics Commission ensuring equal access for all to vaccines and therapeutic development to confront COVIS-19; Joint statement of the UNESCO International Bioethics Committee (IBC) and UNESCO World Commission for the Ethics of Science and Technology (COMEST). SHS/IBC-COMEST/COVID-19 Vaccines IP Paris, 21 Sept 2021. https://unesdoc.unesco.org/ark:/48223/pf0000379042.
24. The Lancet COVID-19 Commission. Enhancing global cooperation to end the COVID-19 pandemic. 2021. https://covid19commission.org/enhancing-global-cooperation.
25. A patent waiver on COVID vaccines is right and fair. Nature. 2021;593(7860):478. https://doi.org/10.1038/d41586-021-01242-1. PMID: 34035532.
26. The National Commission for the Protection of Human Subjects of Biomedical and Behavioral Research. The Belmont Report: ethical principles and guidelines for the protection of human subjects of research. 1979. https://www.hhs.gov/ohrp/regulations-and-policy/belmont-report/read-the-belmont-report/index.html.
27. Kloiber O, Shimoda K, Watanabe H, Imamura K, Kurihara C. Declaration of Helsinki: challenges and new trends ahead of us. Clin Eval. 2020;48(1):W9–W27. http://cont.o.oo7.jp/48_1/w9-w27.pdf.
28. World Medical Association. WMA Declaration of Geneva. Adopted in September 1948, last amended in October 2017. https://www.wma.net/policies-post/wma-declaration-of-geneva/.
29. World Medical Association. WMA International Code of Medical Ethics. Adopted in October 1949, last amended in October 2022. https://www.wma.net/policies-post/wma-international-code-of-medical-ethics/.

30. The World Medical Association. WMA Declaration of Lisbon on the Rights of the Patient. Adopted in September/October 1981, last reaffirmed in April 2015. https://www.wma.net/policies-post/wma-declaration-of-lisbon-on-the-rights-of-the-patient/.
31. The World Medical Association. WMA Declaration on Patient Safety. Adopted in October 2002, last revised in Ocboter 2022. https://www.wma.net/policies-post/wma-declaration-on-patient-safety/.
32. The World Medical Association. WMA Declaration of Taipei on ethical considerations regarding health databases and biobanks. Adopted in October 2002, revised in October 2016. https://www.wma.net/policies-post/wma-declaration-of-taipei-on-ethical-considerations-regarding-health-databases-and-biobanks/.
33. International Military Tribunal. Trials of war criminals before the Nuremberg Military Tribunals under Control Council law no. 10 Nuernberg, October 1946-April 1949. Vol. 2, p. 181-2. (The Nuremberg Code).
34. International Committee of the Red Cross. Convention (III) relative to the Treatment of Prisoners of War. Geneva, 12 August 1949. Commentary of 2020 Article 13: Humane treatment of prisoners. 2020. https://ihl-databases.icrc.org/applic/ihl/ihl.nsf/Comment.xsp?action=openDocument&documentId=3DEA78B5A19414AFC1258585004344BD#73.
35. Hoffmann TC, Montori VM, Del Mar C. The connection between evidence-based medicine and shared decision making. JAMA. 2014;312(13):1295–6. https://doi.org/10.1001/jama.2014.10186. PMID: 25268434.
36. World Health Organization. Universal health coverage. https://www.who.int/health-topics/universal-health-coverage#tab=tab_1. Accessed 31 Jul 2022.
37. World Medical Association. WHO and WMA sign memorandum of understanding. 5 Apr 2018. https://www.wma.net/news-post/who-and-wma-sign-memorandum-of-understanding/. Accessed 31 Jul 2022.
38. Tsuboi E, Kurihara C. The 2000 revision of the declaration of Helsinki and its implications in medical ethics: interview with the WMA immediate past president. Clin Eval. 2002;30(1):99–107. Japanese. http://cont.o.oo7.jp/30_1/p99-107.pdf.
39. Kurihara C, Saio T. Expert conference on the revision of the declaration of Helsinki in Tokyo: 2013 revision and perspective toward the 50th anniversary in 2014. Clin Eval. 2013;41(2)e. http://cont.o.oo7.jp/41_2/p337-49eng.pdf.
40. Convention on Biological Diversity. The Cartagena protocol on biosafety. Adopted on 29 January 2000 and entered into force on 11 September 2003. https://bch.cbd.int/protocol/. Accessed 31 Jul 2022.
41. International Bioethics Committee (IBC), United Nations Education, Scientific and Cultural Organization. IBC Report of the International Bioethics Committee (IBC) on the principle of protecting future generations. 2021. SHS/IBC-28/2021/2 Rev. https://unesdoc.unesco.org/ark:/48223/pf0000378723.
42. Council for International Organizations of Medical Sciences. International ethical guidelines for health-related research involving humans. 2016. https://cioms.ch/publications/product/international-ethical-guidelines-for-health-related-research-involving-humans/.
43. Kurihara C, Saio T, Imamura K, Greco D. COVID-19 and bioethics part 3: pandemic and research ethics–democracy, placebo and post-trial access. Clin Eval. 2021;49(38):9–109. http://cont.o.oo7.jp/49sup38/49sup38contents_e.html.
44. Research must do no harm: new guidance addresses all studies relating to people. Nature. 2022;606(7914):434. https://doi.org/10.1038/d41586-022-01607-0. PMID: 35701624.
45. Stewart M, Brown JB, Weston WW, McWhinney IR, McWilliam CL, Freeman TR. Patient-centered medicine: transforming the clinical method. Abingdon: Radcliffe Medical Press; 2013.
46. Greenhalgh T, Hurwitz B, editors. Narrative based medicine. Dialogue and discourse in clinical practice. London: BMJ Books; 1998.
47. DIPEx International. https://dipexinternational.org/. Accessed 31 Jul 2022.

48. Newman SD, Andrews JO, Magwood GS, Jenkins C, Cox MJ, Williamson DC. Community advisory boards in community-based participatory research: a synthesis of best processes. Prev Chronic Dis. 2011;8(3):A70. Epub 2011 Apr 15. PMID: 21477510; PMCID: PMC3103575.
49. UNAIDS. Good participatory practice: guidelines for biomedical HIV prevention trials. 29 JUN 2011. Accessed 31 Jul 2022. https://www.unaids.org/sites/default/files/media_asset/JC1853_GPP_Guidelines_2011_en_0.pdf
50. Prepared by Dr Catherine Hankins for the World Health Organization. Good participatory practice guidelines for trials of emerging (and re-emerging) pathogens that are likely to cause severe outbreaks in the near future and for which few or no medical countermeasures exist (GPP-EP) outcome document of the consultative process. December 2016. Accessed 31 Jul 2022. https://cdn.who.int/media/docs/default-source/blue-print/good-participatory-practice-for-trials-of-(re-)emerging-pathogens-(gpp-ep)_guidelines.pdf?sfvrsn=102e3e70_9&download=true.
51. Council for International Organizations of Medical Sciences (CIOMS). Report of the CIOMS Working Group XI. Patient involvement in the development, regulation and safe use of medicines. Geneva. 2022. https://doi.org/10.56759/iiew8982.
52. International Council for Harmonisation of Technical Requirements for Pharmaceuticals for Human Use. ICH harmonised guideline: general considerations for clinical studies E8(R1). Adopted on 6 October 2021. https://www.ich.org/page/efficacy-guidelines.
53. U.S. Food and Drug Administration. MedWatch: the FDA safety information and adverse event reporting program. https://www.fda.gov/safety/medwatch-fda-safety-information-and-adverse-event-reporting-program. Accessed 31 Jul 2022.
54. Directive 2010/84/EU of the European Parliament and of the Council of 15 December 2010 amending, as regards pharmacovigilance, Directive 2001/83/EC on the Community code relating to medicinal products for human use Text with EEA relevance. OJ L 348, 31.12.2010, p. 74-99 Special edition in Croatian: Chapter 13 Volume 056 p. 162-187.
55. Food and Drug Administration. Incorporating the patient's perspective into drug development and communication: an Ad Hoc Task Force Report of the Patient-Reported Outcomes (PRO) Harmonization Group Meeting at the Food and Drug Administration, February 16, 2001.
56. Food and Drug Administration. Guidance for industry: patient-reported outcome measures: use in medical product development to support labeling claims. 2009.
57. Ziegler E, Hill J, Lieske B, Klein J, Knesebeck OV, Kofahl C. Empowerment in cancer patients: does peer support make a difference? A systematic review. Psychooncology. 2022;31(5):683–704. https://doi.org/10.1002/pon.5869. Epub 2022 Jan 5. PMID: 34981594.
58. Beresford P, Wallcraft J. Psychiatric system survivors and emancipatory research: issues, overlaps and differences. In: Barnes C, Mercer G, editors. Doing disability research. Leeds: The Disability Press; 1997. p. 66–87. https://disability-studies.leeds.ac.uk/wp-content/uploads/sites/40/library/Barnes-Chapter-5.pdf.
59. World Medical Association. Declaration of Helsinki: recommendations guiding doctors in clinical research. Adopted in 1964. Rescinded. Accessed 31 Jul 2022. https://www.wma.net/wp-content/uploads/2018/07/DoH-Jun1964.pdf [II.1 "*the doctor must be free to use a new therapeutic measure, if in his judgment it offers hope of saving life, reestablishing health, or alleviating suffering*".]
60. World Health Organization. Ethical issues related to study design for trials on therapeutics for Ebola virus disease. Report of the WHO Ethics Working Group meeting, 20-21 October. Geneva. https://apps.who.int/iris/handle/10665/137509. Accessed 31 Jul 2022.
61. World Health Organization. Emergency use of unproven clinical interventions outside clinical trials: ethical considerations. 12 Apr 2022. https://www.who.int/publications/i/item/9789240041745. Accessed 31 Jul 2022.
62. Kurihara C, Saio T, Imamura K, Greco D. The mission and experience of layperson members of the IRB/EC in Korea and Japan: participating in discussion and voting as representatives of the citizenry, society, and patients. Clin Eval. 2020;47(3):449–85. http://cont.o.oo7.jp/47_3/47_3contents_e.html.

63. Nudeshima J. The current state of the French Law on Research involving Human Being: as a result of the adjustment of doctrine and practice under the EU regulation. Clin Eval. 2022;50(1):5–10. Japanese.
64. The Health Research Authority. Governance arrangements for Research Ethics Committees. 2021. https://www.hra.nhs.uk/planning-and-improving-research/policies-standards-legislation/governance-arrangement-research-ethics-committees/. Accessed 31 Jul 2022.
65. The Health Research Authority. Integrated research application system. 2021. https://www.hra.nhs.uk/about-us/committees-and-services/integrated-research-application-system/. Accessed 31 Jul 2022.
66. Keiger D, De Pasquale S. Trials & tribulation: a special report. Johns Hopkins Magazine. 2002;54(1):2. https://pages.jh.edu/jhumag/0202web/trials.html.
67. Freire P. Pedagogy of the oppressed. New York: Continuum; 1970.
68. Greco D, Kimura R, Victoria Perottino M, Saio T, Kurihara C. Ethics of international collaborative research: perspectives from Brazil: part 1 selected notes on Paulo Freire: part 2 access, compulsory license, case study. Clin Eval. 2020;48(1):95–123. http://cont.o.oo7.jp/48_1/w95-w123.pdf.
69. Kurihara C, Baroutsou V, Becker S, Brun J, Franke-Bray B, Carlesi R, Chan A, Collia LF, Kleist P, Laranjeira LF, Matsuyama K, Naseem S, Schenk J, Silva H, Kerpel-Fronius S. Linking the Declarations of Helsinki and of Taipei: critical challenges of future-oriented research ethics. Front Pharmacol. 2020;11:579714. https://doi.org/10.3389/fphar.2020.579714.
70. Taichman DB, Sahni P, Pinborg A, Peiperl L, Laine C, James A, et al. Data sharing statements for clinical trials: a requirement of the International Committee of Medical Journal Editors. PLoS Med. 2017;14(6):e1002315.
71. Kurihara C, Mimura M, Oi H, Taruno H, Sato Y, Koike R, Watanabe H. Challenges and proposals for improvement of the implementation of the clinical trial act, based on the comparison of the three research regulations in Japan-report 2: focusing on data-driven research and privacy protection. Clin Eval. 2022;50(1):21–48. Japanese.
72. Wendler D. Broad versus blanket consent for research with human biological samples. Hast Cent Rep. 2013;43(5):3–4. https://doi.org/10.1002/hast.200.
73. Kaye J, Whitley EA, Lund D, Morrison M, Teare H, Melham K. Dynamic consent: a patient interface for twenty-first century research networks. Eur J Hum Genet. 2015;23(2):141–6. https://doi.org/10.1038/ejhg.2014.71. Epub 2014 May 7. PMID: 24801761; PMCID: PMC4130658.
74. Chadwick R, Levitt M, Shickle D, eds. The right to know and the right not to know: genetic privacy and responsibility. Cambridge University Press; 2014.
75. De Angelis C, Drazen JM, Frizelle FA, Haug C, Hoey J, Horton R, Kotzin S, Laine C, Marusic A, Overbeke AJ, Schroeder TV, Sox HC, van der Weyden MB. Clinical trial registration: a statement from the International Committee of Medical Journal Editors. NEJM. 2004;351(12):1250–1.
76. Parsa-Parsi R, Blackmer J, Ehni HJ, Janbu T, Kloiber O, Wiesing U. Reconsidering the declaration of Helsinki. Lancet. 2013;382(9900):1246–7. https://doi.org/10.1016/S0140-6736(13)62094-2. PMID: 24120200.

Part III
Alternative Frameworks for Innovation and Drug Development Strategies

Medicines Development for Global Health: Learning from COVID-19 Vaccines R&D

Varvara Baroutsou

Abstract The concept of this chapter is about biomedical research and development (R&D) in the global public interest. The diversity and disparities in health equity during the COVID-19 pandemic highlight the urgency and feasibility of transforming the R&D ecosystem. Collaborative work, precompetitive common and public funding applied during COVID-19 vaccines development may solve global priority issues for public health, including chronic, life threating, and neglected diseases.

To complement this concept, the chapter refers to unmet needs not served by market interest, open science, and open innovation. In parallel it also presents a holistic view based on author's expertise in research and experimental development activities of global pharmaceutical companies, academic institutions, research institutes, and clinical research centers.

Keywords Global health · Research and development · Public health · Ethics · Unmet medical needs · Innovation

1 Background

This chapter gives an overview of the impact of the COVID-19 pandemic that has exposed the failing lines of national and global healthcare systems, while simultaneously reinforcing the importance of fostering healthcare system resilience. Nowhere has this been more obvious than in health equity and public health, which for decades have been neglected rather than supported in pursuing disease outbreaks preparedness [1].

V. Baroutsou (✉)
International Federation of Associations of Pharmaceutical Physicians and Pharmaceutical Medicine (IFAPP), Woerden, The Netherlands
e-mail: varvara.baroutsou@ifapp.org; secreteriat@ifapp.org

C. Kurihara et al. (eds.), *Ethical Innovation for Global Health*,
https://doi.org/10.1007/978-981-99-6163-4_14

The pandemic has greatly stimulated innovation in areas such as public health surveillance, highlighting some of the many ways in which public and digital health can strengthen and enhance equity in health care planning and delivery. The COVID-19 emergency accelerated research by funding collaborative and well-coordinated public and private efforts to enable safer, higher quality development of life saving vaccines and treatments [2].

The stress test of the pandemic has also brought into focus gaps, challenges, and governance barriers to boost the transition to research data-enabled health care policies that respond to everyday care needs and preempt future pandemics.

Drawing on experience from international research and development (R&D) of COVID-19 vaccines and adopting applied solutions used in the centers of excellence that could apply at scale, preparedness for emerging health threats and/or managing existing significant unmet medical needs may be within reach.

The chapter focuses on global health and medicine research as well as technologies that can relieve society by controlling natural disasters. An invisible divided access to vaccination during the COVID-19 era has also triggered thoughts on transforming the R&D system.

Proactive use of effective solutions is key to progress on the path to safer, higher quality care for all, that is inclusive, global, personalized, and innovative. However, progress will also depend on addressing persistent barriers to transformative change and alignment with overall global health, research and development (R&D), and care strategy stakeholders [3].

2 Build Future Preparedness

2.1 Challenges and Opportunities for Innovation

The COVID-19 experience mandates addressing the increasingly complex scientific, medical, and ethical challenges, quality of sources and types of clinical data, right study design and methodologies, including digital technologies, in our mission to derive knowledge to support public health needs beyond pharma industry interest.

A new global R&D and regulatory, public and private partnership perspective is needed to ensure more comprehensive evidence, more transparency, more patient-centered therapeutic innovation, for targeted, effective, safer, and affordable medicines-vaccines to improve the future of medicine and healthcare [4].

With the current dominant notion that the crisis is behind us, we observe a retrenchment of industry and policy makers to the pre-pandemic funding mechanisms and regulatory models.

Looking to the future with cautious optimism, we wonder how we can retain the innovations that demonstrated a collaborative path to guide global health and societies to control diseases with public funding and exemplary coordination applied by the US "Operation Warp Speed" initiative [5].

2.2 Operation Warp Speed

As part of the U.S. vaccine effort, on May 15, 2020, the federal government announced Operation Warp Speed (OWS), a partnership between the Department of Defense (DOD) and Department of Health and Human Services (HHS). The goal was to produce 300 million doses of COVID-19 vaccines, with initial doses available by January 2021. DOD and HHS have obligated approximately $13 billion as of December 31, 2020, to support the development, manufacture, and distribution of vaccines to help achieve this goal. The OWS was an interagency program that additionally included components of the Centers for Disease Control and Prevention, Food and Drug Administration, the National Institutes of Health, and the Biomedical Advanced Research and Development Authority (BARDA); private firms; and other federal agencies [5].

As of January 30, 2021, five of the six OWS vaccine candidates have entered phase 3 clinical trials, two of which—Moderna's and Pfizer/BioNTech's vaccines—have received an emergency use authorization (EUA) from the Food and Drug Administration (FDA). For vaccines that received EUA, additional data on vaccine effectiveness were generated from further follow-up of participants in clinical trials already underway before the EUA was issued. U.S. Government Accountability Office 's(GAO) analysis of the OWS vaccine candidates' technology readiness levels (TRL) showed that COVID-19 vaccine development under OWS generally followed traditional practices, with some adaptations. FDA issued specific guidance that identified ways that vaccine development may be accelerated during the pandemic. To meet OWS compressed timelines, some vaccine companies relied on data from other vaccines using the same platforms, where available, or conducted certain animal studies at the same time as clinical trials. However, as is done in a non-pandemic environment, all vaccine companies gathered initial safety and antibody response data with a small number of participants before proceeding into large-scale human studies (e.g., phase 3 clinical trials). The two Emergency Use Approvals (EUAs) issued in December 2020 were based on analyses of clinical trial participants and showed about 95 percent efficacy for each vaccine. These analyses included assessments of efficacy after individuals were given two doses of vaccine and after they were monitored for about 2 months for adverse events [5].

As of January 2021, five of the six OWS vaccine companies had started commercial scale manufacturing. OWS officials reported that as of January 31, 2021, companies had released 63.7 million doses—about 32 percent of the 200 million doses that, according to OWS, companies with EUAs have been contracted to provide by March 31, 2021. Vaccine companies faced a number of challenges in scaling up manufacturing to produce hundreds of millions of doses under OWS's accelerated timelines. DOD and HHS worked with vaccine companies to help mitigate manufacturing challenges, by identifying additional manufacturing partners to increase production. Additionally, the U.S. Army Corps of Engineers was overseeing construction projects to expand capacity at vaccine manufacturing facilities. Vaccine companies and DOD and HHS have undertaken several efforts to address possible manufacturing disruptions and mitigate supply chain challenges. These efforts included federal assistance to (1) expedite procurement and delivery of critical manufacturing equipment, (2) develop a list of critical supplies that are common across the six OWS vaccine candidates, and (3) expedite the delivery of necessary equipment and goods coming into the USA. Furthermore, OWS officials have worked with the Department of State to expedite visa approval for key technical personnel, including technicians and engineers to assist with installing, testing, and certifying critical equipment manufactured overseas. OWS supported vaccine companies hiring and training personnel with the specialized skills needed to run vaccine manufacturing processes [5].

"Operation Warp Speed" seemed to signal a new dawn, that demonstrated how prior research [6] and standard practices with some adaptations, additional resources,collaboration and coordination turned into lifesaving vaccines in a matter of months [5].

2.3 Next Generation COVID-19 Vaccines

Despite the availability of safe and effective vaccines, new SARS-COV-2 transmissible variants continue to circulate, due to vaccine hesitancy and access issues and the introduction of the bivalent COVID-19 booster vaccines in the fourth quarter of 2022 seems to be a temporary measure. Currently the next generation COVID-19 vaccines are delayed by technical and financial shortfalls, with ongoing efforts focused on addressing intrinsic development challenges regarding the extent and duration of protection against potential new variants in order to achieve improved long-lasting immunity with reduced disease transmission. A targeted effort such as OWR will be likely required to streamline the development of a new vaccine with the extent and duration needed [7].

The scientific achievement of SARS-CoV-2 vaccine development in less than 1 year, inspired the Coalition for Epidemic Preparedness Innovations (CEPI) to set an aspirational goal for initial vaccine authorization and manufacturing at scale within 100 days against new pandemic pathogens. This proposal has been welcomed by governments and interested vaccine developers worldwide [8].

Stopping the next pandemic by achieving global preparedness through rapid vaccine development matched with the ability to be manufactured throughout the world and be readily available and administered globally, sounds optimistic and ambitious. This is, however, possible to be realized, if governments, policy makers and vaccine developers adopt, fund, execute, and educate the public for this fast development approach by granting population trust. Fair and fast global access to pandemic vaccines require leveraging open science collaborative approaches focusing on vaccines libraries, development platforms, technology transfer supported by regional manufacturing capacity not blocked by tight Intellectual Property rights by single companies [3, 8].

COVID-19 has demonstrated that public funding and stewardship under OWS can reduce R&D costs and risks making affordability and transparency more feasible, sharing knowledge more realistic albeit not guaranteed. A creative combination of regulations, incentives, and persuasion may align private and public interests.

2.4 *Information and Technology Sharing*

Challenges of global equitable access because of restricted access to new medicines and technologies, due to Intellectual Property (IP) rights, high prices, inadequate supply, or production as well as biomedical R&D developments during the pandemic highlighted the urgency for reorienting the system toward public interest [3, 9].

Given that biomedical R&D is becoming increasingly global and expanding rapidly in low- and middle-income countries (LMIC) and networks through application of ethical, sound, transparent and fair practices, provides the opportunity to support the global rapid vaccine development network and achieve the goal set by CEPI [7].

At the Council for Trade-Related Aspects of Intellectual Property Rights (TRIPS) on the sixth of July 2022, World Trade Organisation (WTO) members welcomed the adoption of the TRIPS waiver decision on COVID-19 vaccines at the 12th Ministerial Conference (MC12) and began discussions on a possible extension to cover the production and supply of COVID-19 diagnostics and therapeutics [10]. At MC12, trade ministers adopted the Ministerial Decision on the TRIPS Agreement, which gives members greater scope to take direct action to diversify production of COVID-19 vaccines and to override the exclusive effect of patents through a targeted waiver over the next 5 years. Members of developing countries argued that the waiver on COVID-19 vaccines falls short of their expectation and is not enough to help developing countries comprehensively address current and future health challenges. Equitable access to therapeutics and diagnostics, as pointed out by the World Health Organization (WHO), is critical in helping detect new cases and new variants. They said this waiver extension needs to be discussed with a sense of urgency given the fact that many least developed countries lack access to life saving drugs and testing therapeutics [11].

Under a request on IP and Innovation requested by Australia, Canada, the European Union, Hong Kong, China, Japan, Singapore, Switzerland, Chinese Taipei, the United Kingdom and the USA, the WTO presented their new submission with a focus on IP licensing opportunities (IP/C/W/691,23 June 2022). WTO highlighted several major ways owners of IP assets can secure a broader reach for their products and services through licensing agreements, which enable IP owners to allow the licensee to make or sell the invention during the license period. This includes licensing of patents, copyright, trademarks, and know-how [12].

The COVID-19 crisis has set an unpresented collective focus on public health and increased awareness of the power of information to shape patterns of health and well-being from individual to societal levels [8].

COVID-19 has accelerated uptake of digital health and medicine innovations globally, even though barriers and concerns about risks continue to decelerate adoption. The pandemic has shone a light on the need for new models of health care and rapid information flow with broad population access in many countries. The pandemic has simultaneously exposed substantial disparities in care delivery and outcomes, including high-income countries [9].

3 Global Research Priorities in the Public Interest

Nowadays, we need to focus global research on true unmet medical needs, since we lack medicines in areas where market incentives are inadequate. Recent evidence indicates that the failing pace of Alzheimer's disease research, the residual risk of harm from adverse drug reactions, the unmanaged neglected diseases of poverty and emerging infectious diseases, as well as the prevailing antimicrobial resistance and lack of new antibiotics are critical public health underserved needs [3].

Infectious diseases are not a story of the past given the continuing COVID-19 evolving variants, vaccination hesitancy, and low-income countries (LIC) vaccination rates. Furthermore, malaria, tuberculosis, HIV, cancer, and chronic diseases are lately on the rise, largely due to disruptions in diagnosis and treatment caused by the COVID-19 pandemic.

Health emergencies like the recent outbreak and spread of mpox, the reemergence of polio, the Ebola outbreak with lockdown in Uganda and cholera deaths in Lebanon in 2022 require global vigilance.

And if the COVID-19 crisis is not enough to motivate us to ensure global pathogen surveillance and preparedness for saving lives, we also have to manage the increasingly challenging threat of climate change and natural disasters in a world still suffering by the war in Ukraine.

Serious consideration should be given to applying the lessons learned through COVID-19 and eliciting collaborative R&D priorities. These can be based on public health needs, strong ethical and sound research, open science, investment in scientific, technological, and regulatory capacity, timely access to health technologies and therapeutic advances that are safe and effective, to share benefits equitably, by ensuring affordability, availability, suitability, and technology transfer [3, 13, 14].

Moreover, the unfair as well as delayed distribution of COVID-19 vaccines and lack of adequate public health, primary care infrastructure to LIC resulted in excessive mortality, allowed the evolution of the Omicron variants and indicated that each region of the world should be more self-reliant by establishing continental/or national health testing and therapeutic approaches plus health access policies, e.g., compulsory licensing or licensing agreements. The WHO and many global health researchers underline that the disparities during the pandemic highlight the need for regional manufacturing.

4 The COVID-19 Technology Access Pool

Based on the harsh reality of inequities, several countries in the global south, WHO, and other groups are joining efforts in a long-term initiative to build vaccine and drug manufacturing capacity throughout Africa, South America, Asia, and Eastern Europe. WHO director general Tedros Ghebreyesus, on the announcement of the initiative of a vaccine technology transfer collaborative hub in South Africa, stated that "the COVID-19 pandemic has shown that reliance on few companies to supply global public goods is limiting and dangerous." The Medicines Patent Pool (MPP) and the World Health Organization (WHO), Afrigen Biologics (PTY) Limited, the Biologicals and Vaccines Institute of Southern Africa (Biovac), the South African Medical Research Council (SAMRC), and Africa Centres for Disease Control and Prevention (Africa CDC) have signed a letter of intent to address the global imbalance of manufacturing capacity for COVID-19 vaccines. This new Consortium aims to make their own effective mRNA vaccine against COVID-19, before expanding into other diseases that are relevant to their regions,e.g., HIV, Zika, and other. This collaborative hub experiment is testing whether an open science model can build capacity [15].

Despite a plethora of challenges the Africa Centers for Disease Control and Prevention has announced a goal of producing 60% of vaccines used by Africa, in Africa, within the next 20 years. Together with the African Union they launched the Partnership for African Vaccine Manufacturing to scale up pharmaceutical capacity, including African Development Bank Foundation to invest in pharmaceutical manufacturing capabilities and a World Bank funded program to upgrade vaccine making to facilities at the Pasteur Institute of Dakar in Senegal [14, 15].

These initiatives are a clear signal of determination that sufficient investment combined with global political will and legal agreements will promote Africa's nascent mRNA manufacturing industry to protect public health.

The precedent of HIV antiretroviral treatment's introduction through compulsory licensing in LIC with concrete successful outcomes in Brazil, Thailand, and Africa back then, reinforce the development of regional manufacturing of advanced platforms of innovative vaccines or treatments for the public interest. Three factors were critical to this success: legislation for free access to treatment; public sector capacity to manufacture medicines; and strong civil society action to support government initiatives to improve access.

Compulsory licensing or licensing agreements can be instrumental for alleviating insufficient supplies of necessary pharmaceuticals as well as mitigating prohibitively expensive medicines or vaccines prices. While the rewards of patent protection are necessary to support continual innovation, the compulsory licensing exception exists for public health emergencies such as the recent COVID-19 crisis. International organizations can play a key role by providing the legal know-how as well as setting a supportive tone for using licensing agreements. In the process, pharmaceutical companies and G20 countries should not deter or retaliate against developing countries pursuing such public health measures in the time of a pandemic [16].

The agreement that allows the United Nations(UN) -backed Medicines Patent Pool (MPP) to sub-license PF-07321332 for production and distribution by qualified generic medicine manufacturers worldwide, post-regulatory authorization and approval was a landmark licensing agreement between Pfizer and MPP. "This was the first license allowing generic manufacturing of this drug," as stated by Hervé Verhoosel, spokesperson of the global health agency that created the MPP a decade ago. It was an important first step to help ensure that the latest tools for fighting COVID-19 become available in low- and middle-income countries, covering around 53 per cent of the global population, at the same time as they became available in the wealthiest nations [17].

The pharmaceutical companies that produced COVID-19 vaccines have benefited greatly from huge sums of public funding for research and development and advance purchase commitments, amounting to between US$2·2 billion and $4·1 billion (by Feb 1, 2021) from Germany, the UK, and North America combined. Unfortunately, these governments did not make their support conditional on measures that would enable more vaccine to be produced through, for example, patent pools (e.g., the COVID-19 Technology Access Pool) or non-exclusive licensing, which would allow pharmaceutical companies with spare manufacturing capacity to increase supply. So far, most effort has gone into increasing production capacity in the vaccine developers' own facilities or through subcontracts and licensing arrangements with other developers, such as AstraZeneca's agreement with the Serum Institute of India, or Sanofi's support in filling and packing bottles of Pfizer–BioNTech's vaccine [18].

Widespread failures during the COVID-19 pandemic at multiple levels worldwide have led to millions of preventable deaths and a reversal in progress toward sustainable development for many countries. The Lancet Commission on lessons for the future from the COVID-19 pandemic makes important recommendations in three key areas of interest, including G20 support for financing research, development, and the production capacities of low-income and middle-income countries [18].

5 Conclusion

It is essential for R&D stakeholders to conduct effective post-COVID-19 reviews in the learning gained. R&D experts, institutions, sponsors, and authorities should articulate a more holistic view and take a relevant action to translate the COVID-19 vaccines development experience to an effective R&D model for global health. The reactivation of OWS would be required to streamline the development of the next generation of COVID-19 vaccines with the extent and duration needed. The slow progress observed in terms of waiving COVID-19 IP rights for COVID-19 vaccines, diagnostics, and therapeutics from WTO needs to be accelerated in order to facilitate LMIC collaborative hubs in Africa and global south to succeed. Building vaccine manufacturing capacity in South Africa is the first step in a broader effort to boost local production to address health emergencies and strengthen regional health security. COVID-19 has demonstrated that public funding and governmental stewardship reduce R&D costs and risks, making affordability and transparency more feasible, as well as sharing knowledge more realistic. A combination of regulations, incentives, and governmental policies can support private and public interest to converge for the common good. The collaboration of key actors, industry, government, academia, funding agencies, to collectively provide an enabling environment for the transformation of biomedical R&D to serve global public interest and particular initiatives for low- and middle-income countries would contribute to equitable access and benefit sharing across the world in accordance with the Universal Declaration of Human Rights to share scientific advancements and its benefits.

Conflict of Interest The author has no competing interests to declare that are relevant to the content of this chapter.

References

1. Frieden TR, McClelland A. Preparing for pandemics and other health threats. Societal approaches to protect and improve health. JAMA. 2022;328:1585.
2. Herzlinger R, Richman BD, Schulman KA. Maintaining health care innovations after the pandemic. JAMA Health Forum. 2023;4:e225404.
3. Swaminathan S, et al. Reboot biomedical R&D in the global public interest. Nature. 2022;602:207–10.
4. Shah SJ, Butler J, Shah SH, Kamphaus TN, Sachdev V. Accelerating therapeutic discoveries for heart failure: a new public–private partnership. Nat Drug Discov. 2022;21:781–2.
5. OPERATION WARP SPEED. Accelerated COVID19 Vaccine Development Status and Efforts to Address Manufacturing Challenges. Report of United States Government Accountability Office, accessible version February 2021. https://www.gao.gov/assets/720/712410.pdf.
6. Su S, Du L, Jiang S. Learning from the past: development of safe and effective COVID-19 vaccines. Nat Rev Microbiol. 2021;19:211–9.
7. Marks PW, Crupusso PA, Adashi EY. Urgent need for next generation COVID vaccines. JAMA. 2023;329(1):19–20.
8. Saville M, et al. Delivering pandemic vaccines in 100 days—what will it take? NEJM. 2022;387:e3.
9. Califf RM. The FDA and the clinical community. JAMA. 2022;328:1043.
10. TRIPS Council welcomes MC12 TRIPS waiver decision, discusses possible extension. https://www.wto.org/english/news_e/news22_e/trip_08jul22_e.htm.
11. Ministerial decision on the TRIPS agreement adopted on 17 June 2022. https://docs.wto.org/dol2fe/Pages/SS/directdoc.aspx?filename=q:/WT/MIN22/30.pdf&Open=True.
12. Intellectual Property and Innovation: IP Licensing Opportunities. Accessed 23 Jun 2022. https://docs.wto.org/dol2fe/Pages/SS/directdoc.aspx?filename=q:/IP/C/W691.pdf&Open=True.
13. Zelmer J et al. Transforming care and outcomes with digital health through and beyond the pandemic. NEJM Catal. 2022. https://doi.org/10.1056/CAT.22.0053.
14. The radical plan for vaccine equity. Amy Maxmen. Nature. 2022;607(7918):226–33.

15. Dutton J. First African-made mRNA vaccine, a test platform for the future. Nature. https://www.nature.com/articles/d44148-022-00151-3. Accessed 26 Oct 2022.
16. Wong H. The case for compulsory licensing during COVID-19. J Glob Health. 2020;10(1):0103584. https://doi.org/10.7189/jogh.10.010358.
17. COVID-19: Licensing agreement for new candidate drug 'an important first step'. https://news.un.org/en/story/2021/11/1105862.
18. Bozorgmehr K, Jahn R, Stuckler D, McKee M. Free licensing of vaccines to end the COVID-19 crisis. Lancet. 2021;397:1261. https://doi.org/10.1016/S0140-6736(21)00467-0.

Development of Portfolios and Pipelines of Drugs for the Treatment, Prevention and Control of Neglected Tropical Diseases

Kiyoshi Kita, Haruki Yamada, Fumiko Hirabayashi, and Simon L. Croft

Abstract Over the past quarter of a century, the extent and impact of neglected tropical diseases (NTDs) on populations worldwide have been determined, and both treatment and control measures have been developed. Product development partnerships (PDPs) have been established to bring forward new tools for these efforts. A huge movement, including international organizations, academia, pharmaceutical and biotech sectors, committed to advancing the treatment, control and elimination of NTDs has grown over this period. The leading role of PDPs has not only been to develop and implement new tools but also acted as an effective catalyst to encourage and enable pharma and academia to engage in these diseases, to establish portfolios to enable choice and to connect to other innovative public–private partnerships that can deliver new and established treatments and control measures to neglected populations.

Keywords Neglected tropical diseases (NTDs) · Product development partnerships (PDPs) · Portfolio · Open innovation · Global health

---

K. Kita
School of Tropical Medicine and Global Health, Nagasaki University, Nagasaki, Japan

Department of Host-Defense Biochemistry, Institute of Tropical Medicine (NEKKEN), Nagasaki University, Sakamoto, Nagasaki, Japan

H. Yamada
Kitasato University, Tokyo, Japan

NPO DNDi Japan, Tokyo, Japan

F. Hirabayashi
School of Tropical Medicine and Global Health, Nagasaki University, Nagasaki, Japan

NPO DNDi Japan, Tokyo, Japan

S. L. Croft (✉)
London School of Hygiene and Tropical Medicine, London, UK
e-mail: Simon.Croft@lshtm.ac.uk

C. Kurihara et al. (eds.), *Ethical Innovation for Global Health*,
https://doi.org/10.1007/978-981-99-6163-4_15

1 Introduction

Neglected tropical diseases (NTDs), although the concept was first described in the 1980s [1], only became an important item on the research agenda at the beginning of the 2000s through the Drugs for Neglected Diseases initiative (DNDi), the advocacy of Molyneux, Hotez et al., and their inclusion by the WHO and their definition of NTDs. DNDi is a not-for-profit organization established in 2003 [2] by Doctors Without Borders (Médecins Sans Frontières: MSF), together with five global public organizations and the World Health Organization (WHO)/Special Programme for Research & Training in Tropical Diseases (TDR) as a permanent observer. Currently, 20 NTDs have been designated by the WHO (Table 1), and 1 billion patients worldwide are waiting for the development and provision of therapeutic and control agents for these diseases. Most of them are infectious diseases that are prevalent in low- and middle-income countries in Africa, Latin America and Asia, for which it has been difficult for pharmaceutical companies to develop products without financial incentive.

Here we provide an overview of the strategies, portfolio and achievements of drug development for neglected diseases, which could provide some perspectives for innovation oriented towards global health.

The 2030 NTD Targets [4]

The WHO 2030 roadmap sets global targets and milestones to prevent, control, eliminate and eradicate 20 NTDs and disease groups. The recent progress report titled "Ending the neglect to attain the Sustainable Development Goals: A road map for neglected tropical diseases 2021–2030" emphasizes that three foundational pillars will support global efforts to achieve the targets:

1. accelerate programmatic action (pillar 1)
2. intensify cross-cutting approaches (pillar 2)
3. change operating models and culture to facilitate country ownership (pillar 3)

Table 1 "Neglected tropical diseases (NTDs)" by WHO

Neglected tropical diseases (NTDs) [3]	
Buruli ulcer	Mycetoma, chromoblastomycosis and other deep mycoses
Chagas disease	Onchocerciasis
Dengue and chikungunya	Rabies
Dracunculiasis	Scabies and other ectoparasitoses
Echinococcosis	Schistosomiasis
Foodborne trematodiases	Soil-transmitted helminthiases
Human African trypanosomiasis	Snakebite envenoming
Leishmaniasis	Taeniasis and cysticercosis
Leprosy	Trachoma
Lymphatic filariasis	Yaws

2 The NTD Movement: Establishment and Recent Trends

Since the concept became recognized as a critical agenda in the world, a huge movement including advocacy, funding [5] and public health has grown around NTDs. DNDi was launched in 2003 to develop new treatments initially for three kinetoplastid diseases—human African trypanosomiasis (HAT, also known as sleeping sickness), visceral leishmaniasis and Chagas disease. Since then, as a result of global R&D efforts by DNDi with its partners from both the South and the North, facilitated by the expanded disease target, 12 new treatments have been developed for six diseases and are now used in endemic countries. This includes two new chemical entities (NCEs)—fexinidazole for HAT and ravidasvir for Hepatitis C—and one novel combination therapy for HAT. Significant advances have also been made for visceral leishmaniasis. Whilst having identified a number of improved combination treatments for different regions, DNDi now has three NCEs in its leishmaniasis R&D portfolio at the clinical stage and two other compounds are expected to advance to clinical trials soon. In addition, a potential single-dose oral treatment for HAT is in late-stage clinical development.

DNDi established many research collaborations with global partners since its establishment. In Japan, a joint research agreement was signed with the Kitasato Institute on July 15, 2005, to discover new types of anti-trypanosomal molecules from natural products for the development of a therapeutic drug for human African trypanosomiasis. Also, a research collaboration on ascofuranone with the University of Tokyo was selected as one of the first projects for DNDi. Research partnerships with Japanese academia and pharma have since expanded and steadily advanced. Most recently, fosravuconazole, an orally bioavailable azole developed by Eisai Ltd, has been in clinical trial in Sudan for eumycetoma and could be available soon as an effective and affordable treatment.

The environment surrounding the development of therapeutic agents for NTDs has changed significantly since the beginning of the twenty-first century. In 2012, representatives from 13 of the world's major pharmaceutical companies, the World Bank, donor countries, donation groups such as the Bill & Melinda Gates Foundation, ministries of health of endemic countries, DNDi and WHO gathered together and issued the London Declaration[1] as a statement for the control of NTDs. The following year, in 2013, the Global Health Innovative Technology (GHIT) Fund [6] was established in Japan as a public–private partnership fund for global health R&D. Unlike any other in the world, the GHIT Fund facilitates and invests in global partnerships for the discovery and development of new health technologies, including drugs, vaccines and diagnostics against malaria, tuberculosis and NTDs. Since

[1] London Declaration. http://www.who.int/neglected_diseases/London_Declaration_NTDs.pdf. In 2012, 13 of the world's major pharmaceutical companies, DNDi, the Bill & Melinda Gates Foundation and the World Bank signed the agreement, cooperating with the WHO roadmap and aiming to eliminate 10 NTDs by 2020, and to provide free medicines and conduct research and development. This was declared as the London Declaration, in front of the governments of the United Kingdom and the United States and the Ministries of Health of NTDs-endemic countries.

then the discovery and development of drugs for the treatment of NTDs by Japanese industry and academia has moved forward. Furthermore, the 2015 Nobel Prize in Physiology and Medicine was awarded to Satoshi Ōmura, Distinguished Emeritus Professor of Kitasato University, the Kitasato Institute, for the discovery and development of ivermectin, a therapeutic and preventative agent for onchocerciasis and filariasis, used in mass drug administration (MDA) programmes for NTDs. This has greatly facilitated the understanding of the development and use of NTDs therapeutic drugs.

Most recently, on June 17, 2022, "Kigali Declaration on Neglected Tropical Diseases (NTDs)" was signed by worldwide partners of academic institutions, non-governmental organizations, pharmaceutical companies, governments and others [7]. It is expected that efforts in this area will build on the progress of the London Declaration, delivering on the recent WHO roadmap 2021–2030 [4], promoting the achievement of Universal Health Coverage (UHC) and Sustainable Development Goals (SDGs) [8].

3 Landscape of Players and Roles

As indicated in Sect. 2, over the past two decades there have been significant changes in the awareness of the importance of NTDs in terms of disease burden, impact on poverty, and the needs for new tools (drugs, diagnostics, vaccines) for treatment, prevention, control and elimination. In addition, there have been major programmes where tools (drugs, insecticides) have been available, for example for many helminth infections, to implement disease control and elimination through both mass drug administration (MDA) and focused drug administration; for onchocerciasis this has been an ongoing process for 4 decades. WHO has now taken a step further and established a road map (2021–2030) (https://www.who.int/teams/control-of-neglected-tropical-diseases/overview) for the control, treatment and prevention of NTDs. This road map focuses on three pillars: (1) reduction of incidence and prevalence, (2) integration of interventions and (3) facilitation of endemic country ownership. Several organizations work with WHO and are a key part of the effort, for example the Uniting to Combat NTDs alliance (https://unitingtocombatntds.org/ntds/), which has galvanized many efforts. Another key change over the past two decades has been the level of funding for research and development of methods to diagnose, prevent, treat and control NTDs. The level of funding has been monitored each year by G-Finder (https://www.policycuresresearch.org/g-finder/). Of relevance to this discussion, the 2021 G-Finder reported that in the decade 2011–2021 $1484 billion was spent on R&D for the three kinetoplastid diseases—leishmaniasis, Chagas disease and human African trypanosomiasis (HAT); this is a significant sum.

In relation to this chapter and the needs for controlling NTDs, the focus will be on preventive chemotherapy and transmission control (PCT) for NTDs[2]. This chapter will also consider some other diseases of poverty (TB, HIV, malaria), in particular malaria (https://www.who.int/westernpacific/activities/guiding-research-on-infectious-diseases-of-poverty).

3.1 Product Development Partnerships (PDPs) as a Catalyst

A main driver in the research and development of new tools has been the PDPs. PDPs are not-for-profit organizations that develop new medicines, vaccines and diagnostics by building partnerships between government, private sector, academia and philanthropic organizations. Over the past two decades, DNDi, established in 2003, the Medicines for Malaria Venture (MMV), established in 1999, the Foundation for Innovative New Diagnostics (FIND), founded in 2003, and the TB Alliance for the discovery and development of new drugs to treat tuberculosis, established in 2000, have been exemplars of this approach. PDPs do not stand alone; all must place their strategic objectives and practices within the landscape of funders, of stakeholders, of actors, of evaluators, of academic groups and others (see Fig. 1). Within this context, PDPs have taken a central role in building interactions and taking a leadership in re-building the R&D portfolios and pipelines of drugs and diagnostics for NTDs, malaria and TB. The WHO has played an additional essential role to enable engagement with endemic countries. It takes time and effort to build relationships within this landscape, especially to ensure mutual understanding about aims and objectives. For example, the private sector, which has a profit motive, is a key partner in development, and to work with this sector effectively requires PDPs to adapt their culture and vice versa (Fig. 1).

A clear demonstration of the impact of PDPs can be seen from the current portfolios of discovery and development projects, which have grown over the past two decades (see Sect. 3.2). Initially, as part of the process of establishing credibility, the PDPs took on projects which were "low-hanging fruits", for example, drug combinations and repurposing drugs from one indication to another, whilst they started to commit fully to the search for novel chemical entities and radically new treatments. During this evolution PDPs, like MMV and DNDi, have in addition acted as a catalyst to others to participate by: (1) demonstrating that the model can work, (2)

[2] NTDs can be considered as those requiring.

1. Preventive chemotherapy and transmission control (PCT): This intervention focuses on the availability of safe and effective drugs, which make it feasible to implement large-scale preventive chemotherapy.
2. Innovative and intensified disease management (IDM): The goal of IDM interventions is to manage diseases within primary healthcare systems, and ultimately to eliminate those diseases as public health problems. This intervention is used when cost-effective control tools do not exist and where large-scale use of tools is limited.

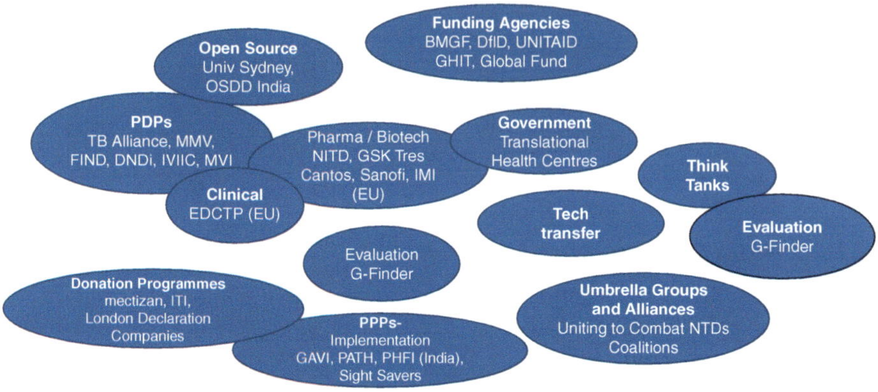

Fig. 1 Academic–private sector–public partnerships need to understand the landscape of players. This figure is a reproduction with a small update from the previous publication in Clinical Evaluation. 2018; 46(2): 197-235. http://cont.0.007.jp/46_2/w55-w73.pdf

through the encouragement of academic groups to reshape themselves to be involved in the drug discovery and development process, and most importantly, (3) through the encouragement of the biotech and pharma sectors to re-engage in R&D for neglected and diseases of poverty. At the same time, this development of partnerships across sectors has led to additional funding from governments, like the UK government, and philanthropic organizations (Bill & Melinda Gates Foundation: BMGF, Wellcome Trust) and shown the way for new organizations like the GHIT Fund. This role of PDPs as a catalyst has not been given sufficient recognition.

3.2 PDP Portfolios and Clinical Targets

The scale of efforts and the impacts of the PDPs could not have been envisaged 20 years ago. One particular area of success, as mentioned above, has been the building of portfolios. Taking MMV as an example, as this PDP is focused on one disease, has received extensive funding, has a clarity around strategy and objectives and from an early stage engaged with the private sector. The portfolio of anti-malarial projects (https://www.mmv.org/mmv-pipeline-antimalarial-drugs) and the number of projects now in the pipeline are exceptional; never in the history of drug discovery and drug development has there been such a focus and concentration of funding and effort to enable such a comprehensive portfolio to be built [9]. The DNDi portfolio is more complex as it encompasses several parasitic, fungal and viral diseases in their approach. The importance of having these large portfolios is that it enables MMV and DNDi to choose which projects to progress and which ones to drop; it encourages decision making, which is supported through target product profiles and clinical candidate profiles (Fig. 2). Any review of the MMV portfolios from 2000 onwards will reveal how many projects have been dropped and how few have progressed to clinical stages of development, especially NCE projects. The removal of weak projects enables the PDP to focus funding on those projects more likely to prove successful.

Another key feature of the PDP approach has been the adoption and development of target product profiles (https://www.who.int/observatories/global-observatory-on-health-research-and-development/analyses-and-syntheses/target-product-profile/links-to-who-tpps-and-ppcs), a tool used previously and extensively by the pharmaceutical industry as a guideline for selection of compounds and drugs to progress in the pipeline. It requires a clear picture of the use of potential drugs in defined (diseased) human populations. This approach has also now been used extensively by MMV and DNDi. MMV has taken the full complexity of the *Plasmodium* life cycle and the different species that cause malaria into account in the discovery and development process for a new anti-malarial drug [10], reflecting the different requirements of drugs for short course oral treatment, for prophylaxis as well as treatment, for *Plasmodium vivax* in addition to *P. falciparum* and for malaria eradication where drugs that also target gametocytes responsible for parasite transmission are needed (https://www.mmv.org/research-development/information-scientists/target-product-profiles-target-candidate-profiles). DNDi has also shown similar detailed approach (https://dndi.org/research-and-development/target-product-profiles/), in, for example, consideration of the demands for new

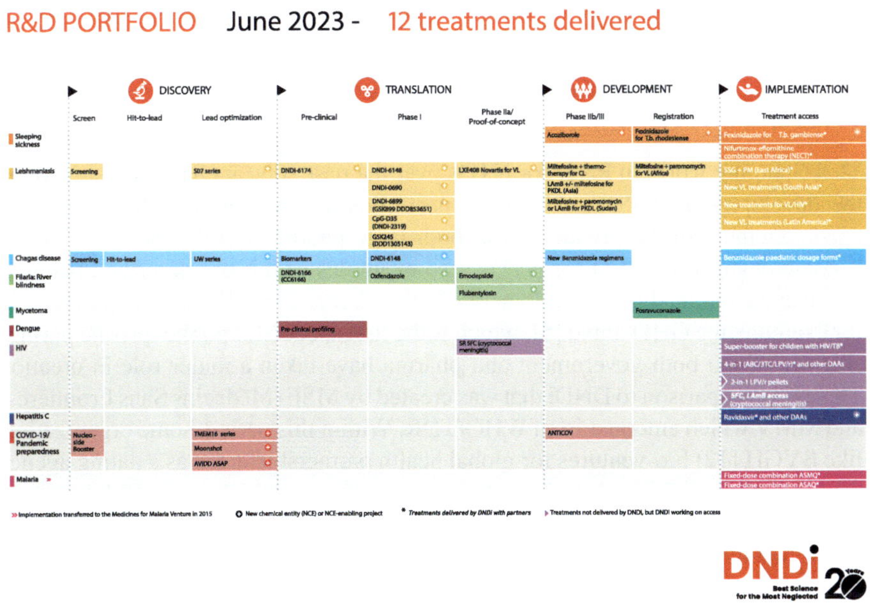

Fig. 2 DNDi R&D Portfolio. The latest DNDi R&D portfolio (as of June 2023) reproduced from the website, with DNDi's permission, updated after completion of this manuscript. The R&D portfolio is updated regularly and can be accessed from https://dndi.org/research-development/portfolio/

treatments for leishmaniasis—to include both visceral and cutaneous forms of the disease, as well as HIV-visceral leishmaniasis co-infections and PKDL (post kala-azar dermal leishmaniasis). For Chagas disease, the PDP has to reflect the need for a drug that is not just for treatment or cure but is also effective in the prevention of disease progression, from the indeterminate asymptomatic stage to the chronic debilitating stage. This sophisticated approach relating to disease complexity has to be considered within the structure of the portfolio, so the choice and selection of compounds for progression or removal from the portfolio need to consider for which TPP a compound is suitable or not. This extends also to the consideration and selection of appropriate drug combinations that are required for specific forms of diseases, for shorter courses of treatment, for prevention of the development of resistance, and roles in disease control, elimination and eradication.

DNDi is working on over 40 projects, including more than 20 new chemical entities. It is also running over 20 clinical trials.

The approach has also been adopted in earlier parts of the R&D process, for example, in defining requirements for the selection of individual compounds for status as clinical candidates. This clinical candidate profile supports the key decision making for the large investment required to take a compound forward into clinical development. Clinical candidate properties have been defined for malaria by MMV [10] and for DNDi in projects; for example the benzoxaborole for VL, DNDi-6148 (c). The earlier stages of selection for lead identification and lead optimization processes adopted by PDPs have been described by Katsuno et al. [11].

3.3 New Models for Partnership and R&D

This changing world in R&D for NTDs over the past 20 years has not just involved the development of PDPs with their portfolios and pipelines. It has also given encouragement to the private sector, both large pharma and the biotech sector to become engaged in R&D for NTDs; two examples of this are described below. It has also encouraged the development of other forms of public–private partnership, for example, the GHIT Fund [5], which is the first successful public–private partnership for which both government and pharma have taken a major role in creation. This is in comparison to DNDi that was created by MSF (Médecins Sans Frontières), and MMV which emerged from WHO/TDR. Within this period, some organizations like BVGH [12] bio-ventures for global health partnership acted as a dating agency, bringing pharma, biotech and academia together around compound libraries and disease models; BVGH's success was reflected in the over 100 collaboration agreements that they had arranged by 2017. Individuals, with knowledge and experience, have built other not-for-profit organizations, for example, Medicines Development for Global Health (https://www.medicinesdevelopment.com/), which under the leadership of Mark Sullivan has taken forward the development and registration of moxidectin for onchocerciasis—a drug that also has potential for another NTD, scabies. Other new models for earlier discovery stage research include the GSK

Open Lab established at their Tres Cantos site to enable interaction with academia and PDPs for screening, novel targets, model elaboration and the progression of compounds within a precompetitive setting (https://www.openlabfoundation.org/). TB drug development remains a big challenge despite the establishment in 2000 of the PDP, the TB Alliance [13]. To further support R&D for this disease of poverty, the Bill & Melinda Gates Foundation has funded TB drug Accelerator, led by Ken Duncan (ex-Glaxo Wellcome), to bring together key researchers from biotech, pharma, academic and clinicians from endemic countries, to accelerate this process. A similar "accelerator" model has successfully been adopted by DNDi to aid novel discovery. A more recent and bigger pharma-driven venture with a broader strategy has been the establishment of the Merck Global Health Institute (2017) (https://www.merckglobalhealthinstitute.com/), which encompasses not just R&D for NTDs, with a particular focus on schistosomiasis, but also has within its mission capacity building to strengthen local health systems and ensure an access path for products for sustained market availability and affordability. The importance of the role of NTDs experts from endemic countries in the R&D process has recently been included as one of the three pillars of the WHO NTDs road map (*ibid*). This inclusion and support for endemic country expertise through platforms has long been the strategy of DNDi, initially through the establishment of the Leishmaniasis East Africa Platform (LEAP) in 2004 [14], a model subsequently adopted for HAT, Chagas disease and cutaneous leishmaniasis.

3.4 PDP Pharma Biotech Partnership and Impact

One major criticism of big pharma 25 years ago was the limited access given to their vast compound libraries. This changed over a decade ago when GSK screened 2 million compounds against *Plasmodium falciparum*, not only publishing the data [15] but also providing open access to the 13,000 hits, allowing MMV and other groups in academia to pursue as novel anti-malarial leads. Other companies have also screened their compound libraries. Novartis, through their Institute for Tropical Diseases in Singapore, identified promising anti-malarial candidates, two of which are now in clinical trials; the development of one (KF156, ganaplacide) has been recently reviewed [16]. This new approach has cascaded into a multi-partner Open Access screening platform with availability of compounds from the Malaria Box to be tested against NTDs [17].

A similar pattern of change has been seen for leishmaniases and trypanosomiasis with GSK again leading the way in screening and publishing the data [18] and giving access to compounds for PDPs and academia to take further [19]. Working closely with Dundee University (UK), two compounds have advanced to clinical candidates. At the same time Novartis (GNF, San Diego), with funding from the Wellcome Trust, screened 1.5 million compounds from the company library against these same parasites and identified several leads, one with a novel mechanism of action, inhibition of proteasome function [20]; one of which, LXE408 [21], is being further developed in a Novartis-DNDi partnership and is now in clinical trials as an oral drug for visceral leishmaniasis. The screening of large compound libraries utilizes high-throughput and content screening (HTS/HCS) technologies; DNDi played a key role in initiating this approach with the Pasteur Institute in Seoul in 2006 [22].

In another approach DNDi, working in close partnership with SCYNEXIS, discovered and developed a series of novel compounds, benzoxaboroles, for the treatment of human African trypanosomiasis, focusing on compounds able to cross the blood–brain barrier [23]. One of these, acoziborole, is now in Phase III clinical trials for human African trypanosomiasis in the Democratic Republic of the Congo (DRC) and has the potential to be a single-dose treatment for this disease (https://www.clinicaltrials.gov/ct2/show/NCT05256017). This series of oxaboroles have wide antimicrobial activity and came from Anacor, a company previously based in Palo Alto. The initial contact between DNDi and Anacor came about through the close interactions between DNDi and MMV in the period from 2004 to 2007, with MMV providing the initial contacts with that company. Regular exchange, both formal and informal, between PDPs enhances interaction between organizations to the benefit of all.

3.5 Rules, Guidelines, Understandings and Structures for Partnerships

The interactions between PDPs, pharma, biotech and academia—interactions that have proved so essential to drug R&D and the identification of several potential new drugs and treatments for NTDs and other diseases of poverty—require understanding and time to build relationships. The different sectors have different priorities and staff requirements in addition to the common goal of delivering a new treatment. After two decades scientists from academia, the private sector and PDPs understand the rules of collaboration, of agreements, recognizing the importance of concepts of licensing, the needs for confidentiality, the importance of when the time is right for publication and the need for advocacy. This has been a part of the real change in the interactions that are normally formalized with legal agreements, royalty-free license agreements, etc. There are some organizations, for example the European Union and the WHO, where bureaucratic processes take time to ensure country inclusion. Also, surprisingly, there have been academic institutions that do not understand the need for not-for-profit for NTDs, a matter better understood by the private sector. The review processes of pharma, despite a clear endpoint, can seem excessive but meeting safety and regulatory requirements is essential. The PDP–academia–private sector understanding has been greatly helped by the recruitment of ex-pharmaceutical industry researchers and managers as companies like Pfizer and Astra Zeneca have closed their infectious disease units. These highly skilled ex-pharma scientists have been a bonus for PDPs, including DNDi, and for academia, an example being within a biomedical research institute at the University of Dundee, UK, where their experience has played a key role in the development of one anti-malarial candidate and two anti-leishmanial candidates.

Another need for mutual understanding is linked to the concept of "innovation". Innovation has become one of the keywords of the past decade; it is a driver for funding and high profiles in research. However, the concept of innovation can mean different things to different organizations, and understanding the "language of innovation" is important. For governments that provide funding and country support, innovation is normally linked to economic development and value for money. For the WHO, innovation is linked to solving a public health problem that will help to lift people out of poverty. On the contrary, for industry and the private sector, with R&D within the precompetitive space, it is important to understand that innovation can lead to major investment and an outcome; this requires a profile and impact that includes recognition for the companies concerned.

4 Outstanding Issues and the Future

4.1 Sustaining the Expertise

We are not at an end by any means. The main challenge is funding because success can mean, like for human African trypanosomiasis, a tick in a box that the disease is now controlled and no more research is needed. Around the year 2000, leprosy was so-called eliminated, a tick was put in the box, and funds for research were withdrawn. Yet today we have increasing transmission of leprosy in India, Brazil, the Philippines. If you fail to pay sufficient attention to an infectious disease when the prevalence is low, then you know it will come back. In India and Bangladesh where there is a likelihood that VL will be eliminated, there is a possibility that there will be no new generation of scientists working on Kala Azar (visceral leishmaniasis). The real issue for elimination is not just sustaining the funding but also sustaining the expertise, sustaining the momentum of working across the pharma–biotech–academia sectors.

There is another important challenge, namely ensuring the involvement of endemic countries in drug R&D. There are some good examples, India and Brazil, where translational health institutes are established. ASEAN countries have developed a programme for NTDs but this has yet to be enacted. DNDi has set up a drug discovery accelerator in Latin America centred in Brazil. Prof. Kelly Chibale, a Zambian chemist, founded and developed a centre (H3D) in Cape Town, South Africa, undertaking drug R&D for TB and malaria; his group is responsible for developing the first anti-malarial drug to come out of an African laboratory.

4.2 *Connect Development to Implementation*

The last point is that scientists working in R&D for NTDs have to always remember and understand that all the drugs in development have to be used in the settings where there is poverty or broken health systems, for example, Afghanistan, Syria and South Sudan. The common goal for scientists is to try to develop efficacious drugs—the focus is on efficacy and safety. The difference between efficacy and effectiveness has to be clearly understood—what works not in the clinical trials Phase III and Phase IV, but what works in the village and is acceptable to the affected communities. This difference was clearly considered in documents considering the eradication of malaria [24].

There have been examples from India with the lymphatic filariasis elimination programme. GSK donates 700 million tablets of albendazole each year to endemic countries for MDA. However, when it came to the analysis of lymphatic filariasis areas in an eastern state of India, only 40% of the villages were receiving the drug due to health systems issues. There were also additional challenges of compliance and adherence, with not all those who received the drug in the target population actually taking the tablets. It is always important to consider that the endpoint is not just efficacious treatments but to make new drugs that are acceptable and effective, similarly for new vaccines and new diagnostics, for the populations in endemic regions. This is linked back to the target product profile discussed above, but also calls for early discussions with health systems.

5 Conclusion

Within the past 20 years, there has been a sea change in both the way NTDs are perceived and the way new tools (drugs, diagnostics, vaccines) are (1) developed and (2) new and old tools are used, in the treatment, control and elimination of NTDs. PDPs, the pharmaceutical industry, biotech, academia, governments, charities and NGOs have all played a vital part in this change. DNDi, the PDP which has focused on the development and implementation of drugs for NTDs, has had success in both repurposing and extending indications of treatments as well as developing two new drugs for human African trypanosomiasis. It has in addition built a portfolio of compounds, some in clinical trial, with potential to treat leishmaniasis and Chagas disease; in particular for visceral leishmaniasis the success is shown with three clinical candidates developed in collaboration with large pharma and academia. However, two major challenges remain. Firstly, sustainability—will governments and funders maintain interest as patient numbers decrease? Will the commitment be sufficient to achieve the goal of elimination of several NTDs? Secondly—how to improve integration with health systems so that they are prepared for implementation of these new tools and to ensure target populations are aware of the benefits? How will the health systems ensure suitable surveillance systems are in place to monitor decreasing numbers to the point of elimination and identify outbreaks? The challenges are clear for the next decade; fortunately some endemic countries have been exemplary in demonstrating how to meet these challenges.

Acknowledgements This manuscript is an update of the round-table discussion published in Clinical Evaluation, 2018; 46(2): 197-235 (Japanese translation with some edition) as well as original English version in the same issue, page 407–410 (http://cont.0.007.jp/46_2/w55-w73.pdf), under the authorizations of publishers of previous and this publications.

We thank Chieko Kurihara for her enthusiastic support for this previous publication and, as one of the editors, for this chapter. The authors are also grateful to James Arkinstall, Joelle Tanguy and Julie Archer of DNDi Geneva, and Kaori Nakatani, Yoko Noda and the team members of DNDi Japan for all their assistance.

Conflict of Interests Kiyoshi Kita was a Scientific Advisory Committee (SAC) member of DNDi from 2013 to 2018 and was also Chair of Selection Committee member of GHIT from 2013 to 2022.

Haruki Yamada was a member of the DNDi SAC from 2003 to 2011 and was the Chair of Board of DNDi Japan from 2009 to 2023.

Fumiko Hirabayashi was Japan Representative of DNDi from 2004 and 2016 and is a board member of DNDi Japan since 2009.

Simon Croft was the R&D Director of DNDi from 2004 and 2007 and was a member of the DNDi SAC from 2011 to 2021.

References

1. Keating C. Ken Warren and the Rockefeller foundation's great neglected diseases network, 1978–1988: the transformation of tropical and global medicine. Mol Med. 2014;20(Suppl 1):S24–30.
2. Drugs for Neglected Diseases initiative. [cited 2023 April 30]. Available from: https://dndi.org/.
3. World Health Organization. Neglected tropical diseases. [cited 2023 April 29]. Available from https://www.who.int/news-room/questions-and-answers/item/neglected-tropical-diseases#:~:text=NTDs%20include%3A%20Buruli%20ulcer%3B%20Chagas,rabies%3B%20scabies%20and%20other%20ectoparasitoses%3B.
4. World Health Organization. Ending the neglect to attain the sustainable development goals; a road map for neglected tropical diseases 2021–2030. Available from https://www.who.int/teams/control-of-neglected-tropical-diseases/global-report-on-neglected-tropical-diseases-2023.
5. Policy Cures Research. G-Finder 2022 neglected disease research and development: the status quo won't get us there. [cited 30 April 30]. Available from https://policy-cures-website-assets.s3.ap-southeast-2.amazonaws.com/wp-content/uploads/2023/02/09104948/2022-G-FINDER-Neglected-Disease-report.pdf.
6. GHIT Fund. [cited 2023 April 20]. https://www.ghitfund.org/en.
7. Uniting to Combat NTDs. Kigali Declaration. [cited 2023 June 15]. Available from https://unitingtocombatntds.org/es/the-kigali-declaration/.
8. United Nations. Resolution adopted by the general assembly on 25 September 2015- transforming our world: the 2030 Agenda for Sustainable Development. 21 October 2015. [cited 2023 June 15]. Available from https://www.un.org/en/development/desa/population/migration/generalassembly/docs/globalcompact/A_RES_70_1_E.pdf.

9. MMV. MMV-supported projects. [cited 2023 April 30]. https://view.officeapps.live.com/op/view.aspx?src=https%3A%2F%2Fwww.mmv.org%2Fsites%2Fdefault%2Ffiles%2Fcontent%2Fdocument%2FMMV_Portfolio_Dec_2022.pptx&wdOrigin=BROWSELINK.
10. Burrows JN, et al. New developments in anti-malarial target candidate and product profiles. Malar J. 2017;16:26.
11. Katsuno K, et al. Hit and lead criteria in drug discovery for infectious diseases of the developing world. Nat Rev Drug Discov. 2015;14(11):751–8.
12. BIO ventures for global health. [cited 2023 April 29]. https://bvgh.org/.
13. TB Alliance. [cited 2023 April 29]. https://www.tballiance.org/.
14. Wasunna M, et al. The Leishmaniasis East Africa Platform (LEAP): strengthening clinical trial capacity in resource-limited countries to deliver new treatments for visceral leishmaniasis. Trans R Soc Trop Med Hyg. 2016;110(6):321–3.
15. Gamo FJ, et al. Thousands of chemical starting points for antimalarial lead identification. Nature. 2010;465(7296):305–10.
16. Koller R, et al. The early preclinical and clinical development of ganaplacide (KAF156), a novel antimalarial compound. Expert Opin Investig Drugs. 2018;27:803–10.
17. Van Voorhis WC, et al. Open source drug discovery with the malaria box compound collection for neglected diseases and beyond. PLoS Pathog. 2016;12(7):e1005763.
18. Pena I, et al. New compound sets identified from high throughput phenotypic screening against three kinetoplastid parasites: an open resource. Sci Rep. 2015;5:8771.
19. Olias-Molero A, et al. Antileishmanial drug discovery and development: time to reset the model? Microorganisms. 2021;9(12):2500.
20. Khare S, et al. Nature. 2016;537(7691):229–3.
21. Nagle A, et al. J Med Chem. 2020;63(19):10773–81.
22. Siqueira-Neto JL, et al. PLoS One. 2012;6(6):e1671.
23. Wring S, et al. Parasitology. 2014;141:104–18.
24. The malERA Consultative Group on Health Systems and Operational Research. A research agenda for malaria eradication: health systems and operational research. PLoS Med. 2011;8(1):e1000397. https://doi.org/10.1371/journal.pmed.1000397.

Patient and Public Involvement (PPI) and Pharmaceutical Development Through Open Innovation Processes: Recent Activities

Kotone Matsuyama, Naoki Tsutsumi, Keiko Inoue, Noriko Iwaya, and Kyoko Imamura

Abstract In recent decades, efforts to involve patients/citizens as important stakeholders in the drug development process have evolved. Originally, patient involvement activity started in the AIDS drug development programs. Reflecting the accumulated efforts of internationally active NPOs and other organizations, patient voices are now expanded in the drug development process. During the COVID-19 pandemic, new approaches, such as decentralized clinical trials (DTC) utilizing digital tools, were substantially facilitated. In such an environment, participants should have relatively high digital literacy. There were some findings that some of the recently performed DCTs were questionable from the perspective of patients' diversity. Also, barriers still exist in educating patients about clinical trials and drug development procedures, which is necessary for them to participate in the sponsor's activities of PPI and in obtaining educational opportunities. From the perspective of access to vaccines and therapeutics trials during the pandemic situation, the goal of the patient involvement process should essentially be the benefit sharing the society. Ensuring fair treatment opportunities for patients is also very important in the drug development process. This chapter introduces actual patient–public engagement activities in the drug development process as a social co-creation resulting from cooperation among relevant stakeholders.

K. Matsuyama (✉)
Department of Health Policy and Management, Nippon Medical School, Bunkyo-ku, Tokyo, Japan
e-mail: m-kotone@nms.ac.jp

N. Tsutsumi
Japanese Institute for Public Engagement (Ji4pe), Midori-ku, Yokohama, Japan

K. Inoue
Association for Medical Malpractice Victims, Higashimurayama, Tokyo, Japan

N. Iwaya
Intractable Disease Society Peer Support Familia Yamaguchi, Yamaguchi, Japan

K. Imamura
Ji4pe, Midori-ku, Yokohama, Japan

C. Kurihara et al. (eds.), *Ethical Innovation for Global Health*,
https://doi.org/10.1007/978-981-99-6163-4_16

Keywords Patient and public involvement · Patient-centered medicine · COVID-19 pandemic · Decentralized clinical trial · Emancipation · Social co-creation

1 Introduction: Pharmaceutical Medicines Development and Patient and Public Involvement

Conventional drug development has been primarily led by pharmaceutical companies. Historically, many of the existing product lines with pharmacological actions and molecular mechanisms accepted as drugs to date have been developed preferentially for infectious diseases, lifestyle-related diseases, and major malignancies, for which the market size is relatively large. As a result, standard treatments have been established for many such disease areas at a very impressive rate since World War II. When a drug that might surpass the standard therapy is found, clinical trials are conducted, and the process of innovation of the standard therapy has continued for a long time. However, the establishment of standard therapies and the achievement of improved treatment satisfaction provided momentum for the development of drugs for more intractable or rare diseases. It is worth noting that 54%, or more than half, or 20 of the 37 new drugs launched in the United States (US) in 2022 are for rare diseases [1].

In the development of drugs for rare and intractable diseases, Patient and Public Involvement (PPI) activities, in which patients and the public are actively involved, are currently the focus of attention. Pharmaceutical companies, mainly in Europe and the US, are also strongly engaged in activities such as patient interviews, patient questionnaires, and other methods to obtain feedback from patients and form patient advisory boards to incorporate their opinions directly. In order to promote these PPI activities that enhance the social value of pharmaceuticals, it is a prerequisite to provide educational opportunities for patients and citizens to understand the process, methodologies, and related regulations of drug development. It is important to make better use of their insights and experiences gained from good collaborative communications, as well as to increase the value of the input of patients and citizens as representatives of our society. To accelerate PPI activities, it is an urgent task to deepen mutual understanding among stakeholders so that patients and citizens can properly understand and participate in drug development.

2 History of Patient Engagement in the United States, Europe and International Initiatives

Patient and public involvement have started with demonstrations of people living with AIDS, and the US and European Union (EU) regulators have explicitly stated their commitment to patient engagement. Here is a history of the evolution of such activities in the Western world.

2.1 *Government Actions in the US*

In the US, collaboration with HIV/AIDS patient groups and US government agencies began in 1988 after the demonstration activities of AIDS patient groups [2]. One of the groups was known as ACT UP, or AIDS Coalition to Unleash Power, which was founded in New York City in 1987 as a political action group in response to the AIDS crisis. The group's first action, in the spring of 1987, was a march on Wall Street to protest the high cost and lack of availability of HIV treatment [3]. They activate the connections with the stakeholders such as the US Food and Drug Administration (FDA), the National Institutes of Health (NIH) and the Centers for Disease Control (CDC) and related institutions [4]. Due to their activism or increasing advocacy, in 1991, patient representatives participated for the first time in an advisory committee for an AIDS drug, and patient voice was reflected in the drug development process. In addition, US FDA Safety and Innovation Act (FDASIA) was set in 2012, a working group within the FDA was established legally, and patient involvement in various drug development has been implemented. Currently, according to the "Patient-Focused Drug Development Initiative" (PFDD), meetings are held to systematically incorporate the experiences, perspectives, and needs of patients in the drug development process in each disease area, and these meetings are used for review and approval processes. The PFDD is a group of researchers who meet regularly in each disease area to discuss experiences, perspectives, and needs from the patient's point of view. In particular, the FDA published "Guidance Series for Enhancing the Incorporation of the Patient's Voice in Medical Product Development and Regulatory Decision Making" describing four different methodologies called Patient Engagement-Based Drug Development Practices [5–7].

2.2 *European Initiatives with Patient–Public Involvement*

In Europe, the European Medicines Agency (EMA) was established in 1995. Soon after the establishment, a dialogue between the people living with HIV and the EMA was started in 1996 at the request of those patients. This was followed by patient participation in the Committee for Orphan Medicinal Products (COMP) in

2000, the establishment of a working group with patients in 2003, and the Patients and Consumers Working Party (PCWP) in 2006. Since the establishment of PCWP, these PPI activities have matured organizationally [8].

2.3 International Initiatives

The International Council for Harmonisation of Technical Requirements for Pharmaceuticals for Human Use (ICH), which aims for international harmonization of drug regulations, and its E8R1 Guideline reached the Step 4 agreement (now transited to the Step 5 finalized document). According to the E8R1 Guideline, Reflection Paper on Patient-Focused Drug Development, issued in 2021, patient and public involvement activities were considered an important process in drug development, and other related guidelines, such as the E6R3 guideline, are now being developed [9]. As described in the E8R1 section "2.3 Patient Input into Drug Development," the views of patients can be valuable throughout all phases of drug development.

And the International Council of Medical Societies (CIOMS) has added a statement on stakeholder engagement in its 2016 research ethics guidelines [10], and it is essential to conduct drug development and health-related research with considering various stakeholders, including patients. Recently, CIOMS produced the report titled "Patient involvement in the development, regulation, and safe use of medicines [11]." This report describes the importance of systematically involving patients throughout the lifecycle of medicine's development and clinical use.

2.4 Successful Examples

A good example of a European initiative to educate patient groups is the work of the European Patients' Academy on Therapeutic Innovation (EUPATI), which develops training and educational materials for patients and patient representatives in pharmaceutical research and development and other related areas, with the aim of improving patient knowledge of pharmaceutical research and development (R&D). EUPATI provides a good example of patient empowerment through accessible, innovative, and inclusive education to enhance patient involvement. The educational materials and training provided by EUPATI are developed to help patients and the public participate in pharmaceutical R&D and are widely used in Europe [12].

Another example of patient–public engagement, which is by pharmaceutical companies, is described below. One of the factors that are barriers for patients to participate in clinical trials is the insufficiency of their knowledge and awareness of clinical trials. To reduce patients' fears about clinical trials and to spread correct knowledge about drug development, in 2020, the German company Boehringer Ingelheim (BI) launched the Clinical Trial Ambassador Project with several patient organizations in the Austrian area [13]. Well-educated patients (e.g., patient organization representatives) acted as clinical trial ambassadors to raise awareness of clinical trials, including responses to patients' concerns and questions about clinical trials. In addition, in the project, opinions among the clinical trial ambassadors were exchanged as one of the education for patients about clinical trials.

Education to increase the capacity and capability of patient organization representatives is conducted by expert trainers affiliated with EUPATI and is based on the slogan "Knowledge Is Power," which conveys the knowledge necessary to act as clinical trial ambassadors. As a Clinical Trial Ambassador, one will be consulted by other patients about clinical trials and communicate about them in an easy-to-understand manner. People like clinical trial ambassadors, who can bridge between patients and the pharmaceutical industry and work toward maximizing the social value of pharmaceuticals and other products, are indispensable for the social co-creation of pharmaceutical development. Collaboration between an industry and a patient group may cause a financial conflict of interest. Responding to such concern, European regulators, industry associations, and patient groups developed "Guiding Principles on Reasonable Agreements between Patient Advocates and Pharmaceutical Companies" in 2018 [14], which has been continuously undated. This principle and other toolkits provide important suggestions for formulating agreements between industry and patient groups, avoiding the bias of conflict of interest, keeping confidentiality mutually, as well as facilitating fair management of intellectual property rights.

3 Patient and Public Involvement During the COVID-19 Pandemic

The global spread of new coronavirus infections has drastically changed the healthcare environment worldwide. Many patients had to stay at home, and the depletion of medical resources became an issue. During the global pandemic, the US FDA was one of the first to discuss how clinical trials should be conducted under COVID-19, and in March 2020, the FDA issued a guidance "Conduct of Clinical Trials of Medical Products During a COVID-19 Public Health Emergency [15]." The FDA required that even when access to clinical trials was restricted from the perspective of logistics and medical resources, clinical trials must be conducted in a manner that protects the safety of participants and their well-being. The guidance also included an example of how to deal with clinical trial participants and stated that it was acceptable to deliver investigational drugs to the participant's home and to use video and telephone to communicate with patients remotely. Another FDA guidance mentioned that "*Decentralized Clinical Trials (DCT) hold promise to reduce patient and sponsor burden and increase accrual and retention of a more diverse trial population*" [16].

An especially important issue regarding DCTs using electronic tools is how to ensure the protection of the safety of trial participants and simultaneously establish validated methods of electronic data transmission, including data reliability and personal data protection. The Japanese Ministry of Health, Labour and Welfare (MHLW) also issued a guideline on March 30, 2023, titled "Points to Consider Regarding Explanation and Consent Using Electromagnetic Methods in Electronic Clinical Trials and Post-Marketing Clinical Trials" to explain how clinical trials should be conducted while ensuring GCP compliance and system validation [17].

During the COVID pandemic, data was accumulated through the use of electronic Patient Reported Outcome (ePRO), which means that participants entered their therapeutic outcome information into the database on their own. Not only in the case of clinical trials, but observational, data-driven research utilizing digital tools is also particularly important during health crisis situations. As an example, a joint study was conducted in 2022 by Toyonaka City, Osaka University, and a private Med-Tech company to examine the use of PHRs utilizing an application for collecting personal information from patients and others as a community survey on the aftereffects of coronavirus infection, led by Osaka University. The study has so far yielded approximately 4000 valid data [18]. The main results are summarized as follows:

3.1 Resident Survey Summary

- Target population: Patients affected by COVID-19 by the end of March 2022; reported date July 2022.
- Number of valid shipments: 26,880 (excluding undeliverable items, etc.)
- Number of responses.

  Written 2492.
  Apps 1555.
  Total 4047.

- Response rate 15.1.

  Summary of results:

- 4047 Toyonaka City residents with a history of COVID-19 infection responded to a questionnaire about post-infection sequelae.
- Reflecting the pandemic situation, 77% of patients were infected after the Omicron strain outbreak.
- 75% of respondents had been vaccinated at least twice at the time of response.
- 4% were on oxygen, 1% on ventilatory management, and 0.03% on Extracorporeal Membrane Oxygenation (ECMO).
- One month after infection, the most frequent symptoms were fatigue, cough, and alopecia, with 1.6% of infected individuals experiencing difficulty with daily living.

- 47.7% of all respondents experienced post-illness symptoms.
- Women were more likely than men to experience sequelae.
- Frequency of sequelae did not change by age, treatment, or vaccination history.
- Those with underlying disease were shown to be less likely to experience sequelae, and those with severe disease were more likely to experience sequelae.
- Compared to viruses in the early stages of the pandemic, the frequency of sequelae was higher when infected with alpha, delta, and omicron strains in their dominant period.

In the clinical study described above, a population study conducted during the COVID-19 pandemic is unique in that it used a combination of documents and mobile phone-accessible applications. It is a critical perspective that cooperation among the government, industry, and academia allows clinical research to be conducted quickly and maintains scientific and ethical integrity to make results valuable for diagnosis and treatment in clinical practice.

4 Changes in the Clinical Trial Environment During the COVID-19 Pandemic

The COVID-19 pandemic has brought a dramatic digital transformation to the clinical trial environment. Prior to the pandemic, the clinical trial environment was dominated by in-person visits to medical institutions, limited use of electromagnetic records, and in-person research review committee meetings. But in the pandemic phase, a substantial volume and variety of digital tools come to be applied to support clinical trial management and data collection.

In a survey conducted by Tsutsumi et al. in May 2020 regarding the clinical trial review environment in Japan immediately after the declaration of a state of emergency due to the pandemic, changes were observed, such as conducting clinical trial reviews remotely and using digitization tools [19]. The results of the survey are as follows. A total of 97 institutional review boards (IRBs) participated in the survey. The review meeting was affected in all clinical trials, with a greater impact on "Ongoing" trials than "New" trials. The most common countermeasure was the introduction of a digital communication system. Thirty-four of the 38 IRBs (89%) installed a digital system after the outbreak. The primary concern regarding the use of a digital communication system in the IRB review and the committee meeting process was "facilities" (in medical institutions for the environment of digital use), followed by "operation of the system" and " document delivery for review." The IRB meeting is one of the opportunities for the patients and the public to participate as "laypersons," but it is important for them to be well-qualified regarding the knowledge of methodologies and regulations of clinical trials and research involving humans. Moreover, the COVID-19 pandemic forced IRB members to use digital tools.

Similarly, one of the challenges for patients to participate in a clinical trial in a DCT setting is the high level of digital literacy requirement. Some of the reports of DCTs performed recently suggested questionable situations from the point of view of patient diversity. For example, in a fully virtual DCT, the DeTAP trial (Decentralized Trial in Afib Patients, NCT04471623), they used a combination of remote technology and dedicated remote clinical trial coordinators to inform and educate trial participants virtually in order to increase compliance with the clinical trial protocol. On the other hand, it was found that most of the participants were white, resided in urban areas, and nearly 70% had a bachelor's degree or higher education [20]. It is clear that the results of this study cannot be easily generalized. This example indicated that some level of digital environment required for the participants and the need to complete patient education about the trial in question might lead to a loss of diversity in the study population. To avoid such deficiency, it is paramount to provide equitable opportunities for education for a variety of populations to keep the diversity of trial participants. Such problems associated with the use of digital tools could be resolved by means of involving patients and the public in the early stages of the development program as well as in the ethics committee review.

Digitalization has to be accomplished to bring benefits to patients. "PatientsLikeMe," a patient social networking site originally founded in 2005 as a community for patients with amyotrophic lateral sclerosis (ALS), has become the world's largest social networking community with 850,000 participants in 2800 conditions. It is said to have contributed to the success of more than 100 clinical trials [21]. However, to connect these English-speaking patient communities to non-English-speaking patients, one must overcome differences in language and cultural backgrounds. Despite such continuing efforts, we still have to recognize the barriers that patients of non-English speakers often confront. Since the medical environment varies from country to country, it is important to create a place where patients and their families can communicate in their native language and about the medical environment in their home country, even if it is a small community. Nevertheless, it is becoming possible to communicate using automatic translation and other means in a text-based Social Networking Service.

5 Patient Engagement from a Global Overview; Working Toward Localization

The acquisition of correct and sufficient knowledge about medicine's development by the patients and public involvement and their empowerment in the decision-making process are also important factors in the evolution of PPI activities. It is necessary to provide an appropriate adequate pool of knowledge and skills for PPI activities, considering the different cultural backgrounds in each country, the differences in legal frameworks for drug licensing, as well as health systems, including insurance coverage. However, there are few examples of appropriate localization of Western industry/regulators-oriented standardized methods to be practicable in Asia, Africa, South America, and Oceania. But it must be emphasized that some areas of these regions have substantial experiences to harness local knowledge to change the situation of inequity and unbalance of sharing of benefits resulting from research [22–24]. From the viewpoint of benefit sharing, the insufficiency of diversity in clinical trials in industrialized countries must be proactively corrected to improve global health. Therefore, how to implement appropriate localization to promote PPI activities remains an issue of great international importance.

Since PPI in drug development is deeply related to local elements in each country, it is necessary to localize the methodologies that have been systematically established in English-speaking regions, and the advanced efforts in low- and middle-income countries (LMICs) must be more considered. Hence, the paramount importance is to promote PPI activities that are tailored to the health system and cultural characteristics of each region. True patient-centered medicine cannot be realized if pharmaceutical companies and researchers simply take the initiative to involve patients from their own perspectives. They must not utilize patients and the public as means to achieve their own purpose. The basis of localized and systematic PPI cannot be found without well-constructed patient education, participation, and empowerment. Not only do experts help provide education to patients and the public, but also it is a prerequisite that educated patients/citizens take the initiative to realize emancipation [25] by themselves. It will create a PPI ecosystem embedded in pharmaceutical lifecycle management that contributes to patient-centered medicine.

With this goal in mind, the authors have collaborated with experts, patients, and citizens to create an educational system. Below, we introduce the authors' efforts in Japan.

6 Need to Establish PPI Activity in Japan

For example, in Japan, there is not an abundance of materials introducing PPI activities. First, there is the issue of language: most publications on PPI activities are written in English, so appropriate translation and local development of the needed publications is essential to make the materials available in Japan. Currently, well-recognized materials in Japan include the "Patient and Public Involvement (PPI) Guidebook" developed by a research funding agency named the Japan Agency for Medical Research and Development (AMED) [26]; "Patient and Public Involvement in Clinical Trial Implementation Planning Tools and Supporting Materials for Gathering Patient Insight and Feedback on Clinical Trial Experience" developed by TransCelerate [27]; and "Guidance for Implementing Patient Involvement" developed by the Clinical Evaluation Committee of the Japan Pharmaceutical Manufacturers Association [28].

On the other hand, it is questionable whether there are enough leaders in patient organizations who can become leaders in patient engagement. In the "Third Survey of Awareness and Activities of Patient Organizations" (postal mail and Internet survey) conducted by the Japan Pharmaceutical Manufacturers Association in October 2020, more than 90% of the 144 patient organizations that participated in the survey cited peer interaction and support as an initiative of patient organizations, while 26% of the organizations cited activities related to clinical trials [29]. The number of activities related to clinical trials is still low, and there is a clear difference among each organization's activities of social interaction, which are the typical activities of patient groups.

Of course, it is undeniable that these factors may be due to still insufficient educational system to enhance knowledge of drug development and clinical trials among patients and the public, as well as negative information and images of clinical trials, such as risks of side effects, and data falsification by academic researchers influenced by pharmaceutical companies [30–32]. Inherently, the data and results of clinical trials must be treated as public goods, and furthermore, considering our experience of the COVID-19 pandemic, they should be regarded as "global public good." And their achievements must be widely returned to our society, most prioritizing the needs of patients and the public. Drug development, with the participation of patients and citizens, must correctly meet the needs of society. Research and development that responds precisely to the local community's health needs is the only way to respond to the health needs that exist in each region of the world and contribute to global health. Localization is thus a prerequisite. To this end, from the perspective of localization, it is very important to encourage patients and the public by providing knowledge of drug development and clinical trials in a way that takes into account the local cultural background and to create a platform for drug development in such a way that it is correctly created as a public good that can be equitably shared among the society, globally.

From this perspective, we describe in the next sections the efforts in Japan as endorsed by the author's group.

7 Good Examples of Adapted PPI Activity

Although this is one example of activity in Japan, we would like to introduce our experience of how to apply an international standard for excellent patient education and patient involvement in a culturally sensitive manner.

7.1 Construction of the Educational Course

Establishment of the Japanese Institute for Public Engagement (Ji4pe) and its study courses:

Ji4pe was established in June 2020 as a non-profit general incorporated association with the aim of providing continuous learning courses in which patients, citizens, industry, and academia can all participate as a part of social co-creation.

Initiatives of Ji4pe are categorized into four basic areas: (1) operation of public engagement (PE) learning courses; (2) consultation on various aspects of patient care; (3) proposal of learning materials for children as a basis for the patient–public engagement activities; and (4) research and survey related to drug development and medical application [33]. The educational activity mentioned above (1) is described below.

The PE courses include the following steps:

- Course A: a short course on drug development and related regulations;
- Course B: an intermediate-term course including organizational leadership development, improving health literacy, and participation and leadership in clinical trials;
- Course E: developing lay reviewers for ethics review committees, which are primarily for patients and the public;
- Course C: annual long-term course from drug discovery to post-marketing; and.
- Course D: workplace-based competency assessment course to obtain international certification as a Specialist in Medicines Development (SMD), which are mainly for industry and academic staff.

C and D courses are based on the syllabus of the International Certification Program in Pharmaceutical Medicine (based on the International Federation of Associations of Pharmaceutical Physicians and Pharmaceutical Medicine (IFAPP) and PharmaTrain Federation's standards). These courses have been launched sequentially since October 2020.

Although the lectures and exercises are in an online format instead of face-to-face lectures, due to the COVID-19 pandemic, the following efforts were made to promote sustainable learning. For example, study materials for each lecture are distributed in advance, and preparation is recommended. Lectures are facilitated for questions and discussions. Online examinations are used to check the level of understanding after the course, module-based exams and final exams are required, and participants who fulfilled the specified requirements are awarded certificates upon course completion. Before and after each lecture, an online survey is conducted to collect participants' expectations, and feedback on the lecture, and ongoing program maintenance is conducted based on participants' feedback. Support is also provided for absentees and a review environment for participants by sharing archives of recorded lecture content, questions, and discussions with participants. The implementation of appropriate and phased support for learners through these and other initiatives is well-received by both lecturers and participants. An increasing number of patients and the public who have completed the basic pharmaceutical medicine course have continued in challenged to higher-level courses for further learning, thus laying the groundwork for collaboration between patients and the public, and the pharmaceutical industry. Based on these efforts to achieve and maintain high standards in education, Ji4pe is now recognized as a Centre of Excellence by the PharmaTrain Federation [34].

The experiences of two patients/citizens who are co-authors of this chapter are described as follows. The first case is an example that the values of DTC and patient education using digital tools were recognized in a pandemic situation. The next one is an example where patient education has resulted not only in the patient's benefit but also in giving back to society.

7.2 Testimonials from Course Participants

7.2.1 Noriko Iwaya, Intractable Disease Society Support Familia Yamaguchi

My son has congenital thrombotic thrombocytopenic purpura. The researcher approached us in November 2019 to participate in an international clinical trial (as the fifth case in Japan). We confirmed the trial schedule and the investigational medical center, but we declined to participate because the trial schedule was difficult to follow in our daily life. It was before the COVID-19 pandemic. In the first place, the difference between a "clinical trial according to Good Clinical Practice (GCP) Ordinance under the Japanese Act on Securing Quality, Efficacy, and Safety of Products Including Pharmaceuticals and Medical Devices" and a "Non-GCP clinical study according to the Japanese Ethical Guidelines for Medical and Biological Research Involving Human Subjects" was unclear to us. If the trial could accommodate my son's schedule, he might have been able to participate. Moreover, even if the investigational medical center was far away, the possibility of participation might have expanded with the DCT functions. It was frustrating for a family who had difficulty understanding the researcher's explanations. To resolve these questions and gain more knowledge, I thought I should learn PPI.

In May 2020, I first started learning about PPI in an online seminar hosted by the Department of IT Healthcare Social Collaboration, Graduate School of Pharmaceutical Sciences, University of Tokyo, collaborating with Ji4pe. This experience led me to take the A and C courses organized by Ji4pe, which was established after the university seminar.

The major concern for my son and other patients with the same disease during the COVID-19 pandemic was not only the measures to avoid infection but also the situation of new vaccine development. Feeling the need to share this information and sympathy, our patient association was established in September 2020, with members consisting of patients and families spread across Japan, including Hokkaido, Miyagi, Osaka, and Yamaguchi Prefectures. To ensure a common understanding of "clinical trials" and "vaccines" among the members, a scholarship system was introduced to support members taking the A course of Ji4pe.

In addition, as the organizer of an "Intractable Disease Café" beyond disease boundaries in Yamaguchi Prefecture, I felt that although we wanted to promote PPI in our region, the people around me seemed to have little interest in drug development. As "clinical trial" have become familiar language due to the daily media coverage of the COVID-19 vaccine development, I started to spend more time in PPI promotion to encourage its public relations (as an advocator) and to bridge the learning gap between the metropolitan area and our region, utilizing what I have learned, and I want to continue practicing PPI.

7.2.2 Keiko Inoue, A Board Member of the Association for Medical Malpractice Victims

I am currently a member of some ethics review committees, and as I gain review experience, I have come to want to learn systematically about the process from drug development to marketing approval and post-marketing surveillance. In addition, from my experience as a patient and another experience with medical malpractice, I also felt the need for patients and the public to actively participate in medical care in order to realize safe and high-quality medical care. For these reasons, I first took Ji4pe's A course, which provides knowledge about pharmaceutical medicine, and after that, B course, where I could learn about organizational management.

What I learned led me to the next step. For example, the knowledge I gained here gave me the opportunity to participate in the creation of a common Informed Consent Form template for clinical trials. In addition, as a new activity of our association, a survey was conducted among its members to clarify the problems and difficulties faced by medical malpractice victims, which were presented at the poster session of the Japan Society of Medical Safety. As it was the first time that a presentation by a medical malpractice victim was provided at the Society meeting, it attracted a lot of attention. Furthermore, I am currently in charge of one of the Ji4pe courses as a lecturer on the theme of communication in medical care.

These experiences made me realize that it is important to connect what I have learned to the next step and expand and change the world. I would like to take the first step to change society by returning the knowledge I have gained to society at large rather than making it my own.

8 Conclusion: Patient Involvement as Social Co-Creation

As described above, it will become increasingly important for social co-creation to take place in the field of drug development through patient and public involvement. Patients or the public who can contribute to such a positive cycle can be defined as follows.

The candidate should have: (1) sufficient and up-to-date knowledge of drug development, including clinical trials; (2) the ability to communicate constructively and developmentally with relevant stakeholders; (3) the ability to effectively communicate information and knowledge as a patient/public representative to an unspecified audience in patient/citizen's organizations; and (4) the insight, ideas, analysis, judgment, and agility to further enhance the above aspects.

Such patients and the public would be expected to accelerate drug development in a more rapid and positive manner. When patients and the public are properly informed about clinical trials and become aware that their participation will lead to innovation and access to new treatment options, it will be possible to increase the number of positive experiences participating in clinical trials. If such a positive cycle is generated, even more patients and the public will enjoy the opportunity to participate in clinical trials and contribute to the development of future treatment options, thereby realizing "social co-creation" in drug development.

It should also be noted that the digital transformation of the clinical trial environment during the pandemic could be a factor in enhancing diversity in clinical trial participation, including participation in review committees as a first step of social co-creation. Also, DCTs must bring changes by equitably developing a digitally accessible environment among various groups of patients and the public, providing high digital literacy to them through education, in order to maintain the diversity of clinical trials. In addition to patient knowledge education on clinical trials, it will be necessary to keep in mind that improving literacy for using digital tools will also be a requisite factor in patient and public involvement in the future.

The stakeholders of medicines development, such as industry, government, academia, and the public, including patients, will help make the right decisions regarding treatment options and also help to identify unmet medical needs and provide benefit sharing among the global community. It is a prerequisite that professionals not only provide education to patients and citizens but also that emancipated patients/citizens take the initiative in creating a better society. To this end, it is important that all the stakeholders work hand in hand to create the ecosystem as a social co-creation that will continue the future improvement of global public health, and it could help overcome the unacceptable inequalities in access to developed products.

Acknowledgments The authors would like to express our deep gratitude to the clinical trial ambassador project 2022 members, especially Mr. Yuta Kasai (Astellas Pharma Inc.), Mr. Kenma Nozaki (Nippon Boehringer Ingelheim Co. Ltd.), Mr. Shintaro Omuro (Astellas Pharma Inc.), Ms. Keiko Omomo (CMIC HealthCare Institute Co., Ltd.) and Mr. Nobutaka Yagi (Nippon Boehringer Ingelheim Co. Ltd.).

Conflict of Interest Naoki Tsutsumi, Keiko Inoue, Noriko Iwaya, and Kyoko Imamura have no conflict of interest to be declared related to this manuscript. Kotone Matsuyama received honoraria from Nippon Kayaku Co. Ltd., Astellas Pharma Inc., and Nippon Boehringer Ingelheim Co. Ltd.

References

1. Advancing Health Through Innovation: New Drug Therapy Approvals 2022, CDER, US FDA, 2023 [Internet]. https://www.fda.gov/drugs/new-drugs-fda-cders-new-molecular-entities-and-new-therapeutic-biological-products/new-drug-therapy- approvals-2022. Accessed 31 Mar 2023.
2. Learn more about Patient Engagement [Internet]. https://www.fda.gov/patients/learn-more-about-patient-engagement. Accessed 31 Mar 2023.
3. ACT UP [Internet]. https://actupny.com. Accessed 3 Jul 2023.
4. Jim Eigo J, Harrington M, McCarthy M, Spinella S, Sugden R. FDA action handbook. New York: ACT UP; 1988. https://actupny.org/documents/FDAhandbook1.html. Accessed 3 Jul 2023.
5. 21st Century Cures Act.
6. The Food and Drug Administration Reauthorization Act of 2017 Title I.
7. FDA Patient-Focused Drug Development Guidance Series for Enhancing the Incorporation of the Patient's Voice in Medical Product Development and Regulatory Decision Making [Internet]. https://www.fda.gov/drugs/development-approval-process-drugs/fda-patient-focused-drug-development-guidance-series-enhancing- incorporation-patients-voice-medical. Accessed 31 Mar 2023.
8. https://www.ema.europa.eu/en/partners-networks/patients-consumers.
9. ICH reflection paper - proposed ICH guideline work to advance Patient Focused Drug Development (PFDD) [Internet]. https://www.ema.europa.eu/en/documents/scientific-guideline/ich-reflection-paper-proposed-ich-guideline-work-advance-patient-focused-drug-development-pfdd_en.pdf. Accessed 31 Mar 2023.
10. 2016 International ethical guidelines for health-related research involving humans, CIOMS[Internet]. Accessed 31 May 2023. https://doi.org/10.56759/rgxl7405
11. Patient involvement in the development, regulation and safe use of medicines, CIOMS, 2022 [Internet]. Accessed 31 Mar 2023. https://doi.org/10.56759/iiew8982
12. EUPATI [Internet]. https://eupati.eu. Accessed 31 Mar 2023.
13. The Clinical Trial Ambassadors [Internet]. https://www.clinicaltrialambassadors.com/. Accessed 31 Mar 2023.
14. Workgroup of European Cancer Patient Advocacy Networks, Myeloma Patients Europe, Patient Focused Medicines Development. Guiding Principles on Rea-sonable Agreements between Patient Advocates and Pharmaceutical Companies. Accessed 23 Oct 2018.
15. FDA Guidance on Conduct of Clinical Trials of Medical Products During the COVID-19 Public Health Emergency [Internet]. https://www.fda.gov/regulatory-information/search-fda-guidance-documents/fda-guidance-conduct-clinical-trials-medical-products-during-covid-19-public-health-emergency. Accessed 26 Jun 2023.
16. Advancing Oncology Decentralized Trials [Internet]. https://www.fda.gov/about-fda/oncology-center-excellence/advancing-oncology-decentralized-trials. Accessed 30 Apr 2023.
17. Explanation and Consent Using Electromagnetic Methods in Clinical Trials and Post-Marketing Clinical Trials and Consent for Clinical Trials and Post-Marketing Clinical Trials Using Electromagnetic Means, MHLW [Internet]. https://www.mhlw.go.jp/hourei/doc/tsuchi/T230331I0140.pdf. Accessed 30 Apr 2023.
18. Aftereffects survey of COVID-19 infection, City of Toyonaka [Internet]. https://www.city.toyonaka.osaka.jp/kenko/covid19_support/koronakouisyousien.files/tyousakekka3.pdf. Accessed 30 Apr 2023.
19. Tsutsumi N, Kurihara C, Matsuyama K, Tsukahara K, Yoshida K, Tsuda T, Oi H, Tokai Y, Nagao N, Imamura K. Utilization of information and communication technology in IRBs as a countermeasure against the COVID-19 outbreak and future challenges. Clin Eval. 2020;48(2):355–66.

20. Sarraju A, Seninger C, Parameswaran V, et al. Pandemic-proof recruitment and engagement in a fully decentralized trial in atrial fibrillation patients (DeTAP). NPJ Digit Med. 2022;5:80. https://doi.org/10.1038/s41746-022-00622-9.
21. PatientsLikeMe [Internet]. https://www.patientslikeme.com. Accessed 5 May 2023.
22. Newman SD, Andrews JO, Magwood GS, Jenkins C, Cox MJ, Williamson DC. Community advisory boards in community-based participatory research: a synthesis of best processes. Prev Chronic Dis. 2011;8(3):A70. Epub 2011 Apr 15. PMID: 21477510; PMCID: PMC3103575.
23. UNAIDS. Good participatory practice: Guidelines for biomedical HIV prevention trials. 2011. [Internet]. https://www.unaids.org/sites/default/files/media_asset/JC1853_GPP_Guidelines_2011_en_0.pdf. Accessed 26 Jun 2023.
24. Prepared by Dr Catherine Hankins for the World Health Organization. Good participatory practice guidelines for trials of emerging (and re-emerging) pathogens that are likely to cause severe outbreaks in the near future and for which few or no medical countermeasures exist (GPP-EP) Outcome document of the consultative process. 2016. [Internet]. https://cdn.who.int/media/docs/default-source/blue-print/good-participatory-practice-for-trials-of-(re-)emerging-pathogens-(gpp-ep)_guidelines.pdf?sfvrsn=102e3e70_9&download=true. Accessed 26 Jun 2023.
25. Freire P. Pedagogy of the oppressed. New York: Continuum; 1970.
26. Patient and Public Involvement (PPI) Guidebook - A first step toward patient and researcher collaboration [Internet]. 2023. https://www.amed.go.jp/ppi/guidebook.html. Accessed 31 Mar 2023.
27. Navigate to the Patient Experience Toolkits [Internet]. https://www.transceleratebiopharmainc.com/initiatives/patient-experience/. Accessed 26 Jun 2023.
28. Clinical Evaluation Subcommittee, Committee on Drug Evaluation, Japan Pharmaceutical Manufacturers Association. Guidebook for Pharmaceutical Companies to Implement Activities Based on Patient Centricity-Drug Development Utilizing Patients' Voices. September 2019 [Internet]. https://www.jpma.or.jp/information/evaluation/results/allotment/patient_centricity3.html. Accessed 26 Jun 2023.
29. Japan Pharmaceutical Manufacturers Association. Report on the results of the third survey on awareness and activities of patient groups. 2021.
30. Emoto H, Nimiya K, Nishimura Y. Questionnaire survey on patients' image of clinical trials analysis and subsequent social awareness. In: Proceedings of the Annual Conference of the Society for Research and Innovation, vol. 35; 2020. p. 770–3.
31. Aizawa A, Yanagida S, Kato S, Yokoi T, Owaki T, Onita K, Takahashi Y, et al. Current status and future of information provision for the purpose of promoting clinical trials. Clin Med. 2004;20(10):1025–43.
32. Events and feelings after completion of clinical trial participation [Internet]. https://www.dipex-j.org/clinical-trial/topic/after/image. Accessed 26 Jun 2023.
33. Imamura K, Tsutsumi N. Future of drug development and human resource development supporting social coexistence. Clin Eval. 2020;48(3):643–8.
34. Pharma Train Course Providers List [Internet]. https://www.pharmatrain.eu/pharmatrain-course-providers. Accessed 26 Jun 2023.

GPSR Compliance

The European Union's (EU) General Product Safety Regulation (GPSR) is a set of rules that requires consumer products to be safe and our obligations to ensure this.

If you have any concerns about our products, you can contact us on ProductSafety@springernature.com

In case Publisher is established outside the EU, the EU authorized representative is:

Springer Nature Customer Service Center GmbH
Europaplatz 3
69115 Heidelberg, Germany

Batch number: 10732090

Printed by Printforce, the Netherlands